Electric Motors and Drives

Electric Motors and Drives
Fundamentals, Types and Applications

Third edition

Austin Hughes
*Senior Fellow, School of Electronic and Electrical Engineering,
University of Leeds*

AMSTERDAM • BOSTON • HEIDELBERG • LONDON • NEW YORK • OXFORD
PARIS • SAN DIEGO • SAN FRANCISCO • SINGAPORE • SYDNEY • TOKYO
Newnes is an imprint of Elsevier

Newnes is an imprint of Elsevier
The Boulevard, Langford Lane, Kidlington, Oxford, OX5 1GB
30 Corporate Drive, Suite 400, Burlington, MA 01803, USA

First edition 1990
Second edition 1993
Third edition 2006
Reprinted 2006, 2007, 2008 (twice), 2009

Notice
No responsibility is assumed by the publisher for any injury and/or damage to persons
or property as a matter of products liability, negligence or otherwise, or from any use
or operation of any methods, products, instructions or ideas contained in the material
herein. Because of rapid advances in the medical sciences, in particular, independent
verification of diagnoses and drug dosages should be made

British Library Cataloguing in Publication Data
A catalogue record for this book is available from the British Library

Library of Congress Cataloging-in-Publication Data
A catalog record for this book is available from the Library of Congress

ISBN: 978-0-7506-4718-2

For information on all Newnes publications
visit our website at www.elsevierdirect.com

Transferred to Digital Printing in 2009

Working together to grow
libraries in developing countries

www.elsevier.com | www.bookaid.org | www.sabre.org

ELSEVIER   BOOK AID
            International   Sabre Foundation

CONTENTS

3 CONVENTIONAL D.C. MOTORS

PREFACE

Like its predecessors, the third edition of this book is intended primarily for non-specialist users and students of electric motors and drives. My original aim was to bridge the gap between specialist textbooks (which are pitched at a level too academic for the average user) and the more prosaic 'handbooks', which are full of useful detail but provide little opportunity for the development of any real insight or understanding. The fact that the second edition was reprinted ten times indicated that there had indeed been a gap in the market, and that a third edition would be worthwhile. It was also gratifying to learn that although the original book was not intended as yet another undergraduate textbook, teachers and students had welcomed the book as a gentle introduction to the subject.

The aim throughout is to provide the reader with an understanding of how each motor and drive system works, in the belief that it is only by knowing what should happen that informed judgements and sound comparisons can be made. Given that the book is aimed at readers from a range of disciplines, introductory material on motors and power electronics is clearly necessary, and this is presented in the first two chapters. Many of these basic ideas crop up frequently throughout the book, so unless the reader is well-versed in the fundamentals it would be wise to absorb the first two chapters before tackling the later material. In addition, an awareness of the basic ideas underlying feedback and closed-loop control is necessary in order to follow the sections dealing with drives, and this has now been provided as an Appendix.

The book explores most of the widely used modern types of motors and drives, including conventional and brushless d.c., induction motors (mains and inverter-fed), stepping motors, synchronous motors (mains and converter-fed) and reluctance motors. The d.c. motor drive and the induction motor drive are given most importance, reflecting their dominant position in terms of numbers. Understanding the d.c. drive is particularly important because it is still widely used as a frame of reference for other drives: those who develop a good grasp of the d.c. drive will find their know-how invaluable in dealing with all other types, particularly if they can establish a firm grip on the philosophy of the control scheme.

Younger readers may be unaware of the radical changes that have taken place over the past 40 years, so perhaps a couple of paragraphs are appropriate to put the current scene into perspective. For more than a century, many different types of motors were developed, and each became closely associated with a particular application. Traction, for example, was seen as the exclusive preserve of the series d.c. motor, whereas the shunt d.c. motor, though outwardly indistinguishable, was seen as being quite unsuited to traction applications. The cage induction motor was (and still is) the most widely used but was judged as being suited only for applications that called for constant speed. The reason for the plethora of motor types was that there was no easy way of varying the supply voltage and/or frequency to obtain speed control, and designers were therefore forced to seek ways of providing speed control within the motor itself. All sorts of ingenious arrangements and interconnections of motor windings were invented, but even the best motors had a limited range of operating characteristics, and all of them required bulky control equipment gear-control, which was manually or electromechanically operated, making it difficult to arrange automatic or remote control.

All this changed from the early 1960s when power electronics began to make an impact. The first major breakthrough came with the thyristor, which provided a relatively cheap, compact and easily controlled variable-speed drive using the d.c. motor. In the 1970s, the second major breakthrough resulted from the development of power-electronic inverters, providing a three-phase variable-frequency supply for the cage induction motor and thereby enabling its speed to be controlled.

These major developments resulted in the demise of many of the special motors, leaving the majority of applications in the hands of comparatively few types, and the emphasis has now shifted from complexity inside the motor to sophistication in supply and control arrangements.

From the user's point of view this is a mixed blessing. Greater flexibility and superior levels of performance are available, and there are fewer motor types to consider. But if anything more than constant speed is called for, the user will be faced with the purchase of a complete drive system, consisting of a motor together with its associated power electronics package. To choose wisely requires not only some knowledge of motors, but also the associated power-electronics and the control options that are normally provided.

Development in the world of electrical machines tends to be steady rather than spectacular, which means that updating the second edition has called for only modest revision of the material covering the how and why of motors, though in most areas explanations have been extended, especially where feedback indicated that more clarity was called for. After much weighing the pros and cons I decided to add a chapter on the equivalent circuit of the induction motor, because familiarity with the terminology of the equivalent circuit is necessary in order to engage in serious dialogue with induction motor suppliers or experts. However those who find the circuit emphasis not to their liking can be reassured that they can skip Chapter 7 without prejudicing their ability to tackle the subsequent chapter on induction motor drives.

The power electronics area has matured since the 1993 edition of the book, but although voltage and current ratings of individual switching devices continue to improve, and there is greater integration of drive electronics and power devices, there has been no strategic shift that would call for a radical revision of the material in the second edition. The majority of drive converters now use IGBT or MOSFET devices, but the old-fashioned bipolar transistor symbol has been retained in most of the diagrams because it has the virtue of showing the direction of current flow, and is therefore helpful in understanding circuit operation.

The style of the book reflects my own preference for an informal approach, in which the difficulty of coming to grips with new ideas is not disguised. Deciding on the level at which to pitch the material was originally a headache, but experience suggested that a mainly descriptive approach with physical explanations would be most appropriate, with mathematics kept to a minimum to assist digestion. The most important concepts (such as the inherent e.m.f. feedback in motors, or the need for a switching strategy in converters) are deliberately reiterated to reinforce understanding, but should not prove too tiresome for readers who have already 'got the message'. I had hoped to continue without numbered headings, as this always seems to me to make the material seem lighter, but cross referencing is so cumbersome without numbering that in the end I had to give in.

I have deliberately not included any computed magnetic field plots, nor any results from the excellent motor simulation packages that are now available because experience suggests that simplified diagrams are actually better as learning vehicles. All of the diagrams have been redrawn, and many new ones have been added.

Review questions have been added at the end of each chapter. The number of questions broadly reflects my judgement of the relative importance of each chapter, and they are intended to help build confidence and to be used selectively. A drives user might well not bother with the basic machine-design questions in the first two chapters, but could benefit by tackling the applications-related questions in subsequent chapters. Judicious approximations are called for in most of the questions, and in some cases there is either insufficient explicit information or redundant data: this is deliberate and designed to reflect reality.

Answers to the numerical questions are printed in the book, with fully worked and commented solutions on the accompanying website http://books.elsevier.com/companions/0750647183. The best way to learn is to make an unaided attempt before consulting a worked solution, so the extra effort in consulting the website will perhaps encourage best practice. In any event, my model solution may not be the best!

Austin Hughes

1

ELECTRIC MOTORS

INTRODUCTION

Electric motors are so much a part of everyday life that we seldom give them a second thought. When we switch on an electric drill, for example, we confidently expect it to run rapidly up to the correct speed, and we do not question how it knows what speed to run at, or how it is that once enough energy has been drawn from the supply to bring it up to speed, the power drawn falls to a very low level. When we put the drill to work it draws more power, and when we finish the power drawn from the mains reduces automatically, without intervention on our part.

The humble motor, consisting of nothing more than an arrangement of copper coils and steel laminations, is clearly rather a clever energy converter, which warrants serious consideration. By gaining a basic understanding of how the motor works, we will be able to appreciate its potential and its limitations, and (in later chapters) see how its already remarkable performance can be further enhanced by the addition of external electronic controls.

This chapter deals with the basic mechanisms of motor operation, so readers who are already familiar with such matters as magnetic flux, magnetic and electric circuits, torque, and motional e.m.f can probably afford to skim over much of it. In the course of the discussion, however, several very important general principles and guidelines emerge. These apply to all types of motors and are summarised in Section 1.8. Experience shows that anyone who has a good grasp of these basic principles will be well equipped to weigh the pros and cons of the different types of motor, so all readers are urged to absorb them before tackling other parts of the book.

PRODUCING ROTATION

Nearly all motors exploit the force which is exerted on a current-carrying conductor placed in a magnetic field. The force can be demonstrated by placing a bar magnet near a wire carrying current (Figure 1.1), but anyone trying the experiment will probably be disappointed to discover how feeble the force is, and will doubtless be left wondering how such an unpromising effect can be used to make effective motors.

We will see that in order to make the most of the mechanism, we need to arrange a very strong magnetic field, and make it interact with many conductors, each carrying as much current as possible. We will also see later that although the magnetic field (or 'excitation') is essential to the working of the motor, it acts only as a catalyst, and all of the mechanical output power comes from the electrical supply to the conductors on which the force is developed. It will emerge later that in some motors the parts of the machine responsible for the excitation and for the energy converting functions are distinct and self-evident. In the d.c. motor, for example, the excitation is provided either by permanent magnets or by field coils wrapped around clearly defined projecting field poles on the stationary part, while the conductors on which force is developed are on the rotor and supplied with current via sliding brushes. In many motors, however, there is no such clear-cut physical distinction between the 'excitation' and the 'energy-converting' parts of the machine, and a single stationary winding serves both purposes. Nevertheless, we will find that identifying and separating the excitation and energy-converting functions is always helpful in understanding how motors of all types operate.

Returning to the matter of force on a single conductor, we will first look at what determines the magnitude and direction of the force,

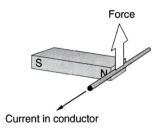

Force

S   N

Current in conductor

Figure 1.1   *Mechanical force produced on a current-carrying wire in a magnetic field*

before turning to ways in which the mechanism is exploited to produce rotation. The concept of the magnetic circuit will have to be explored, since this is central to understanding why motors have the shapes they do. A brief introduction to magnetic field, magnetic flux, and flux density is included before that for those who are not familiar with the ideas involved.

Magnetic field and magnetic flux

When a current-carrying conductor is placed in a magnetic field, it experiences a force. Experiment shows that the magnitude of the force depends directly on the current in the wire, and the strength of the magnetic field, and that the force is greatest when the magnetic field is perpendicular to the conductor.

In the set-up shown in Figure 1.1, the source of the magnetic field is a bar magnet, which produces a magnetic field as shown in Figure 1.2.

The notion of a 'magnetic field' surrounding a magnet is an abstract idea that helps us to come to grips with the mysterious phenomenon of

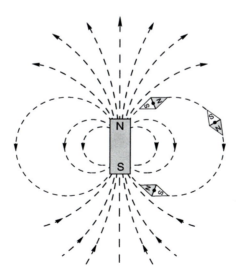

Figure 1.2   *Magnetic flux lines produced by a permanent magnet*

magnetism: it not only provides us with a convenient pictorial way of picturing the directional effects, but it also allows us to quantify the 'strength' of the magnetism and hence permits us to predict the various effects produced by it.

The dotted lines in Figure 1.2 are referred to as magnetic flux lines, or simply flux lines. They indicate the direction along which iron filings (or small steel pins) would align themselves when placed in the field of the bar magnet. Steel pins have no initial magnetic field of their own, so there is no reason why one end or the other of the pins should point to a particular pole of the bar magnet.

However, when we put a compass needle (which is itself a permanent magnet) in the field we find that it aligns itself as shown in Figure 1.2. In the upper half of the figure, the S end of the diamond-shaped compass settles closest to the N pole of the magnet, while in the lower half of the figure, the N end of the compass seeks the S of the magnet. This immediately suggests that there is a direction associated with the lines of flux, as shown by the arrows on the flux lines, which conventionally are taken as positively directed from the N to the S pole of the bar magnet.

The sketch in Figure 1.2 might suggest that there is a 'source' near the top of the bar magnet, from which flux lines emanate before making their way to a corresponding 'sink' at the bottom. However, if we were to look at the flux lines inside the magnet, we would find that they were continuous, with no 'start' or 'finish'. (In Figure 1.2 the internal flux lines have been omitted for the sake of clarity, but a very similar field pattern is produced by a circular coil of wire carrying a d.c. See Figure 1.6 where the continuity of the flux lines is clear.). Magnetic flux lines always form closed paths, as we will see when we look at the 'magnetic circuit', and draw a parallel with the electric circuit, in which the current is also a continuous quantity. (There must be a 'cause' of the magnetic flux, of course, and in a permanent magnet this is usually pictured in terms of atomic-level circulating currents within the magnet material. Fortunately, discussion at this physical level is not necessary for our purpose.)

Magnetic flux density

Along with showing direction, the flux plots also convey information about the intensity of the magnetic field. To achieve this, we introduce the idea that between every pair of flux lines (and for a given depth into the paper) there is a same 'quantity' of magnetic flux. Some people have no difficulty with such a concept, while others find that the notion of quantifying something so abstract represents a serious intellectual challenge. But whether the approach seems obvious or not, there is no denying of the practical utility of quantifying the mysterious stuff we call magnetic flux, and it leads us next to the very important idea of magnetic flux density (B).

When the flux lines are close together, the 'tube' of flux is squashed into a smaller space, whereas when the lines are further apart the same tube of flux has more breathing space. The flux density (B) is simply the flux in the 'tube' (Φ) divided by the cross sectional area (A) of the tube, i.e.

$$B = \frac{\Phi}{A} \qquad (1.1)$$

The flux density is a vector quantity, and is therefore often written in bold type: its magnitude is given by equation (1.1), and its direction is that of the prevailing flux lines at each point. Near the top of the magnet in Figure 1.2, for example, the flux density will be large (because the flux is squashed into a small area), and pointing upwards, whereas on the equator and far out from the body of the magnet the flux density will be small and directed downwards.

It will be seen later that in order to create high flux densities in motors, the flux spends most of its life inside well-defined 'magnetic circuits' made of iron or steel, within which the flux lines spread out uniformly to take full advantage of the available area. In the case shown in Figure 1.3, for example, the cross-sectional area at bb′ is twice that at aa′, but the flux is constant so the flux density at bb′ is half that at aa′.

It remains to specify units for quantity of flux, and flux density. In the SI system, the unit of magnetic flux is the weber (Wb). If one weber of flux is distributed uniformly across an area of 1m² perpendicular to the flux, the flux density is clearly one weber per square metre

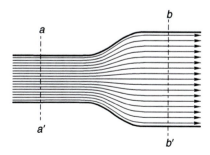

Figure 1.3   *Magnetic flux lines inside part of an iron magnetic circuit*

(Wb/m^2). This was the unit of magnetic flux density until about 40 years ago, when it was decided that one weber per square meter would henceforth be known as one tesla (T), in honour of Nikola Tesla who is generally credited with inventing the induction motor. The widespread use of B (measured in tesla) in the design stage of all types of electro-magnetic apparatus means that we are constantly reminded of the importance of tesla; but at the same time one has to acknowledge that the outdated unit did have the advantage of conveying directly what flux density is, i.e. flux divided by area.

In the motor world we are unlikely to encounter more than a few milliwebers of flux, and a small bar magnet would probably only produce a few microwebers. On the other hand, values of flux density are typically around 1 T in most motors, which is a reflection of the fact that although the quantity of flux is small, it is also spread over a small area.

Force on a conductor

We now return to the production of force on a current-carrying wire placed in a magnetic field, as revealed by the setup shown in Figure 1.1.

The direction of the force is shown in Figure 1.1: it is at right angles to both the current and the magnetic flux density. With the flux density horizontal and to the right, and the current flowing out of the paper, the force is vertically upward. If either the field or the current is reversed, the force acts downwards, and if both are reversed, the force will remain upward.

We find by experiment that if we double either the current or the flux density, we double the force, while doubling both causes the force to increase by a factor of four. But how about quantifying the force? We need to express the force in terms of the product of the current and the magnetic flux density, and this turns out to be very straightforward when we work in SI units.

The force on a wire of length l, carrying a current I and exposed to a uniform magnetic flux density B throughout its length is given by the simple expression

$$F = BIl \qquad (1.2)$$

where F is in newtons when B is in tesla, I in amperes, and l in metres.

This is a delightfully simple formula, and it may come as a surprise to some readers that there are no constants of proportionality involved in equation 1.2. The simplicity is not a coincidence, but stems from the fact that the unit of current (the ampere) is actually defined in terms of force.

Strictly, equation 1.2 only applies when the current is perpendicular to the field. If this condition is not met, the force on the conductor will be less; and in the extreme case where the current was in the same direction as the field, the force would fall to zero. However, every sensible motor designer knows that to get the best out of the magnetic field it has to be perpendicular to the conductors, and so it is safe to assume in the subsequent discussion that B and I are always perpendicular. In the remainder of this book, it will be assumed that the flux density and current are mutually perpendicular, and this is why, although B is a vector quantity (and would usually be denoted by bold type), we can drop the bold notation because the direction is implicit and we are only interested in the magnitude.

The reason for the very low force detected in the experiment with the bar magnet is revealed by equation 1.2. To obtain a high force, we must have a high flux density, and a lot of current. The flux density at the ends of a bar magnet is low, perhaps 0.1 tesla, so a wire carrying 1 amp will experience a force of only 0.1 N/m (approximately 100 gm wt). Since the flux density will be confined to perhaps 1 cm across the end face of the magnet, the total force on the wire will be only 1 gm. This would be barely detectable, and is too low to be of any use in a decent motor. So how is more force obtained?

The first step is to obtain the highest possible flux density. This is achieved by designing a 'good' magnetic circuit, and is discussed next. Secondly, as many conductors as possible must be packed in the space where the magnetic field exists, and each conductor must carry as much current as it can without heating up to a dangerous temperature. In this way, impressive forces can be obtained from modestly sized devices, as anyone who has tried to stop an electric drill by grasping the chuck will testify.

MAGNETIC CIRCUITS

So far we have assumed that the source of the magnetic field is a permanent magnet. This is a convenient starting point as all of us are familiar with magnets, even if only of the fridge-door variety. But in the majority of motors, the working magnetic field is produced by coils of wire carrying current, so it is appropriate that we spend some time looking at how we arrange the coils and their associated iron 'magnetic circuit' so as to produce high magnetic fields which then interact with other current-carrying conductors to produce force, and hence rotation.

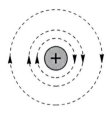

Figure 1.4   *Magnetic flux lines produced by a straight, current-carrying wire*

First, we look at the simplest possible case of the magnetic field surrounding an isolated long straight wire carrying a steady current (Figure 1.4). (In the figure, the + sign indicates that current is flowing into the paper, while a dot is used to signify current out of the paper: these symbols can perhaps be remembered by picturing an arrow or dart, with the cross being the rear view of the fletch, and the dot being the approaching point.) The flux lines form circles concentric with the wire, the field strength being greatest close to the wire. As might be expected, the field strength at any point is directly proportional to the current. The convention for determining the direction of the field is that the positive direction is taken to be the direction that a right-handed corkscrew must be rotated to move in the direction of the current.

Figure 1.4 is somewhat artificial as current can only flow in a complete circuit, so there must always be a return path. If we imagine a parallel 'go' and 'return' circuit, for example, the field can be obtained by superimposing the field produced by the positive current in the go side with the field produced by the negative current in the return side, as shown in Figure 1.5.

We note how the field is increased in the region between the conductors, and reduced in the regions outside. Although Figure 1.5 strictly only applies to an infinitely long pair of straight conductors, it will probably not come as a surprise to learn that the field produced by a single turn of wire of rectangular, square or round form is very much the same as that shown in Figure 1.5. This enables us to build up a picture of the field

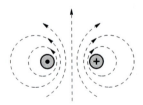

Figure 1.5   *Magnetic flux lines produced by current in a parallel go and return circuit*

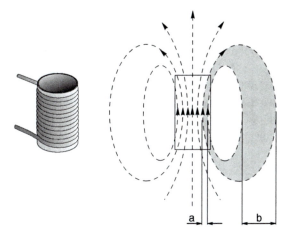

Figure 1.6   *Multi-turn cylindrical coil and pattern of magnetic flux produced by current in the coil. (For the sake of clarity, only the outline of the coil is shown on the right.)*

that would be produced in air, by the sort of coils used in motors, which typically have many turns, as shown for example in Figure 1.6.

The coil itself is shown on the left in Figure 1.6 while the flux pattern produced is shown on the right. Each turn in the coil produces a field pattern, and when all the individual field components are superimposed we see that the field inside the coil is substantially increased and that the closed flux paths closely resemble those of the bar magnet we looked at earlier. The air surrounding the sources of the field offers a homogeneous path for the flux, so once the tubes of flux escape from the concentrating influence of the source, they are free to spread out into the whole of the surrounding space. Recalling that between each pair of flux lines there is an equal amount of flux, we see that because the flux lines spread out as they leave the confines of the coil, the flux density is much lower outside than inside: for example, if the distance 'b' is say four times 'a' the flux density B_b is a quarter of B_a.

Although the flux density inside the coil is higher than outside, we would find that the flux densities which we could achieve are still too low to be of use in a motor. What is needed firstly is a way of increasing the flux density, and secondly a means for concentrating the flux and preventing it from spreading out into the surrounding space.

Magnetomotive force (MMF)

One obvious way to increase the flux density is to increase the current in the coil, or to add more turns. We find that if we double the current, or the number of turns, we double the total flux, thereby doubling the flux density everywhere.

We quantify the ability of the coil to produce flux in terms of its magnetomotive force (MMF). The MMF of the coil is simply the product of the number of turns (N) and the current (I), and is thus expressed in ampere-turns. A given MMF can be obtained with a large number of turns of thin wire carrying a low current, or a few turns of thick wire carrying a high current: as long as the product NI is constant, the MMF is the same.

Electric circuit analogy

We have seen that the magnetic flux which is set up is proportional to the MMF driving it. This points to a parallel with the electric circuit, where the current (amps) that flows is proportional to the EMF (volts) driving it.

In the electric circuit, current and EMF are related by Ohm's Law, which is

$$\text{Current} = \frac{\text{EMF}}{\text{Resistance}} \quad \text{i.e.} \quad I = \frac{V}{R} \tag{1.3}$$

For a given source EMF (volts), the current depends on the resistance of the circuit, so to obtain more current, the resistance of the circuit has to be reduced.

We can make use of an equivalent 'magnetic Ohm's law' by introducing the idea of reluctance ($\mathcal{R}$). The reluctance gives a measure of how difficult it is for the magnetic flux to complete its circuit, in the same way that resistance indicates how much opposition the current encounters in the electric circuit. The magnetic Ohm's law is then

$$\text{Flux} = \frac{\text{MMF}}{\text{Reluctance}} \quad \text{i.e.} \quad \Phi = \frac{NI}{\mathcal{R}} \tag{1.4}$$

We see from equation 1.4 that to increase the flux for a given MMF, we need to reduce the reluctance of the magnetic circuit. In the case of the example (Figure 1.6), this means we must replace as much as possible of the air path (which is a 'poor' magnetic material, and therefore constitutes a high reluctance) with a 'good' magnetic material, thereby reducing the reluctance and resulting in a higher flux for a given MMF.

The material which we choose is good quality magnetic steel, which for historical reasons is usually referred to as 'iron'. This brings several very dramatic and desirable benefits, as shown in Figure 1.7.

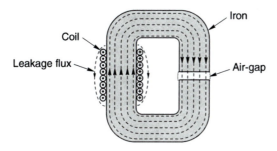

Figure 1.7   *Flux lines inside low-reluctance magnetic circuit with air-gap*

Firstly, the reluctance of the iron paths is very much less than the air paths which they have replaced, so the total flux produced for a given MMF is very much greater. (Strictly speaking therefore, if the MMFs and cross-sections of the coils in Figures 1.6 and 1.7 are the same, many more flux lines should be shown in Figure 1.7 than in Figure 1.6, but for the sake of clarity similar numbers are indicated.) Secondly, almost all the flux is confined within the iron, rather than spreading out into the surrounding air. We can therefore shape the iron parts of the magnetic circuit as shown in Figure 1.7 in order to guide the flux to wherever it is needed. And finally, we see that inside the iron, the flux density remains uniform over the whole cross-section, there being so little reluctance that there is no noticeable tendency for the flux to crowd to one side or another.

Before moving on to the matter of the air-gap, we should note that a question which is often asked is whether it is important for the coils to be wound tightly onto the magnetic circuit, and whether, if there is a multi-layer winding, the outer turns are as effective as the inner ones. The answer, happily, is that the total MMF is determined solely by the number of turns and the current, and therefore every complete turn makes the same contribution to the total MMF, regardless of whether it happens to be tightly or loosely wound. Of course it does make sense for the coils to be wound as tightly as is practicable, since this not only minimises the resistance of the coil (and thereby reduces the heat loss) but also makes it easier for the heat generated to be conducted away to the frame of the machine.

The air-gap

In motors, we intend to use the high flux density to develop force on current-carrying conductors. We have now seen how to create a high flux density inside the iron parts of a magnetic circuit, but, of course, it is physically impossible to put current-carrying conductors inside the iron. We therefore arrange for an air-gap in the magnetic circuit, as shown in Figure 1.7. We will see shortly that the conductors on which the force is to be produced will be placed in this air-gap region.

If the air-gap is relatively small, as in motors, we find that the flux jumps across the air-gap as shown in Figure 1.7, with very little tendency to balloon out into the surrounding air. With most of the flux lines going straight across the air-gap, the flux density in the gap region has the same high value as it does inside the iron.

In the majority of magnetic circuits consisting of iron parts and one or more air-gaps, the reluctance of the iron parts is very much less than the reluctance of the gaps. At first sight this can seem surprising, since the distance across the gap is so much less than the rest of the path through the iron. The fact that the air-gap dominates the reluctance is simply a reflection of how poor air is as a magnetic medium, compared to iron. To put the comparison in perspective, if we calculate the reluctances of two paths of equal length and cross-sectional area, one being in iron and the other in air, the reluctance of the air path will typically be 1000 times greater than the reluctance of the iron path. (The calculation of reluctance will be discussed in Section 1.3.4.)

Returning to the analogy with the electric circuit, the role of the iron parts of the magnetic circuit can be likened to that of the copper wires in the electric circuit. Both offer little opposition to flow (so that a negligible fraction of the driving force (MMF or EMF) is wasted in conveying the flow to where it is usefully exploited) and both can be shaped to guide the flow to its destination. There is one important difference, however. In the electric circuit, no current will flow until the circuit is completed, after which all the current is confined inside the wires. With an iron magnetic circuit, some flux can flow (in the surrounding air) even before the iron is installed. And although most of the flux will subsequently take the easy route through the iron, some will still leak into the air, as shown in Figure 1.7. We will not pursue leakage flux here, though it is sometimes important, as will be seen later.

Reluctance and air-gap flux densities

If we neglect the reluctance of the iron parts of a magnetic circuit, it is easy to estimate the flux density in the air-gap. Since the iron parts are then in effect 'perfect conductors' of flux, none of the source MMF (NI) is used in driving the flux through the iron parts, and all of it is available to push the flux across the air-gap. The situation depicted in Figure 1.7

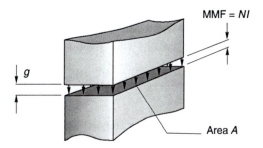

Figure 1.8 *Air-gap region, with MMF acting across opposing pole-faces*

therefore reduces to that shown in Figure 1.8, where an MMF of NI is applied directly across an air-gap of length g.

To determine how much flux will cross the gap, we need to know its reluctance. As might be expected, the reluctance of any part of the magnetic circuit depends on its dimensions, and on its magnetic properties, and the reluctance of a rectangular 'prism' of air, of cross-sectional area A and length g as in Figure 1.8 is given by

$$\mathcal{R}_g = \frac{g}{A\mu_0} \qquad (1.5)$$

where μ_0 is the so-called 'primary magnetic constant' or 'permeability of free space'. Strictly, as its name implies, μ_0 quantifies the magnetic properties of a vacuum, but for all engineering purposes the permeability of air is also μ_0. The value of the primary magnetic constant (μ_o) in the SI system is $4\pi \times 10^{-7}$ H/m; rather surprisingly, there is no name for the unit of reluctance.

In passing, we should note that if we want to include the reluctance of the iron part of the magnetic circuit in our calculation, its reluctance would be given by

$$\mathcal{R}_{fe} = \frac{l_{fe}}{A\mu_{fe}}$$

and we would have to add this to the reluctance of the air-gap to obtain the total reluctance. However, because the permeability of iron (μ_{fe}) is so much higher than μ_0, the iron reluctance will be very much less than the gap reluctance, despite the path length l being considerably longer than the path length (g) in the air.

Equation 1.5 reveals the expected result that doubling the air-gap would double the reluctance (because the flux has twice as far to go), while doubling the area would halve the reluctance (because the flux has two equally appealing paths in parallel). To calculate the flux, Φ, we use the magnetic Ohm's law (equation 1.4), which gives

$$\Phi = \frac{\text{MMF}}{\mathcal{R}} = \frac{NI\,A\mu_0}{g} \qquad (1.6)$$

We are usually interested in the flux density in the gap, rather than the total flux, so we use equation 1.1 to yield

$$B = \frac{\Phi}{A} = \frac{\mu_0\,NI}{g} \qquad (1.7)$$

Equation 1.7 is delightfully simple, and from it we can calculate the air-gap flux density once we know the MMF of the coil (NI) and the length of the gap (g). We do not need to know the details of the coil-winding as long as we know the product of the turns and the current, nor do we need to know the cross-sectional area of the magnetic circuit in order to obtain the flux density (though we do if we want to know the total flux, see equation 1.6).

For example, suppose the magnetising coil has 250 turns, the current is 2 A, and the gap is 1 mm. The flux density is then given by

$$B = \frac{4\pi \times 10^{-7} \times 250 \times 2}{1 \times 10^{-3}} = 0.63 \text{ tesla}$$

(We could of course obtain the same result using an exciting coil of 50 turns carrying a current of 10 A, or any other combination of turns and current giving an MMF of 500 ampere-turns.)

If the cross-sectional area of the iron was constant at all points, the flux density would be 0.63 T everywhere. Sometimes, as has already been mentioned, the cross-section of the iron reduces at points away from the air-gap, as shown for example in Figure 1.3. Because the flux is compressed in the narrower sections, the flux density is higher, and in Figure 1.3 if the flux density at the air-gap and in the adjacent pole-faces is once again taken to be 0.63 T, then at the section aa' (where the area is only half that at the air-gap) the flux density will be $2 \times 0.63 = 1.26\,\text{T}$.

Saturation

It would be reasonable to ask whether there is any limit to the flux density at which the iron can be operated. We can anticipate that there must be a limit, or else it would be possible to squash the flux into a

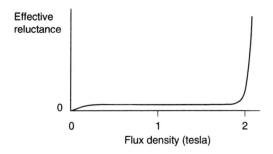

Figure 1.9   *Sketch showing how the effective reluctance of iron increases rapidly as the flux density approaches saturation*

vanishingly small cross-section, which we know from experience is not the case. In fact there is a limit, though not a very sharply defined one.

Earlier we noted that the iron has almost no reluctance, at least not in comparison with air. Unfortunately this happy state of affairs is only true as long as the flux density remains below about 1.6 – 1.8 T, depending on the particular steel in question. If we try to work the iron at higher flux densities, it begins to exhibit significant reluctance, and no longer behaves like an ideal conductor of flux. At these higher flux densities, a significant proportion of the source MMF is used in driving the flux through the iron. This situation is obviously undesirable, since less MMF remains to drive the flux across the air-gap. So just as we would not recommend the use of high-resistance supply leads to the load in an electric circuit, we must avoid overloading the iron parts of the magnetic circuit.

The emergence of significant reluctance as the flux density is raised is illustrated qualitatively in Figure 1.9.

When the reluctance begins to be appreciable, the iron is said to be beginning to 'saturate'. The term is apt, because if we continue increasing the MMF, or reducing the area of the iron, we will eventually reach an almost constant flux density, typically around 2 T. To avoid the undesirable effects of saturation, the size of the iron parts of the magnetic circuit are usually chosen so that the flux density does not exceed about 1.5 T. At this level of flux density, the reluctance of the iron parts will be small in comparison with the air-gap.

Magnetic circuits in motors

The reader may be wondering why so much attention has been focused on the gapped C-core magnetic circuit, when it appears to bear little

Figure 1.10   *Evolution of d.c. motor magnetic circuit from gapped C-core*

resemblance to the magnetic circuits found in motors. We will now see that it is actually a short step from the C-core to a magnetic motor circuit, and that no fundamentally new ideas are involved.

The evolution from C-core to motor geometry is shown in Figure 1.10, which should be largely self-explanatory, and relates to the field system of a d.c. motor.

We note that the first stage of evolution (Figure 1.10, left) results in the original single gap of length g being split into two gaps of length $g/2$, reflecting the requirement for the rotor to be able to turn. At the same time the single magnetising coil is split into two to preserve symmetry. (Relocating the magnetising coil at a different position around the magnetic circuit is of course in order, just as a battery can be placed anywhere in an electric circuit.) Next, (Figure 1.10, centre) the single magnetic path is split into two parallel paths of half the original cross-section, each of which carries half of the flux: and finally (Figure 1.10, right), the flux paths and pole-faces are curved to match the rotor. The coil now has several layers in order to fit the available space, but as discussed earlier this has no adverse effect on the MMF. The air-gap is still small, so the flux crosses radially to the rotor.

TORQUE PRODUCTION

Having designed the magnetic circuit to give a high flux density under the poles, we must obtain maximum benefit from it. We therefore need to arrange a set of conductors, fixed to the rotor, as shown in Figure 1.11, and to ensure that conductors under a N-pole (at the top of Figure 1.11) carry positive current (into the paper), while those under the S-pole carry negative current. The tangential electromagnetic ('*BIl*') force (see equation 1.2) on all the positive conductors will be to the left, while the force on the negative ones will be to the right. A nett couple, or torque will therefore be exerted on the rotor, which will be caused to rotate.

(The observant reader spotting that some of the conductors appear to have no current in them will find the explanation later, in Chapter 3.)

Figure 1.11   *Current-carrying conductors on rotor, positioned to maximise torque. (The source of the magnetic flux lines (arrowed) is not shown.)*

At this point we should pause and address three questions that often crop up when these ideas are being developed. The first is to ask why we have made no reference to the magnetic field produced by the current-carrying conductors on the rotor. Surely they too will produce a magnetic field, which will presumably interfere with the original field in the air-gap, in which case perhaps the expression used to calculate the force on the conductor will no longer be valid.

The answer to this very perceptive question is that the field produced by the current-carrying conductors on the rotor certainly will modify the original field (i.e. the field that was present when there was no current in the rotor conductors.) But in the majority of motors, the force on the conductor can be calculated correctly from the product of the current and the 'original' field. This is very fortunate from the point of view of calculating the force, but also has a logical feel to it. For example in Figure 1.1, we would not expect any force on the current-carrying conductor if there was no externally applied field, even though the current in the conductor will produce its own field (upwards on one side of the conductor and downwards on the other). So it seems right that since we only obtain a force when there is an external field, all of the force must be due to that field alone.

The second question arises when we think about the action and reaction principle. When there is a torque on the rotor, there is presumably an equal and opposite torque on the stator; and therefore we might wonder if the mechanism of torque production could be pictured using the same ideas as we used for obtaining the rotor torque. The answer is yes; there is always an equal and opposite torque on the stator, which is why it is usually important to bolt a motor down securely. In some machines (e.g. the induction motor) it is easy to see that torque is produced on the stator by the interaction of the air-gap flux density and the stator currents, in exactly the same way that the flux density interacts with the rotor currents to produce torque on the rotor. In other motors, (e.g. the d.c. motor we have been looking at), there is no simple physical argument which can be advanced to derive the torque on the stator, but nevertheless it is equal and opposite to the torque on the rotor.

The final question relates to the similarity between the set-up shown in Figure 1.10 and the field patterns produced for example by the electromagnets used to lift car bodies in a scrap yard. From what we know of the large force of attraction that lifting magnets can produce, might not we expect a large radial force between the stator pole and the iron body of the rotor? And if there is, what is to prevent the rotor from being pulled across to the stator?

Again the affirmative answer is that there is indeed a radial force due to magnetic attraction, exactly as in a lifting magnet or relay, although the mechanism whereby the magnetic field exerts a pull as it enters iron or steel is entirely different from the 'BIl' force we have been looking at so far.

It turns out that the force of attraction per unit area of pole-face is proportional to the square of the radial flux density, and with typical air-gap flux densities of up to 1 T in motors, the force per unit area of rotor surface works out to be about $40\,\mathrm{N/cm}^2$. This indicates that the total radial force can be very large: for example the force of attraction on a small pole-face of only $5 \times 10\,\mathrm{cm}$ is 2000 N, or about 200 Kg. This force contributes nothing to the torque of the motor, and is merely an unwelcome by-product of the 'BIl' mechanism we employ to produce tangential force on the rotor conductors.

In most machines the radial magnetic force under each pole is actually a good deal bigger than the tangential electromagnetic force on the rotor conductors, and as the question implies, it tends to pull the rotor onto the pole. However, the majority of motors are constructed with an even number of poles equally spaced around the rotor, and the flux density in each pole is the same, so that − in theory at least − the resultant force on the complete rotor is zero. In practice, even a small eccentricity will cause the field to be stronger under the poles where the air-gap is smaller, and this will give rise to an unbalanced pull, resulting in noisy running and rapid bearing wear.

Magnitude of torque

Returning to our original discussion, the force on each conductor is given by equation 1.2, and it follows that the total tangential force F depends on the flux density produced by the field winding, the number of conductors on the rotor, the current in each, and the length of the rotor. The resultant torque or couple[1] (T) depends on the radius of the rotor (r), and is given by

$$T = Fr \qquad (1.8)$$

We will develop this further in Section 1.5, after we examine the remarkable benefits gained by putting the conductors into slots.

The beauty of slotting

If the conductors were mounted on the surface of the rotor iron, as in Figure 1.11, the air-gap would have to be at least equal to the wire diameter, and the conductors would have to be secured to the rotor in order to transmit their turning force to it. The earliest motors were made like this, with string or tape to bind the conductors to the rotor.

Unfortunately, a large air-gap results in an unwelcome high-reluctance in the magnetic circuit, and the field winding therefore needs many turns and a high current to produce the desired flux density in the air-gap. This means that the field winding becomes very bulky and consumes a lot of power. The early (Nineteenth-century) pioneers soon hit upon the idea of partially sinking the conductors on the rotor into grooves machined parallel to the shaft, the intention being to allow the air-gap to be reduced so that the exciting windings could be smaller. This worked extremely well as it also provided a more positive location for the rotor conductors, and thus allowed the force on them to be transmitted to the body of the rotor. Before long the conductors began to be recessed into ever deeper slots until finally (see Figure 1.12) they no longer stood proud of the rotor surface and the air-gap could be made as small as was consistent with the need for mechanical clearances between the rotor and the stator. The new 'slotted' machines worked very well, and their pragmatic makers were unconcerned by rumblings of discontent from sceptical theorists.

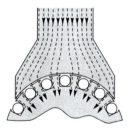

Figure 1.12   *Influence on flux paths when the rotor is slotted to accommodate conductors*

---

[1] Older readers will probably have learned the terms *Couple* and *Moment (of a force)* long before realising that they mean the same as torque.

The theorists of the time accepted that sinking conductors into slots allowed the air-gap to be made small, but argued that, as can be seen from Figure 1.12, almost all the flux would now pass down the attractive low-reluctance path through the teeth, leaving the conductors exposed to the very low leakage flux density in the slots. Surely, they argued, little or no '*BIl*' force would be developed on the conductors, since they would only be exposed to a very low flux density.

The sceptics were right in that the flux does indeed flow down the teeth; but there was no denying that motors with slotted rotors produced the same torque as those with the conductors in the air-gap, provided that the average flux densities at the rotor surface were the same. So what could explain this seemingly too good to be true situation?

The search for an explanation preoccupied some of the leading thinkers long after slotting became the norm, but finally it became possible to verify theoretically that the total force remains the same as it would have been if the conductors were actually in the flux, but almost all of the tangential force now acts on the rotor teeth, rather than on the conductors themselves.

This is remarkably good news. By putting the conductors in slots, we simultaneously enable the reluctance of the magnetic circuit to be reduced, and transfer the force from the conductors themselves to the sides of the iron teeth, which are robust and well able to transmit the resulting torque to the shaft. A further benefit is that the insulation around the conductors no longer has to transmit the tangential forces to the rotor, and its mechanical properties are thus less critical. Seldom can tentative experiments with one aim have yielded rewarding outcomes in almost every other relevant direction.

There are some snags, however. To maximise the torque, we will want as much current as possible in the rotor conductors. Naturally we will work the copper at the highest practicable current density (typically between 2 and 8 A/mm^2), but we will also want to maximise the cross-sectional area of the slots to accommodate as much copper as possible. This will push us in the direction of wide slots, and hence narrow teeth. But we recall that the flux has to pass radially down the teeth, so if we make the teeth too narrow, the iron in the teeth will saturate, and lead to a poor magnetic circuit. There is also the possibility of increasing the depth of the slots, but this cannot be taken too far or the centre region of the rotor iron – which has to carry the flux from one pole to another – will become so depleted that it too will saturate.

In the next section we look at what determines the torque that can be obtained from a rotor of a given size, and see how speed plays a key role in determining the power output.

SPECIFIC LOADINGS AND SPECIFIC OUTPUT

Specific loadings

A design compromise is inevitable in the crucial air-gap region, and designers constantly have to exercise their skills to achieve the best balance between the conflicting demands on space made by the flux (radial) and the current (axial).

As in most engineering design, guidelines emerge as to what can be achieved in relation to particular sizes and types of machines, and motor designers usually work in terms of two parameters, the specific magnetic loading, and the specific electric loading. These parameters have a direct bearing on the output of the motor, as we will now see.

The specific magnetic loading $(\overline{B})$ is the average of the magnitude of the radial flux density over the entire cylindrical surface of the rotor. Because of the slotting, the average flux density is always less than the flux density in the teeth, but in order to calculate the magnetic loading we picture the rotor as being smooth, and calculate the average flux

Plate 1.1 *Totally enclosed fan-ventilated (TEFV) cage induction motor. This particular example is rated at 200 W (0.27 h.p.) at 1450 rev/min, and is at the lower end of the power range for 3-phase versions. The case is of cast aluminium, with cooling air provided by the covered fan at the non-drive end. Note the provision for alternative mounting. (Photograph by courtesy of Brook Crompton)*

density by dividing the total radial flux from each 'pole' by the surface area under the pole.

The specific electric loading (usually denoted by the symbol $(\overline{A})$, the A standing for Amperes) is the axial current per metre of circumference on the rotor. In a slotted rotor, the axial current is concentrated in the conductors within each slot, but to calculate $\overline{A}$ we picture the total current to be spread uniformly over the circumference (in a manner similar to that shown in Figure 1.12, but with the individual conductors under each pole being represented by a uniformly distributed 'current sheet'). For example, if under a pole with a circumferential width of 10 cm we find that there are five slots, each carrying a current of 40 A, the electric loading is $\dfrac{5 \times 40}{0.1} = 2000\,\text{A/m}$.

Many factors influence the values which can be employed in motor design, but in essence the specific magnetic and electric loadings are limited by the properties of the materials (iron for the flux, and copper for the current), and by the cooling system employed to remove heat losses.

The specific magnetic loading does not vary greatly from one machine to another, because the saturation properties of most core steels are similar. On the other hand, quite wide variations occur in the specific electric loadings, depending on the type of cooling used.

Despite the low resistivity of the copper conductors, heat is generated by the flow of current, and the current must therefore be limited to a value such that the insulation is not damaged by an excessive temperature rise. The more effective the cooling system, the higher the electric loading can be. For example, if the motor is totally enclosed and has no internal fan, the current density in the copper has to be much lower than in a similar motor which has a fan to provide a continuous flow of ventilating air. Similarly, windings which are fully impregnated with varnish can be worked much harder than those which are surrounded by air, because the solid body of encapsulating varnish provides a much better thermal path along which the heat can flow to the stator body. Overall size also plays a part in determining permissible electric loading, with large motors generally having higher values than small ones.

In practice, the important point to be borne in mind is that unless an exotic cooling system is employed, most motors (induction, d.c. etc.) of a particular size have more or less the same specific loadings, regardless of type. As we will now see, this in turn means that motors of similar size have similar torque capabilities. This fact is not widely appreciated by users, but is always worth bearing in mind.

Torque and motor volume

In the light of the earlier discussion, we can obtain the total tangential force by first considering an area of the rotor surface of width w and length L. The axial current flowing in the width w is given by $I = w\overline{A}$, and on average all of this current is exposed to radial flux density $\overline{B}$, so the tangential force is given (from equation 1.2) by $\overline{B} \times w\overline{A} \times L$. The area of the surface is wL, so the force per unit area is $\overline{B} \times \overline{A}$. We see that the product of the two specific loadings expresses the average tangential stress over the rotor surface.

To obtain the total tangential force we must multiply by the area of the curved surface of the rotor, and to obtain the total torque we multiply the total force by the radius of the rotor. Hence for a rotor of diameter D and length L, the total torque is given by

$$T = (\overline{B}\,\overline{A}) \times (\pi DL) \times D/2 = \frac{\pi}{2}(\overline{B}\,\overline{A})D^2 L \qquad (1.9)$$

This equation is extremely important. The term $D^2 L$ is proportional to the rotor volume, so we see that for given values of the specific magnetic and electric loadings, the torque from any motor is proportional to the rotor volume. We are at liberty to choose a long thin rotor or a short fat one, but once the rotor volume and specific loadings are specified, we have effectively determined the torque.

It is worth stressing that we have not focused on any particular type of motor, but have approached the question of torque production from a completely general viewpoint. In essence our conclusions reflect the fact that all motors are made from iron and copper, and differ only in the way these materials are disposed. We should also acknowledge that in practice it is the overall volume of the motor which is important, rather than the volume of the rotor. But again we find that, regardless of the type of motor, there is a fairly close relationship between the overall volume and the rotor volume, for motors of similar torque. We can therefore make the bold but generally accurate statement that the overall volume of a motor is determined by the torque it has to produce. There are of course exceptions to this rule, but as a general guideline for motor selection, it is extremely useful.

Having seen that torque depends on rotor volume, we must now turn our attention to the question of power output.

Specific output power – importance of speed

Before deriving an expression for power, a brief digression may be helpful for those who are more familiar with linear rather than rotary systems.

In the SI system, the unit of work or energy is the Joule (J). One joule represents the work done by a force of 1 newton moving 1 metre in its own direction. Hence the work done (W) by a force F which moves a distance d is given by

$$W = F \times d$$

With F in newtons and d in metres, W is clearly in newton-metres (Nm), from which we see that a newton-metre is the same as a joule.

In rotary systems, it is more convenient to work in terms of torque and angular distance, rather than force and linear distance, but these are closely linked as we can see by considering what happens when a tangential force F is applied at a radius r from the centre of rotation. The torque is simply given by

$$T = F \times r.$$

Now suppose that the arm turns through an angle θ, so that the circumferential distance travelled by the force is $r \times \theta$. The work done by the force is then given by

$$W = F \times (r \times \theta) = (F \times r) \times \theta = T \times \theta \qquad (1.10)$$

We note that whereas in a linear system work is force times distance, in rotary terms work is torque times angle. The units of torque are newton-metres, and the angle is measured in radians (which is dimensionless), so the units of work done are Nm, or Joules, as expected. (The fact that torque and work (or energy) are measured in the same units does not seem self-evident to this author!)

To find the power, or the rate of working, we divide the work done by the time taken. In a linear system, and assuming that the velocity remains constant, power is therefore given by

$$P = \frac{W}{t} = \frac{F \times d}{t} = F \times v \qquad (1.11)$$

where v is the linear velocity. The angular equivalent of this is given by

$$P = \frac{W}{t} = \frac{T \times \theta}{t} = T \times \omega \qquad (1.12)$$

where ω is the (constant) angular velocity, in radians per second.

We can now express the power output in terms of the rotor dimensions and the specific loadings, using equation 1.9 which yields

$$P = T\omega = \frac{\pi}{2}(\overline{B}\,\overline{A})D^2L\omega \qquad (1.13)$$

Equation 1.13 emphasises the importance of speed in determining power output. For given specific and magnetic loadings, if we want a motor of a given power we can choose between a large (and therefore expensive) low-speed motor or a small (and cheaper) high-speed one. The latter choice is preferred for most applications, even if some form of speed reduction (using belts or gears) is needed, because the smaller motor is cheaper. Familiar examples include portable electric tools, where rotor speeds of 12 000 rev/min or more allow powers of hundreds of watts to be obtained, and electric traction: wherein both cases the high motor speed is geared down for the final drive. In these examples, where volume and weight are at a premium, a direct drive would be out of the question.

The significance of speed is underlined when we rearrange equation 1.13 to obtain an expression for the specific power output (power per unit rotor volume), Q, given by

$$Q = \overline{B}\,\overline{A}\frac{\omega}{2} \qquad (1.14)$$

To obtain the highest possible specific output for given values of the specific magnetic and electric loadings, we must clearly operate the motor at the highest practicable speed. The one obvious disadvantage of a small high-speed motor and gearbox is that the acoustic noise (both from the motor itself and the from the power transmission) is higher than it would be from a larger direct drive motor. When noise must be minimised (for example in ceiling fans), a direct drive motor is therefore preferred, despite its larger size.

ENERGY CONVERSION – MOTIONAL EMF

We now turn away from considerations of what determines the overall capability of a motor to what is almost the other extreme, by examining the behaviour of a primitive linear machine which, despite its obvious simplicity, encapsulates all the key electromagnetic energy conversion processes that take place in electric motors. We will see how the process of conversion of energy from electrical to mechanical form is elegantly represented in an 'equivalent circuit' from which all the key aspects of motor behaviour can be predicted. This circuit will provide answers to such questions as 'how does the motor automatically draw in more power when it is required to work', and 'what determines the steady speed and current'. Central to such questions is the matter of motional e.m.f., which is explored next.

We have already seen that force (and hence torque) is produced on current-carrying conductors exposed to a magnetic field. The force is given by equation 1.2, which shows that as long as the flux density and current remain constant, the force will be constant. In particular, we see that the force does not depend on whether the conductor is stationary or moving. On the other hand, relative movement is an essential requirement in the production of mechanical output power (as distinct from torque), and we have seen that output power is given by the equation $P = T\omega$. We will now see that the presence of relative motion between the conductors and the field always brings 'motional e.m.f.' into play; and we will see that this motional e.m.f. plays a key role in quantifying the energy conversion process.

Elementary motor – stationary conditions

The primitive linear machine is shown pictorially in Figure 1.13 and in diagrammatic form in Figure 1.14.

It consists of a conductor of active[2] length *l* which can move horizontally perpendicular to a magnetic flux density *B*.

It is assumed that the conductor has a resistance (*R*), that it carries a d.c. current (*I*), and that it moves with a velocity (*v*) in a direction perpendicular to the field and the current (see Figure 1.14). Attached to the conductor is a string which passes over a pulley and supports a weight: the tension in the string acting as a mechanical 'load' on the rod. Friction is assumed to be zero.

We need not worry about the many difficult practicalities of making such a machine, not least how we might manage to maintain electrical connections to a moving conductor. The important point is that although

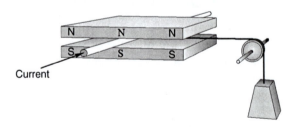

Figure 1.13   *Primitive linear d.c. motor*

---

[2] The active length is that part of the conductor exposed to the magnetic flux density – in most motors this corresponds to the length of the rotor and stator iron cores.

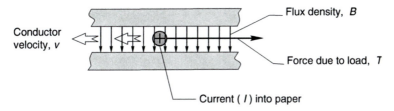

Figure 1.14   *Diagrammatic sketch of primitive linear d.c. motor*

this is a hypothetical set-up, it represents what happens in a real motor, and it allows us to gain a clear understanding of how real machines behave before we come to grips with much more complex structures.

We begin by considering the electrical input power with the conductor stationary (i.e. $v = 0$). For the purpose of this discussion we can suppose that the magnetic field (B) is provided by permanent magnets. Once the field has been established (when the magnet was first magnetised and placed in position), no further energy will be needed to sustain the field, which is just as well since it is obvious that an inert magnet is incapable of continuously supplying energy. It follows that when we obtain mechanical output from this primitive 'motor', none of the energy involved comes from the magnet. This is an extremely important point: the field system, whether provided from permanent magnets or 'exciting' windings, acts only as a catalyst in the energy conversion process, and contributes nothing to the mechanical output power.

When the conductor is held stationary the force produced on it (BIl) does no work, so there is no mechanical output power, and the only electrical input power required is that needed to drive the current through the conductor. The resistance of the conductor is R, the current through it is I, so the voltage which must be applied to the ends of the rod from an external source will be given by $V_1 = IR$, and the electrical input power will be $V_1 I$ or $I^2 R$. Under these conditions, all the electrical input power will appear as heat inside the conductor, and the power balance can be expressed by the equation

electrical input power ($V_1 I$) =

rate of production of heat in conductor ($I^2 R$).   (1.15)

Although no work is being done because there is no movement, the stationary condition can only be sustained if there is equilibrium of forces. The tension in the string (T) must equal the gravitational force on the mass (mg), and this in turn must be balanced by the electromagnetic force on the conductor (*BIl*). Hence under stationary conditions the current must be given by

$$T = mg = BIl, \text{ or } I = \frac{mg}{Bl} \tag{1.16}$$

This is our first indication of the link between the mechanical and electric worlds, because we see that in order to maintain the stationary condition, the current in the conductor is determined by the mass of the mechanical load. We will return to this link later.

Power relationships – conductor moving at constant speed

Now let us imagine the situation where the conductor is moving at a constant velocity (*v*) in the direction of the electromagnetic force that is propelling it. What current must there be in the conductor, and what voltage will have to be applied across its ends?

We start by recognising that constant velocity of the conductor means that the mass (*m*) is moving upwards at a constant speed, i.e. it is not accelerating. Hence from Newton's law, there must be no resultant force acting on the mass, so the tension in the string (*T*) must equal the weight (*mg*). Similarly, the conductor is not accelerating, so its nett force must also be zero. The string is exerting a braking force (*T*), so the electromagnetic force (*BIl*) must be equal to *T*. Combining these conditions yields

$$T = mg = BIl, \text{ or } I = \frac{mg}{Bl} \tag{1.17}$$

This is exactly the same equation that we obtained under stationary conditions, and it underlines the fact that the steady-state current is determined by the mechanical load. When we develop the equivalent circuit, we will have to get used to the idea that in the steady-state one of the electrical variables (the current) is determined by the mechanical load.

With the mass rising at a constant rate, mechanical work is being done because the potential energy of the mass is increasing. This work is coming from the moving conductor. The mechanical output power is equal to the rate of work, i.e. the force (*T* = *BIl*) times the velocity (*v*). The power lost as heat in the conductor is the same as it was when stationary, since it has the same resistance, and the same current. The electrical input power supplied to the conductor must continue to furnish this heat loss, but in addition it must now supply the mechanical output power. As yet we do not know what voltage will have to be applied, so we will denote it by V_2. The power-balance equation now becomes

electrical input power $(V_2 I)$ = rate of production of heat in conductor

+ mechanical output power

$$= I^2 R + (BIl)v$$

$$(1.18)$$

We note that the first term on the right hand side of equation 1.18 represent the heating effect, which is the same as when the conductor was stationary, while the second term represents the additional power that must be supplied to provide the mechanical output. Since the current is the same but the input power is now greater, the new voltage V_2 must be higher than V_1. By subtracting equation 1.15 from equation 1.18 we obtain

$$V_2 I - V_1 I = (BIl)v$$

and thus

$$V_2 - V_1 = Blv = E \qquad (1.19)$$

Equation 1.19 quantifies the extra voltage to be provided by the source to keep the current constant when the conductor is moving. This increase in source voltage is a reflection of the fact that whenever a conductor moves through a magnetic field, an e.m.f. (E) is induced in it.

We see from equation 1.19 that the e.m.f. is directly proportional to the flux density, to the velocity of the conductor relative to the flux, and to the active length of the conductor. The source voltage has to overcome this additional voltage in order to keep the same current flowing: if the source voltage is not increased, the current would fall as soon as the conductor begins to move because of the opposing effect of the induced e.m.f.

We have deduced that there must be an e.m.f. caused by the motion, and have derived an expression for it by using the principle of the conservation of energy, but the result we have obtained, i.e.

$$E = Blv \qquad (1.20)$$

is often introduced as the 'flux-cutting' form of Faraday's law, which states that when a conductor moves through a magnetic field an e.m.f.

given by equation 1.20 is induced in it. Because motion is an essential part of this mechanism, the e.m.f. induced is referred to as a 'motional e.m.f.'. The 'flux-cutting' terminology arises from attributing the origin of the e.m.f. to the cutting or slicing of the lines of flux by the passage of the conductor. This is a useful mental picture, though it must not be pushed too far: the flux lines are after all merely inventions which we find helpful in coming to grips with magnetic matters.

Before turning to the equivalent circuit of the primitive motor, two general points are worth noting. Firstly, whenever energy is being converted from electrical to mechanical form, as here, the induced e.m.f. always acts in opposition to the applied (source) voltage. This is reflected in the use of the term 'back e.m.f.' to describe motional e.m.f. in motors. Secondly, although we have discussed a particular situation in which the conductor carries current, it is certainly not necessary for any current to be flowing in order to produce an e.m.f.: all that is needed is relative motion between the conductor and the magnetic field.

EQUIVALENT CIRCUIT

We can represent the electrical relationships in the primitive machine in an equivalent circuit as shown in Figure 1.15.

The resistance of the conductor and the motional e.m.f. together represent in circuit terms what is happening in the conductor (though in reality the e.m.f. and the resistance are distributed, not lumped as separate items). The externally applied source that drives the current is represented by the voltage V on the left (the old-fashioned battery symbol being deliberately used to differentiate the applied voltage V from the induced e.m.f. E). We note that the induced motional e.m.f. is shown as opposing the applied voltage, which applies in the 'motoring' condition we have been discussing. Applying Kirchoff's law we obtain the voltage equation as

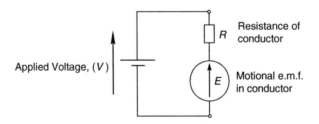

Figure 1.15   *Equivalent circuit of primitive d.c. motor*

$$V = E + IR \text{ or } I = \frac{V - E}{R} \tag{1.21}$$

Multiplying equation 1.21 by the current gives the power equation as

electrical input power $(VI) =$ mechanical output power (EI)

$\qquad\qquad + $ copper loss $(I^2 R)$. $\tag{1.22}$

(Note that the term 'copper loss' used in equation 1.22 refers to the heat generated by the current in the windings: all such losses in electric motors are referred to in this way, even when the conductors are made of aluminium or bronze!)

It is worth seeing what can be learned from these equations because, as noted earlier, this simple elementary 'motor' encapsulates all the essential features of real motors. Lessons which emerge at this stage will be invaluable later, when we look at the way actual motors behave.

If the e.m.f. E is less than the applied voltage V, the current will be positive, and electrical power will flow from the source, resulting in motoring action. On the other hand if E is larger than V, the current will flow back to the source, and the conductor will be acting as a generator. This inherent ability to switch from motoring to generating without any interference by the user is an extremely desirable property of electromagnetic energy converters. Our primitive set-up is simply a machine which is equally at home acting as motor or generator.

A further important point to note is that the mechanical power (the first term on the right hand side of equation 1.22) is simply the motional e.m.f. multiplied by the current. This result is again universally applicable, and easily remembered. We may sometimes have to be a bit careful if the e.m.f. and the current are not simple d.c. quantities, but the basic idea will always hold good.

Finally, it is obvious that in a motor we want as much as possible of the electrical input power to be converted to mechanical output power, and as little as possible to be converted to heat in the conductor. Since the output power is EI, and the heat loss is $I^2 R$ we see that ideally we want EI to be much greater than $I^2 R$, or in other words E should be much greater than IR. In the equivalent circuit (Figure 1.15) this means that the majority of the applied voltage V is accounted for by the motional e.m.f. (E), and only a little of the applied voltage is used in overcoming the resistance.

Motoring condition

Motoring implies that the conductor is moving in the same direction as the electromagnetic force (BIl), and at a speed such that the back e.m.f. (Blv) is less than the applied voltage V. In the discussion so far, we have assumed that the load is constant, so that under steady-state conditions the current is the same at all speeds, the voltage being increased with speed to take account of the motional e.m.f. This was a helpful approach to take in order to derive the steady-state power relationships, but is seldom typical of normal operation. We therefore turn to how the moving conductor will behave under conditions where the applied voltage V is constant, since this corresponds more closely with the normal operations of a real motor.

In the next section, matters are inevitably more complicated than we have seen so far because we include consideration of how the motor increases from one speed to another, as well as what happens under steady-state conditions. As in all areas of dynamics, study of the transient behaviour of our primitive linear motor brings into play additional parameters such as the mass of the conductor (equivalent to the inertia of a real rotary motor) which are absent from steady-state considerations.

Behaviour with no mechanical load

In this section we assume that the hanging weight has been removed, and that the only force on the conductor is its own electromagnetically generated one. Our primary interest will be in what determines the steady speed of the primitive motor, but we must begin by considering what happens when we first apply the voltage.

With the conductor stationary when the voltage V is applied, the current will immediately rise to a value of V/R, since there is no motional e.m.f. and the only thing which limits the current is the resistance. (Strictly we should allow for the effect of inductance in delaying the rise of current, but we choose to ignore it here in the interests of simplicity.) The resistance will be small, so the current will be large, and a high force will therefore be developed on the conductor. The conductor will therefore accelerate at a rate equal to the force on it divided by its mass.

As it picks up speed, the motional e.m.f. (equation 1.20) will grow in proportion to the speed. Since the motional e.m.f. opposes the applied voltage, the current will fall (equation 1.21), so the force and hence the acceleration will reduce, though the speed will continue to rise. The

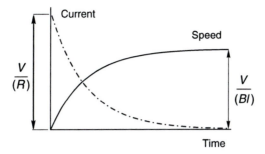

Figure 1.16 *Dynamic (run-up) behaviour of primitive d.c. motor with no mechanical load*

speed will increase as long as there is an accelerating force, i.e. as long as there is current in the conductor. We can see from equation 1.21 that the current will finally fall to zero when the speed reaches a level at which the motional e.m.f. is equal to the applied voltage. The speed and current therefore vary as shown in Figure 1.16: both curves have the exponential shape which characterises the response of systems governed by a first-order differential equation. The fact that the steady-state current is zero is in line with our earlier observation that the mechanical load (in this case zero) determines the steady-state current.

We note that in this idealised situation (in which there is no load applied, and no friction forces), the conductor will continue to travel at a constant speed, because with no nett force acting on it there is no acceleration. Of course, no mechanical power is being produced, since we have assumed that there is no opposing force on the conductor, and there is no input power because the current is zero. This hypothetical situation nevertheless corresponds closely to the so-called 'no-load' condition in a motor, the only difference being that a motor will have some friction (and therefore it will draw a small current), whereas we have assumed no friction in order to simplify the discussion.

Although no power is required to keep the frictionless and unloaded conductor moving once it is up to speed, we should note that during the whole of the acceleration phase the applied voltage was constant and the input current fell progressively, so that the input power was large at first but tapered-off as the speed increased. During this run-up time energy was continually being supplied from the source: some of this energy is wasted as heat in the conductor, but much of it is stored as kinetic energy, and as we will see later, can be recovered.

An elegant self-regulating mechanism is evidently at work here. When the conductor is stationary, it has a high force acting on it, but this force tapers-off as the speed rises to its target value, which corresponds to the back e.m.f. being equal to the applied voltage. Looking back at the expression for motional e.m.f. (equation 1.18), we can obtain an expression for the no-load speed (v_o) by equating the applied voltage and the back e.m.f., which gives

$$E = V = Blv_0, \text{ i.e. } v_0 = \frac{V}{Bl} \qquad (1.23)$$

Equation 1.23 shows that the steady-state no-load speed is directly proportional to the applied voltage, which indicates that speed control can be achieved by means of the applied voltage. We will see later that one of the main reasons why d.c. motors held sway in the speed-control arena for so long is that their speed could be controlled simply by controlling the applied voltage.

Rather more surprisingly, equation 1.23 reveals that the speed is inversely proportional to the magnetic flux density, which means that the weaker the field, the higher the steady-state speed. This result can cause raised eyebrows, and with good reason. Surely, it is argued, since the force is produced by the action of the field, the conductor will not go as fast if the field was weaker. This view is wrong, but understandable. The flaw in the argument is to equate force with speed. When the voltage is first applied, the force on the conductor certainly will be less if the field is weaker, and the initial acceleration will be lower. But in both cases the acceleration will continue until the current has fallen to zero, and this will only happen when the induced e.m.f. has risen to equal the applied voltage. With a weaker field, the speed needed to generate this e.m.f. will be higher than with a strong field: there is 'less flux', so what there is has to be cut at a higher speed to generate a given e.m.f. The matter is summarised in Figure 1.17, which shows how the speed will rise for a given applied voltage, for 'full' and 'half' fields respectively. Note that the initial acceleration (i.e. the slope of the speed-time curve) in the half-

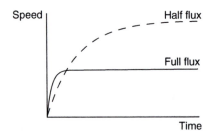

Figure 1.17   *Effect of flux density on the acceleration and steady running speed of primitive d.c. motor with no mechanical load*

flux case is half that of the full-flux case, but the final steady speed is twice as high. In real d.c. motors, the technique of reducing the flux density in order to increase speed is known as 'field weakening'.

Behaviour with a mechanical load

Suppose that, with the primitive linear motor up to its no-load speed we suddenly attach the string carrying the weight, so that we now have a steady force ($T = mg$) opposing the motion of the conductor. At this stage there is no current in the conductor and thus the only force on it will be T. The conductor will therefore begin to decelerate. But as soon as the speed falls, the back e.m.f. will become less than V, and current will begin to flow into the conductor, producing an electromagnetic driving force. The more the speed drops, the bigger the current, and hence the larger the force developed by the conductor. When the force developed by the conductor becomes equal to the load (T), the deceleration will cease, and a new equilibrium condition will be reached. The speed will be lower than at no-load, and the conductor will now be producing continuous mechanical output power, i.e. acting as a motor.

Since the electromagnetic force on the conductor is directly proportional to the current, it follows that the steady-state current is directly proportional to the load which is applied, as we saw earlier. If we were to explore the transient behaviour mathematically, we would find that the drop in speed followed the same first-order exponential response that we saw in the run-up period. Once again the self-regulating property is evident, in that when load is applied the speed drops just enough to allow sufficient current to flow to produce the force required to balance the load. We could hardly wish for anything better in terms of performance, yet the conductor does it without any external intervention on our part. Readers who are familiar with closed-loop control systems will probably recognise that the reason for this excellent performance is that the primitive motor possesses inherent negative speed feedback via the motional e.m.f. This matter is explored more fully in the Appendix.

Returning to equation 1.21, we note that the current depends directly on the difference between V and E, and inversely on the resistance. Hence for a given resistance, the larger the load (and hence the steady-state current), the greater the required difference between V and E, and hence the lower the steady running speed, as shown in Figure 1.18.

We can also see from equation 1.21 that the higher the resistance of the conductor, the more it slows down when a given load is applied. Conversely, the lower the resistance, the more the conductor is able to hold its no-load speed in the face of applied load. This is also illustrated

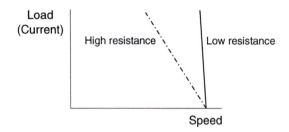

Figure 1.18   *Influence of resistance on the ability of the motor to maintain speed when load is applied*

in Figure 1.18. We can deduce that the only way we could obtain an absolutely constant speed with this type of motor is for the resistance of the conductor to be zero, which is of course not possible. Nevertheless, real d.c. motors generally have resistances which are small, and their speed does not fall much when load is applied – a characteristic which for most applications is highly desirable.

We complete our exploration of the performance when loaded by asking how the flux density influences behaviour. Recalling that the electromagnetic force is proportional to the flux density as well as the current, we can deduce that to develop a given force, the current required will be higher with a weak flux than with a strong one. Hence in view of the fact that there will always be an upper limit to the current which the conductor can safely carry, the maximum force which can be developed will vary in direct proportion to the flux density, with a weak flux leading to a low maximum force and vice-versa. This underlines the importance of operating with maximum flux density whenever possible.

We can also see another disadvantage of having a low flux density by noting that to achieve a given force, the drop in speed will be disproportionately high when we go to a lower flux density. We can see this by imagining that we want a particular force, and considering how we achieve it firstly with full flux, and secondly with half flux. With full flux, there will be a certain drop in speed which causes the motional e.m.f. to fall enough to admit the required current. But with half the flux, for example, twice as much current will be needed to develop the same force. Hence the motional e.m.f. must fall by twice as much as it did with full flux. However, since the flux density is now only half, the drop in speed will have to be four times as great as it was with full flux. The half-flux 'motor' therefore has a load characteristic with a load/speed gradient four times more droopy than the full-flux one. This is shown in Figure 1.19; the applied voltages having been adjusted so that in both

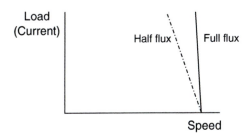

Figure 1.19   *Influence of flux on the drop in steady running speed when load is applied*

cases the no-load speed is the same. The half-flux motor is clearly inferior in terms of its ability to hold the set speed when the load is applied.

We may be tempted to think that the higher speed which we can obtain by reducing the flux somehow makes for better performance, but we can now see that this is not so. By halving the flux, for example, the no-load speed for a given voltage is doubled, but when the load is raised until rated current is flowing in the conductor, the force developed is only half, so the mechanical power is the same. We are in effect trading speed against force, and there is no suggestion of getting something for nothing.

Relative magnitudes of V and E, and efficiency

Invariably we want machines which have high efficiency. From equation 1.20, we see that to achieve high efficiency, the copper loss (I^2R) must be small compared with the mechanical power (EI), which means that the resistive volt-drop in the conductor (IR) must be small compared with either the induced e.m.f. (E) or the applied voltage (V). In other words we want most of the applied voltage to be accounted for by the 'useful' motional e.m.f., rather than the wasteful volt drop in the wire. Since the motional e.m.f. is proportional to speed, and the resistive volt drop depends on the conductor resistance, we see that a good energy converter requires the conductor resistance to be as low as possible, and the speed to be as high as possible.

To provide a feel for the sort of numbers likely to be encountered, we can consider a conductor with resistance of 0.5 Ω, carrying a current of 4 A, and moving at a speed such that the motional e.m.f. is 8 V. From equation 1.19, the supply voltage is given by

$$V = E + IR = 8 + (4 \times 0.5) = 10 \, \text{volts}$$

Hence the electrical input power (VI) is 40 watts, the mechanical output power (EI) is 32 watts, and the copper loss (I^2R) is 8 watts, giving an efficiency of 80%.

If the supply voltage was doubled (i.e. $V = 20$ volts), however, and the resisting force is assumed to remain the same (so that the steady-state current is still 4 A), the motional e.m.f. is given by equation 1.19 as

$$E = 20 - (4 \times 0.5) = 18 \text{ volts}$$

which shows that the speed will have rather more than doubled, as expected. The electrical input power is now 80 watts, the mechanical output power is 72 watts, and the copper loss is still 8 watts. The efficiency has now risen to 90%, underlining the fact that the energy conversion process gets better at higher speeds.

The ideal situation is clearly one where the term IR in equation 1.19 is negligible, so that the back e.m.f. is equal to the applied voltage. We would then have an ideal machine with an efficiency of 100%, in which the steady-state speed would be directly proportional to the applied voltage and independent of the load.

In practice the extent to which we can approach the ideal situation discussed above depends on the size of the machine. Tiny motors, such as those used in wrist-watches, are awful, in that most of the applied voltage is used up in overcoming the resistance of the conductors, and the motional e.m.f. is very small: these motors are much better at producing heat than they are at producing mechanical output power! Small machines, such as those used in hand tools, are a good deal better with the motional e.m.f. accounting for perhaps 70–80% of the applied voltage. Industrial machines are very much better: the largest ones (of many hundreds of kW) use only one or two percent of the applied voltage in overcoming resistance, and therefore have very high efficiencies.

Analysis of primitive motor – conclusions

All of the lessons learned from looking at the primitive motor will find direct parallels in almost all of the motors we look at in the rest of this book, so it is worth reminding ourselves of the key points.

Firstly, we will make frequent reference to the formula for the force (F) on a conductor in a magnetic field, i.e.

$$F = BIl \tag{1.24}$$

and to the formula for the motional induced e.m.f, (E) i.e.

$$E = Blv \qquad (1.25)$$

where B is the magnetic flux density, I is the current, l is the length of conductor and v is the velocity perpendicular to the field. These equations form the theoretical underpinning on which our understanding of motors will rest.

Secondly, we have seen that the *speed* at which the primitive motor runs unloaded is determined by the *applied voltage*, while the *current* that the motor draws is determined by the *mechanical load*. Exactly the same results will hold when we examine real d.c. motors, and very similar relationships will also emerge when we look at the induction motor.

GENERAL PROPERTIES OF ELECTRIC MOTORS

All electric motors are governed by the laws of electromagnetism, and are subject to essentially the same constraints imposed by the materials (copper and iron) from which they are made. We should therefore not be surprised to find that at the fundamental level all motors – regardless of type – have a great deal in common.

These common properties, most of which have been touched on in this chapter, are not usually given prominence. Books tend to concentrate on the differences between types of motors, and manufacturers are usually interested in promoting the virtues of their particular motor at the expense of the competition. This divisive emphasis causes the underlying unity to be obscured, leaving users with little opportunity to absorb the sort of knowledge which will equip them to make informed judgements.

The most useful ideas worth bearing in mind are therefore given below, with brief notes accompanying each. Experience indicates that users who have these basic ideas firmly in mind will find themselves able to understand why one motor seems better than another, and will feel much more confident when faced with the difficult task of weighing the pros and cons of competing types.

Operating temperature and cooling

The cooling arrangement is the single most important factor in determining the output from any given motor.

Plate 1.2   *Steel frame cage induction motor, 150 kW (201 h.p.), 1485 rev/min. The active parts are totally enclosed, and cooling is provided by means of an internal fan which circulates cooling air round the interior of the motor through the hollow ribs, and an external fan which blows air over the case. (Photograph by courtesy of Brook Crompton.)*

Any motor will give out more power if its electric circuit is worked harder (i.e. if the current is allowed to increase). The limiting factor is normally the allowable temperature rise of the windings, which depends on the class of insulation.

For class F insulation (the most widely used) the permissible temperature rise is 100 K, whereas for class H it is 125 K. Thus if the cooling remains the same, more output can be obtained simply by using the higher-grade insulation. Alternatively, with a given insulation the output can be increased if the cooling system is improved. A through-ventilated motor, for example, might give perhaps twice the output power of an otherwise identical but totally enclosed machine.

Torque per unit volume

For motors with similar cooling systems, the rated torque is approximately proportional to the rotor volume, which in turn is roughly proportional to the overall motor volume.

This stems from the fact that for a given cooling arrangement, the specific and magnetic loadings of machines of different types will be more or less the same. The torque per unit length therefore depends first and foremost on the square of the diameter, so motors of roughly the same diameter and length can be expected to produce roughly the same torque.

Power per unit volume – importance of speed

Output power per unit volume is directly proportional to speed.

Low-speed motors are unattractive because they are large, and therefore expensive. It is usually much better to use a high-speed motor with a mechanical speed reduction. For example, a direct drive motor for a portable electric screwdriver would be an absurd proposition.

Size effects – specific torque and efficiency

Large motors have a higher specific torque (torque per unit volume) and are more efficient than small ones.

In large motors the specific electric loading is normally much higher than in small ones, and the specific magnetic loading is somewhat higher. These two factors combine to give the higher specific torque.

Very small motors are inherently very inefficient (e.g. 1% in a wristwatch), whereas motors of over say 100 kW have efficiencies above 95%. The reasons for this scale effect are complex, but stem from the fact that the resistance volt-drop term can be made relatively small in large electromagnetic devices, whereas in small ones the resistance becomes the dominant term.

Efficiency and speed

The efficiency of a motor improves with speed.

For a given torque, power output rises in direct proportion to speed, while electrical losses are – broadly speaking – constant. Under these conditions, efficiency rises with speed.

Rated voltage

A motor can be provided to suit any voltage.

Within limits it is always possible to rewind a motor for a different voltage without affecting its performance. A 200 V, 10 A motor could be rewound for 100 V, 20 A simply by using half as many turns per coil of wire having twice the cross-sectional area. The total amounts of active material, and hence the performance, would be the same.

Short-term overload

Most motors can be overloaded for short periods without damage.

The continuous electric loading (i.e. the current) cannot be exceeded without damaging the insulation, but if the motor has been running with reduced current for some time, it is permissible for the current (and hence the torque) to be much greater than normal for a short period of time. The principal factors which influence the magnitude and duration of the permissible overload are the thermal time-constant (which governs the rate of rise of temperature) and the previous pattern of operation. Thermal time constants range from a few seconds for small motors to many minutes or even hours for large ones. Operating patterns are obviously very variable, so rather than rely on a particular pattern being followed, it is usual for motors to be provided with over-temperature protective devices (e.g. thermistors) which trigger an alarm and/or trip the supply if the safe temperature is exceeded.

REVIEW QUESTIONS

1) The current in a coil with 250 turns is 8 A. Calculate the MMF.

2) The coil in (1) is used in a magnetic circuit with a uniform cross-section made of good-quality magnetic steel and with a 2 mm air-gap. Estimate the flux density in the air-gap, and in the iron. ($\mu_0 = 4\pi \times 10^{-7}$ H/m.)
   How would the answers change if the cross-sectional area of the magnetic circuit was doubled, with all other parameters remaining the same?

3) Calculate the flux in a magnetic circuit that has a cross-sectional area of 18 cm^2 when the flux density is 1.4 T.

4) A magnetic circuit of uniform cross-sectional area has two air-gaps of 0.5 and 1 mm respectively in series. The exciting winding provides an MMF of 1200 Amp-turns. Estimate the MMF across each of the air-gaps, and the flux density.

5) The field winding in a motor consumes 25 W when it produces a flux density of 0.4 T at the pole-face. Estimate the power when the pole-face flux density is 0.8 T.

6) The rotor of a d.c. motor had an original diameter of 30 cm and an air-gap under the poles of 2 mm. During refurbishment the rotor diameter was accidentally reground and was then undersized by 0.5 mm. Estimate by how much the field MMF would have to be

increased to restore normal performance. How might the extra MMF be provided?

7) Estimate the minimum cross-sectional area of a magnetic circuit that has to carry a flux of 5 mWb. (Don't worry if you think that this question cannot be answered without more information – you are right.)

8) Calculate the electromagnetic force on:
    a) a single conductor of length 25 cm, carrying a current of 4 A, exposed to a magnetic flux density of 0.8 T perpendicular to its length.
    b) a coil-side consisting of twenty wires of length 25 cm, each carrying a current of 2 A, exposed to a magnetic flux density of 0.8 T perpendicular to its length.

9) Estimate the torque produced by one of the early machines illustrated in Figure 1.11 given the following:- Mean air-gap flux density under pole-face = 0.4 T; pole-arc as a percentage of total circumference = 75%; active length of rotor = 50 cm; rotor diameter = 30 cm; inner diameter of stator pole = 32 cm; total number of rotor conductors = 120; current in each rotor conductor = 50 A.

10) Motor designers often refer to the 'average flux density over the rotor surface'. What do they really mean? If we want to be really pedantic, what would the average flux density over the (whole) rotor surface be?

11) If the field coils of a motor are rewound to operate from 220 V instead of 110 V, how will the new winding compare with the old in terms of number of turns, wire diameter, power consumption and physical size?

12) A catalogue of DIY power tools indicates that most of them are available in 240 V or 110 V versions. What differences would you expect in terms of appearance, size, weight and performance?

13) Given that the field windings of a motor do not contribute to the mechanical output power, why do they consume power continuously?

14) For a given power, which will be larger, a motor or a generator?

15) Explain briefly why low-speed electrical drives often employ a high-speed motor and some form of mechanical speed reduction, rather than a direct-drive motor.

2

POWER ELECTRONIC CONVERTERS FOR MOTOR DRIVES

INTRODUCTION

In this chapter we look at examples of the power converter circuits which are used with motor drives, providing either d.c. or a.c. outputs, and working from either a d.c. (battery) supply, or from the conventional a.c. mains. The treatment is not intended to be exhaustive, but should serve to highlight the most important aspects which are common to all types of drive converters.

Although there are many different types of converters, all except very low-power ones are based on some form of electronic switching. The need to adopt a switching strategy is emphasised in the first example, where the consequences are explored in some depth. We will see that switching is essential in order to achieve high-efficiency power conversion, but that the resulting waveforms are inevitably less than ideal from the point of view of the motor.

The examples have been chosen to illustrate typical practice, so for each converter the most commonly used switching devices (e.g. thyristor, transistor) are shown. In many cases, several different switching devices may be suitable (see later), so we should not identify a particular circuit as being the exclusive preserve of a particular device.

Before discussing particular circuits, it will be useful to take an overall look at a typical drive system, so that the role of the converter can be seen in its proper context.

General arrangement of drives

A complete drive system is shown in block diagram form in Figure 2.1.

The job of the converter is to draw electrical energy from the mains (at constant voltage and frequency) and supply electrical energy to the motor at whatever voltage and frequency necessary to achieve the desired mechanical output.

Except in the simplest converter (such as a simple diode rectifier), there are usually two distinct parts to the converter. The first is the power stage, through which the energy flows to the motor, and the second is the control section, which regulates the power flow. Control signals, in the form of low-power analogue or digital voltages, tell the converter what it is supposed to be doing, while other low-power feedback signals are used to measure what is actually happening. By comparing the demand and feedback signals, and adjusting the output accordingly, the target output is maintained. The simple arrangement shown in Figure 2.1 has only one input representing the desired speed, and one feedback signal indicating actual speed, but most drives will have extra feedback signals as we will see later. Almost all drives employ closed-loop (feedback) control, so readers who are unfamiliar with the basic principles might find it helpful to read the Appendix at this stage.

A characteristic of power electronic converters which is shared with most electrical systems is that they have very little capacity for storing energy. This means that any sudden change in the power supplied by the converter to the motor must be reflected in a sudden increase in the power drawn from the supply. In most cases this is not a serious problem, but it does have two drawbacks. Firstly, a sudden increase in the current drawn from the supply will cause a momentary drop in the supply voltage, because of the effect of the supply impedance. These voltage 'spikes' will appear as unwelcome distortion to other users on the same supply. And secondly, there may be an enforced delay before the supply can furnish extra power. With a single-phase mains supply, for example, there can be no sudden increase in the power supply from the mains at the instant where the mains voltage is zero, because in-

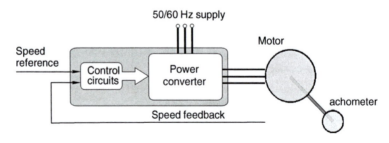

Figure 2.1   *General arrangement of speed-controlled drive*

stantaneous power is necessarily zero at this point in the cycle because the voltage is itself zero.

It would be better if a significant amount of energy could be stored within the converter itself: short-term energy demands could then be met instantly, thereby reducing rapid fluctuations in the power drawn from the mains. But unfortunately this is just not economic: most converters do have a small store of energy in their smoothing inductors and capacitors, but the amount is not sufficient to buffer the supply sufficiently to shield it from anything more than very short-term fluctuations.

VOLTAGE CONTROL – D.C. OUTPUT FROM D.C. SUPPLY

For the sake of simplicity we will begin by exploring the problem of controlling the voltage across a $2\,\Omega$ resistive load, fed from a 12 V constant-voltage source such as a battery. Three different methods are shown in Figure 2.2, in which the circle on the left represents an ideal 12 V d.c. source, the tip of the arrow indicating the positive terminal. Although this setup is not quite the same as if the load was a d.c. motor, the conclusions which we draw are more or less the same.

Method (a) uses a variable resistor (R) to absorb whatever fraction of the battery voltage is not required at the load. It provides smooth (albeit manual) control over the full range from 0 to 12 V, but the snag is that power is wasted in the control resistor. For example, if the load voltage is to be reduced to 6 V, the resistor (R) must be set to $2\,\Omega$, so that half of the battery voltage is dropped across R. The current will be 3 A, the load power will be 18 W, and the power dissipated in R will also be 18 W. In terms of overall power conversion efficiency (i.e. useful power delivered to the load divided by total power from the source) the efficiency is a very poor 50%. If R is increased further, the efficiency falls still lower, approaching zero as the load voltage tends to zero. This method of control is therefore unacceptable for motor control, except perhaps in low-power applications such as model racing cars.

Method (b) is much the same as (a) except that a transistor is used instead of a manually operated variable resistor. The transistor in Figure 2.2(b) is connected with its collector and emitter terminals in series with the voltage source and the load resistor. The transistor is a variable resistor, of course, but a rather special one in which the effective collector–emitter resistance can be controlled over a wide range by means of the base–emitter current. The base–emitter current is usually very small, so it can be varied by means of a low-power electronic circuit (not shown in Figure 2.2) whose losses are negligible in comparison with the power in the main (collector–emitter) circuit.

Method (b) shares the drawback of method (a) above, i.e. the efficiency is very low. But even more seriously as far as power control is concerned, the 'wasted' power (up to a maximum of 18 W in this case) is burned off inside the transistor, which therefore has to be large, well-cooled, and hence expensive. Transistors are never operated in this 'linear' way when used in power electronics, but are widely used as switches, as discussed below.

Switching control

The basic ideas underlying a switching power regulator are shown by the arrangement in Figure 2.2(c), which uses a mechanical switch. By operating the switch repetitively and varying the ratio of on/off time, the average load voltage can be varied continuously between 0 V (switch off all the time) through 6 V (switch on and off for half of each cycle) to 12 V (switch on all the time).

The circuit shown in Figure 2.2(c) is often referred to as a 'chopper', because the battery supply is 'chopped' on and off. When a constant repetition frequency is used, and the width of the on pulse is varied to control the mean output voltage (see the waveform in Figure 2.2), the arrangement is known as 'pulse width modulation' (PWM). An

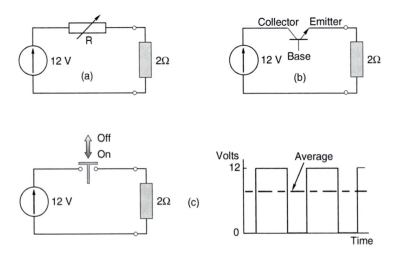

Figure 2.2   *Methods of obtaining a variable-voltage output from a constant-voltage source*

alternative approach is to keep the width of the on pulses constant, but vary their repetition rate: this is known as pulse frequency modulation (PFM).

The main advantage of the chopper circuit is that no power is wasted, and the efficiency is thus 100%. When the switch is on, current flows through it, but the voltage across it is zero because its resistance is negligible. The power dissipated in the switch is therefore zero. Likewise, when the switch is 'off' the current through it is zero, so although the voltage across the switch is 12 V, the power dissipated in it is again zero.

The obvious disadvantage is that by no stretch of imagination could the load voltage be seen as 'good' d.c.: instead it consists of a mean or 'd.c.' level, with a superimposed 'a.c.' component. Bearing in mind that we really want the load to be a d.c. motor, rather than a resistor, we are bound to ask whether the pulsating voltage will be acceptable. Fortunately, the answer is yes, provided that the chopping frequency is high enough. We will see later that the inductance of the motor causes the current to be much smoother than the voltage, which means that the motor torque fluctuates much less than we might suppose, and the mechanical inertia of the motor filters the torque ripples so that the speed remains almost constant, at a value governed by the mean (or d.c.) level of the chopped waveform.

Obviously a mechanical switch would be unsuitable, and could not be expected to last long when pulsed at high frequency. So, an electronic power switch is used instead. The first of many devices to be used for switching was the bipolar junction transistor (BJT), so we will begin by examining how such devices are employed in chopper circuits. If we choose a different device, such as a metal oxide semiconductor field effect transistor (MOSFET) or an insulated gate bipolar transistor (IGBT), the detailed arrangements for turning the device on and off will be different (see Section 2.5), but the main conclusions we draw will be much the same.

Transistor chopper

As noted earlier, a transistor is effectively a controllable resistor, i.e. the resistance between collector and emitter depends on the current in the base–emitter junction. In order to mimic the operation of a mechanical switch, the transistor would have to be able to provide infinite resistance (corresponding to an open switch) or zero resistance (corresponding to a closed switch). Neither of these ideal states can be reached with a real transistor, but both can be closely approximated.

The transistor will be 'off' when the base–emitter current is zero. Viewed from the main (collector–emitter) circuit, its resistance will be very high, as shown by the region Oa in Figure 2.3.

Under this 'cut-off' condition, only a tiny current (I_c) can flow from the collector to the emitter, regardless of the voltage (V_{ce}) between the collector and emitter. The power dissipated in the device will therefore be negligible, giving an excellent approximation to an open switch.

To turn the transistor fully 'on', a base–emitter current must be provided. The base current required will depend on the prospective collector–emitter current, i.e. the current in the load. The aim is to keep the transistor 'saturated' so that it has a very low resistance, corresponding to the region Ob in Figure 2.3. In the example shown in Figure 2.2, if the resistance of the transistor is very low, the current in the circuit will be almost 6 A, so we must make sure that the base–emitter current is sufficiently large to ensure that the transistor remains in the saturated condition when $I_c = 6A$.

Typically in a bipolar transistor (BJT) the base current will need to be around 5–10% of the collector current to keep the transistor in the saturation region: in the example (Figure 2.2), with the full load current of 6 A flowing, the base current might be 400 mA, the collector–emitter voltage might be say 0.33 V, giving an on-state dissipation of 2 W in the transistor when the load power is 72 W. The power conversion efficiency is not 100%, as it would be with an ideal switch, but it is acceptable.

We should note that the on-state base–emitter voltage is very low, which, coupled with the small base current, means that the power

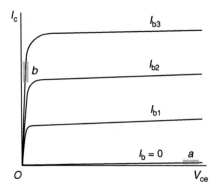

Figure 2.3   *Transistor characteristics showing high-resistance (cut-off) region Oa and low-resistance (saturation) region Ob. Typical 'off' and 'on' operating states are shown by the shaded areas a and b respectively*

required to drive the transistor is very much less than the power being switched in the collector–emitter circuit. Nevertheless, to switch the transistor in the regular pattern as shown in Figure 2.2, we obviously need a base current waveform which goes on and off periodically, and we might wonder how we obtain this 'control' signal. Normally, the base-drive signal will originate from a low-power oscillator (constructed from logic gates, or on a single chip), or perhaps from a microprocessor. Depending on the base circuit power requirements of the main switching transistor, it may be possible to feed it directly from the oscillator; if necessary additional transistors are interposed between the main device and the signal source to provide the required power amplification.

Just as we have to select mechanical switches with regard to their duty, we must be careful to use the right power transistor for the job in hand. In particular, we need to ensure that when the transistor is 'on', we don't exceed the safe current, or else the active semiconductor region of the device will be destroyed by overheating. And we must make sure that the transistor is able to withstand whatever voltage appears across the collector–emitter junction when it is 'off'. If the safe voltage is exceeded, the transistor will breakdown, and be permanently 'on'.

A suitable heatsink will be a necessity. We have already seen that some heat is generated when the transistor is on, and at low switching rates this is the main source of unwanted heat. But at high switching rates, 'switching loss' can also be very important.

Switching loss is the heat generated in the finite time it takes for the transistor to go from on to off or vice versa. The base-drive circuitry will be arranged so that the switching takes place as fast as possible, but in practice it will seldom take less than a few microseconds. During the switch 'on' period, for example, the current will be building up, while the collector–emitter voltage will be falling towards zero. The peak power reached can therefore be large, before falling to the relatively low on-state value. Of course the total energy released as heat each time the device switches is modest because the whole process happens so quickly. Hence if the switching rate is low (say once every second) the switching power loss will be insignificant in comparison with the on-state power. But at high switching rates, when the time taken to complete the switching becomes comparable with the on time, the switching power loss can easily become dominant. In drives, switching rates from hundreds of hertz to the low tens of kilohertz are used: higher frequencies would be desirable from the point of view of smoothness of supply, but cannot be used because the resultant high switching loss becomes unacceptable.

Chopper with inductive load – overvoltage protection

So far we have looked at chopper control of a resistive load, but in a drives context the load will usually mean the winding of a machine, which will invariably be inductive.

Chopper control of inductive loads is much the same as for resistive loads, but we have to be careful to prevent the appearance of dangerously high voltages each time the inductive load is switched 'off'. The root of the problem lies with the energy stored in magnetic field of the inductor. When an inductance L carries a current I, the energy stored in the magnetic field (W) is given by

$$W = \frac{1}{2}LI^2 \tag{2.1}$$

If the inductor is supplied via a mechanical switch, and we open the switch with the intention of reducing the current to zero instantaneously, we are in effect seeking to destroy the stored energy. This is not possible, and what happens is that the energy is dissipated in the form of a spark across the contacts of the switch. This sparking will be familiar to anyone who has pulled off the low-voltage lead to the ignition coil in a car.

The appearance of a spark indicates that there is a very high voltage which is sufficient to breakdown the surrounding air. We can anticipate this by remembering that the voltage and current in an inductance are related by the equation

$$V = L\frac{di}{dt} \tag{2.2}$$

The self-induced voltage is proportional to the rate of change of current, so when we open the switch in order to force the current to zero quickly, a very large voltage is created in the inductance. This voltage appears across the terminals of the switch and if sufficient to breakdown the air, the resulting arc allows the current to continue to flow until the stored magnetic energy is dissipated as heat in the arc.

Sparking across a mechanical switch is unlikely to cause immediate destruction, but when a transistor is used sudden death is certain unless steps are taken to tame the stored energy. The usual remedy lies in the use of a 'freewheel diode' (sometimes called a flywheel diode), as shown in Figure 2.4.

A diode is a one-way valve as far as current is concerned: it offers very little resistance to current flowing from anode to cathode (i.e. in the direction of the broad arrow in the symbol for a diode), but blocks current flow from cathode to anode. Actually, when a power diode conducts in the forward direction, the voltage drop across it is usually not all that dependent on the current flowing through it, so the reference above to the diode 'offering little resistance' is not strictly accurate because it does not obey Ohm's law. In practice the volt-drop of power diodes (most of which are made from silicon) is around 0.7 V, regardless of the current rating.

In the circuit of Figure 2.4(a), when the transistor is on, current (I) flows through the load, but not through the diode, which is said to be reverse-biased (i.e. the applied voltage is trying unsuccessfully to push current down through the diode).

When the transistor is turned off, the current through it and the battery drops very quickly to zero. But the stored energy in the inductance means that its current cannot suddenly disappear. So, since there is no longer a path through the transistor, the current diverts into the only

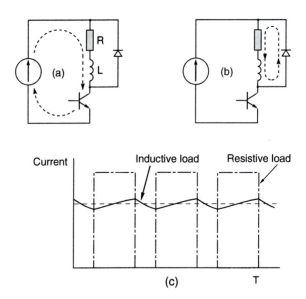

Figure 2.4   *Operation of chopper-type voltage regulator*

other route available, and flows upwards through the low-resistance path offered by the diode, as shown in Figure 2.4(b).

Obviously the current no longer has a battery to drive it, so it cannot continue to flow indefinitely. In fact it will continue to 'freewheel' only until the energy originally stored in the inductance is dissipated as heat, mainly in the load resistance but also in the diode's own (low) resistance. The current waveform during chopping will then be as shown in Figure 2.4(c). Note that the current rises and falls exponentially with a time-constant of L/R, though it never reaches anywhere near its steady-state value in Figure 2.4. The sketch corresponds to the case where the time-constant is much longer than one switching period, in which case the current becomes almost smooth, with only a small ripple. In a d.c. motor drive this is just what we want, since any fluctuation in the current gives rise to torque pulsations and consequent mechanical vibrations. (The current waveform that would be obtained with no inductance is also shown in Figure 2.4: the mean current is the same but the rectangular current waveform is clearly much less desirable, given that ideally we would like constant d.c.)

Finally, we need to check that the freewheel diode prevents any dangerously high voltages from appearing across the transistor. As explained above, when the diode conducts, the forward-bias volt-drop across it is small – typically 0.7 V. Hence while the current is free-wheeling, the voltage at the collector of the transistor is only 0.7 V above the battery voltage. This 'clamping' action therefore limits the voltage across the transistor to a safe value, and allows inductive loads to be switched without damage to the switching element.

Features of power electronic converters

We can draw some important conclusions which are valid for all power electronic converters from this simple example. Firstly, efficient control of voltage (and hence power) is only feasible if a switching strategy is adopted. The load is alternately connected and disconnected from the supply by means of an electronic switch, and any average voltage up to the supply voltage can be obtained by varying the mark/space ratio. Secondly, the output voltage is not smooth d.c., but contains unwanted a.c. components which, though undesirable, are tolerable in motor drives. And finally, the load current waveform will be smoother than the voltage waveform if – as is the case with motor windings – the load is inductive.

D.C. FROM A.C. – CONTROLLED RECTIFICATION

The vast majority of drives of all types draw their power from constant voltage 50 or 60 Hz mains, and in nearly all mains converters the first stage consists of a rectifier which converts the a.c. to a crude form of d.c. Where a constant-voltage (i.e. unvarying average) 'd.c.' output is required, a simple (uncontrolled) diode rectifier is sufficient. But where the mean d.c. voltage has to be controllable (as in a d.c. motor drive to obtain varying speeds), a controlled rectifier is used.

Many different converter configurations based on combinations of diodes and thyristors are possible, but we will focus on 'fully-controlled' converters in which all the rectifying devices are thyristors. These are widely used in modern motor drives.

From the user's viewpoint, interest centres on the following questions:

- How is the output voltage controlled?
- What does the converter output voltage look like? Will there be any problem if the voltage is not pure d.c.?
- How does the range of the output voltage relate to a.c. mains voltage?

We can answer these questions without going too thoroughly into the detailed workings of the converter. This is just as well, because understanding all the ins and outs of converter operation is beyond our scope. On the other hand it is well worth trying to understand the essence of the controlled rectification process, because it assists in understanding the limitations which the converter puts on drive performance (see Chapter 4). Before tackling the questions posed above, however, it is obviously necessary to introduce the thyristor.

The thyristor

The thyristor is an electronic switch, with two main terminals (anode and cathode) and a 'switch-on' terminal (gate), as shown in Figure 2.5. Like a diode, current can only flow in the forward direction, from anode to cathode. But unlike a diode, which will conduct in the forward

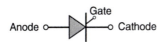

Figure 2.5 *Circuit diagram of thyristor*

direction as soon as forward voltage is applied, the thyristor will continue to block forward current until a small current pulse is injected into the gate-cathode circuit, to turn it on or 'fire' it. After the gate pulse is applied, the main anode–cathode current builds up rapidly, and as soon as it reaches the 'latching' level, the gate pulse can be removed and the device will remain 'on'.

Once established, the anode–cathode current cannot be interrupted by any gate signal. The nonconducting state can only be restored after the anode–cathode current has reduced to zero, and has remained at zero for the turn-off time (typically 100–200 µs in converter-grade thyristors).

When a thyristor is conducting it approximates to a closed switch, with a forward drop of only 1 or 2 V over a wide range of current. Despite the low volt-drop in the 'on' state, heat is dissipated, and heat-sinks must usually be provided, perhaps with fan cooling. Devices must be selected with regard to the voltages to be blocked and the r.m.s and peak currents to be carried. Their overcurrent capability is very limited, and it is usual in drives for devices to have to withstand perhaps twice full-load current for a few seconds only. Special fuses must be fitted to protect against heavy fault currents.

The reader may be wondering why we need the thyristor, since in the previous section we discussed how a transistor could be used as an electronic switch. On the face of it the transistor appears even better than the thyristor because it can be switched off while the current is flowing, whereas the thyristor will remain on until the current through it has been reduced to zero by external means. The primary reason for the use of thyristors is that they are cheaper and their voltage and current ratings extend to higher levels than power transistors. In addition, the circuit configuration in rectifiers is such that there is no need for the thyristor to interrupt the flow of current, so its inability to do so is no disadvantage. Of course there are other circuits (see for example the next section dealing with inverters) where the devices need to be able to switch off on demand, in which case the transistor has the edge over the thyristor.

Single-pulse rectifier

The simplest phase-controlled rectifier circuit is shown in Figure 2.6. When the supply voltage is positive, the thyristor blocks forward current until the gate pulse arrives, and up to this point the voltage across the

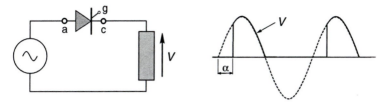

Figure 2.6 *Simple single-pulse thyristor-controlled rectifier, with resistive load and firing angle delay α*

resistive load is zero. As soon as a firing pulse is delivered to the gate-cathode circuit (not shown in Figure 2.6) the device turns on, the voltage across it falls to near zero, and the load voltage becomes equal to the supply voltage. When the supply voltage reaches zero, so does the current. At this point the thyristor regains its blocking ability, and no current flows during the negative half-cycle.

If we neglect the small on-state volt-drop across the thyristor, the load voltage (Figure 2.6) will consist of part of the positive half-cycles of the a.c. supply voltage. It is obviously not smooth, but is 'd.c.' in the sense that it has a positive mean value; and by varying the delay angle (α) of the firing pulses the mean voltage can be controlled.

The arrangement shown in Figure 2.6 gives only one peak in the rectified output for each complete cycle of the mains, and is therefore known as a 'single-pulse' or half-wave circuit. The output voltage (which ideally we would want to be steady d.c.) is so poor that this circuit is never used in drives. Instead, drive converters use four or six thyristors, and produce much superior output waveforms with two or six pulses per cycle, as will be seen in the following sections.

Single-phase fully controlled converter – output voltage and control

The main elements of the converter circuit are shown in Figure 2.7. It comprises four thyristors, connected in bridge formation. (The term 'bridge' stems from early four-arm measuring circuits which presumably suggested a bridge-like structure to their inventors.)

The conventional way of drawing the circuit is shown in Figure 2.7(a), while in Figure 2.7(b) it has been redrawn to assist understanding. The top of the load can be connected (via T1) to terminal A of the mains, or (via T2) to terminal B of the mains, and likewise the bottom of the load can be connected either to A or to B via T3 or T4, respectively.

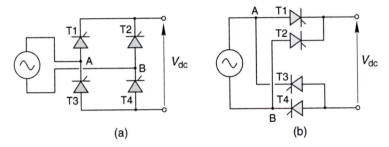

Figure 2.7   *Single-phase 2-pulse (full-wave) fully-controlled rectifier*

We are naturally interested to find what the output voltage waveform on the d.c. side will look like, and in particular to discover how the mean d.c. level can be controlled by varying the firing delay angle (α). This is not as simple as we might have expected, because it turns out that the mean voltage for a given α depends on the nature of the load. We will therefore look first at the case where the load is resistive, and explore the basic mechanism of phase control. Later, we will see how the converter behaves with a typical motor load.

Resistive load

Thyristors T1 and T4 are fired together when terminal A of the supply is positive, while on the other half-cycle, when B is positive, thyristors T2 and T3 are fired simultaneously. The output voltage waveform is shown by the solid line in Figure 2.8. There are two pulses per mains cycle, hence the description 'two-pulse' or full-wave. At every instant the load is either connected to the mains by the pair of switches T1 and T4, or it is connected the other way up by the pair of switches T2 and T3, or it is disconnected. The load voltage therefore consists of rectified

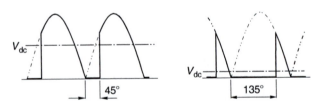

Figure 2.8   *Output voltage waveforms of single-phase fully-controlled rectifier with resistive load, for firing angle delays of 45° and 135°*

chunks of the mains voltage. It is much smoother than in the single-pulse circuit, though again it is far from pure d.c.

The waveforms in Figure 2.8 correspond to $\alpha = 45°$ and $\alpha = 135°$ respectively. The mean value, V_{dc} is shown in each case. It is clear that the larger the delay angle, the lower the output voltage. The maximum output voltage (V_{do}) is obtained with $\alpha = 0°$: this is the same as would be obtained if the thyristors were replaced by diodes, and is given by

$$V_{do} = \frac{2}{\pi} \sqrt{2} V_{rms} \tag{2.3}$$

where V_{rms} is the r.m.s. voltage of the incoming mains. The variation of the mean d.c. voltage with α is given by

$$V_{dc} = \left(\frac{1 + \cos \alpha}{2} \right) V_{do} \tag{2.4}$$

from which we see that with a resistive load the d.c. voltage can be varied from a maximum of V_{do} down to zero by varying α from $0°$ to $180°$.

Inductive (motor) load

As mentioned above, motor loads are inductive, and we have seen earlier that the current cannot change instantaneously in an inductive load. We must therefore expect the behaviour of the converter with an inductive load to differ from that with a resistive load, in which the current can change instantaneously.

The realisation that the mean voltage for a given firing angle might depend on the nature of the load is a most unwelcome prospect. What we would like to say is that, regardless of the load, we can specify the output voltage waveform once we have fixed the delay angle (α). We would then know what value of α to select to achieve any desired mean output voltage. What we find in practice is that once we have fixed α, the mean output voltage with a resistive–inductive load is not the same as with a purely resistive load, and therefore we cannot give a simple general formula for the mean output voltage in terms of α. This is of course very undesirable: if for example we had set the speed of our unloaded d.c. motor to the target value by adjusting the firing angle of the converter to produce the correct mean voltage, the last thing we would want is for the voltage to fall when the load current drawn by the motor increases, as this would cause the speed to fall below the target.

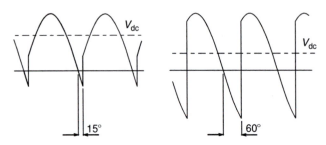

Figure 2.9 *Output voltage waveforms of single-phase fully-controlled rectifier supplying an inductive (motor) load, for firing angle delays of 15° and 60°*

Fortunately however, it turns out that the output voltage waveform for a given α does become independent of the load inductance once there is sufficient inductance to prevent the load current from falling to zero. This condition is known as 'continuous current'; and happily, many motor circuits do have sufficient self-inductance to ensure that we achieve continuous current. Under continuous current conditions, the output voltage waveform only depends on the firing angle, and not on the actual inductance present. This makes things much more straightforward, and typical output voltage waveforms for this continuous current condition are shown in Figure 2.9.

The waveform in Figure 2.9(a) corresponds to $\alpha = 15°$, while Figure 2.9(b) corresponds to $\alpha = 60°$. We see that, as with the resistive load, the larger the delay angle the lower the mean output voltage. However, with the resistive load the output voltage was never negative, whereas we see that for short periods the output voltage can now become negative. This is because the inductance smoothes out the current (see Figure 4.2, for example) so that at no time does it fall to zero. As a result, one pair of thyristors is always conducting, so at every instant the load is connected directly to the mains supply, and the load voltage always consists of chunks of the supply voltage.

It is not immediately obvious why the current switches over (or 'commutates') from the first pair of thyristors to the second pair when the latter are fired, so a brief look at the behaviour of diodes in a similar circuit configuration may be helpful at this point. Consider the setup shown in Figure 2.10, with two voltage sources (each time-varying) supplying a load via two diodes. The question is what determines which diode conducts, and how does this influences the load voltage?

We can consider two instants as shown in the diagram. On the left, V_1 is 250 V, V_2 is 240 V, and D1 is conducting, as shown by the heavy

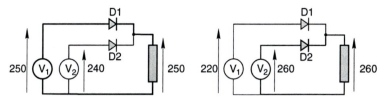

Figure 2.10   *Diagram illustrating commutation between diodes: the current flows through the diode whose anode is at the higher potential*

line. If we ignore the volt-drop across the diode, the load voltage will be 250 V, and the voltage across diode D2 will be $240-250 = -10\,\text{V}$, i.e. it is reverse biased and hence in the nonconducting state. At some other instant (on the right of the diagram), V_1 has fallen to 220 V while V_2 has increased to 260 V: now D2 is conducting instead of D1, again shown by the heavy line, and D1 is reverse-biased by $-40\,\text{V}$. The simple pattern is that the diode with the highest anode potential will conduct, and as soon as it does so it automatically reverse-biases its predecessor. In a single-phase diode bridge, for example, the commutation occurs at the point where the supply voltage passes through zero: at this instant the anode voltage on one pair goes from positive to negative, while on the other pair the anode voltage goes from negative to positive.

The situation in controlled thyristor bridges is very similar, except that before a new device can takeover conduction, it must not only have a higher anode potential, but it must also receive a firing pulse. This allows the changeover to be delayed beyond the point of natural (diode) commutation by the angle α, as shown in Figure 2.9. Note that the maximum mean voltage (V_{do}) is again obtained when α is zero, and is the same as for the resistive load (equation 2.3). It is easy to show that the mean d.c. voltage is now related to α by

$$V_{\text{dc}} = V_{\text{do}} \cos \alpha \qquad (2.5)$$

This equation indicates that we can control the mean output voltage by controlling α, though equation 2.5 shows that the variation of mean voltage with α is different from that for a resistive load (equation 2.4). We also see that when α is greater than 90° the mean output voltage is negative. The fact that we can obtain a nett negative output voltage with an inductive load contrasts sharply with the resistive load case, where the output voltage could never be negative. We will see later that this facility allows the converter to return energy from the load to the supply, and this is important when we want to use the converter with a d.c. motor in the regenerating mode.

It is sometimes suggested (particularly by those with a light-current background) that a capacitor could be used to smooth the output voltage, this being common practice in cheap low-power d.c. supplies. Although this works well at low power, there are two reasons why capacitors are not used with controlled rectifiers supplying motors. Firstly, as will be seen later, it is not necessary for the voltage to be smooth as it is the current which directly determines the torque, and as already pointed out the current is always much smoother than the voltage because of inductance. And secondly, the power levels in most drives are such that in order to store enough energy to smooth out the rectified voltage, very bulky and expensive capacitors would be required.

3-phase fully controlled converter

The main power elements are shown in Figure 2.11. The three-phase bridge has only two more thyristors than the single-phase bridge, but the output voltage waveform is vastly better, as shown in Figure 2.12. There are now six pulses of the output voltage per mains cycle, hence the description 'six-pulse'. The thyristors are again fired in pairs (one in the top half of the bridge and one – from a different leg – in the bottom half), and each thyristor carries the output current for one third of the time. As in the single-phase converter, the delay angle controls the output voltage, but now $\alpha = 0$ corresponds to the point at which the phase voltages are equal (see Figure 2.12).

The enormous improvement in the smoothness of the output voltage waveform is clear when we compare Figure 2.9 and Figure 2.12 , and it indicates the wisdom of choosing a 3-phase converter whenever possible. The very much better voltage waveform also means that the desirable

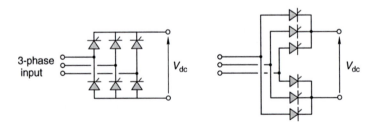

Figure 2.11   *Three-phase fully-controlled thyristor converter*

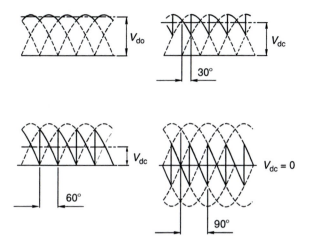

Figure 2.12 *Output voltage waveforms for three-phase fully-controlled thyristor converter supplying an inductive (motor) load, for various firing angles from 0° to 90°. The mean d.c. voltage is shown by the horizontal line, except for α = 90° where the mean d.c. voltage is zero*

'continuous current' condition is much more likely to be met, and the waveforms in Figure 2.12 have therefore been drawn with the assumption that the load current is in fact continuous. Occasionally, even a six-pulse waveform is not sufficiently smooth, and some very large drive converters therefore consist of two six-pulse converters with their outputs in series. A phase-shifting transformer is used to insert a 30° shift between the a.c. supplies to the two 3-phase bridges. The resultant ripple voltage is then 12-pulse.

Returning to the six-pulse converter, the mean output voltage can be shown to be given by

$$V_{dc} = V_{do} \cos \alpha = \frac{3}{\pi} \sqrt{2} \, V_{rms} \cos \alpha \qquad (2.6)$$

We note that we can obtain the full range of output voltages from $+V_{do}$ to $-V_{do}$, so that, as with the single-phase converter, regenerative operation will be possible.

It is probably a good idea at this point to remind the reader that, in the context of this book, our interest in the controlled rectifier is as a supply to the armature of a d.c. motor. When we examine the d.c. motor drive in Chapter 4, we will see that it is the average or mean value of the output voltage from the controlled rectifier that determines the speed, and it is this mean voltage that we refer to when we talk of 'the voltage' from the converter. We must not forget the unwanted a.c. or ripple element, however, as this can be large. For example, we see from Figure 2.12 that to obtain a very low voltage (to make the motor run very slowly) α will be close to zero; but if we were to connect an a.c. voltmeter to the output terminals it could register several hundred volts, depending on the incoming mains voltage!

Output voltage range

In Chapter 4 we will discuss the use of the fully controlled converter to drive a d.c. motor, so it is appropriate at this stage to look briefly at the typical voltages we can expect. Mains a.c. supply voltages obviously vary around the world, but single-phase supplies are usually 220–240 V, and we see from equation 2.3 that the maximum mean d.c. voltage available from a single-phase 240 V supply is 216 V. This is suitable for 180–200 V motors. If a higher voltage is needed (say for a 300 V motor), a transformer must be used to step up the mains.

Turning now to typical three-phase supplies, the lowest three-phase industrial voltages are usually around 380–440 V. (Higher voltages of up to 11 kV are used for large drives, but these will not be discussed here). So with $V_{rms} = 415\,V$ for example, the maximum d.c. output voltage (equation 2.6) is 560 V. After allowances have been made for supply variations and impedance drops, we could not rely on obtaining much more than 520–540 V, and it is usual for the motors used with six-pulse drives fed from 415 V, three-phase supplies to be rated in the range 440–500 V. (Often the motor's field winding will be supplied from single-phase 240 V, and field voltage ratings are then around 180–200 V, to allow a margin in hand from the theoretical maximum of 216 V referred to earlier.)

Firing circuits

Since the gate pulses are only of low power, the gate drive circuitry is simple and cheap. Often a single integrated circuit (chip) contains all the circuitry for generating the gate pulses, and for synchronising them with the appropriate delay angle (α) with respect to the supply voltage. To avoid direct electrical connection between the high voltages in the main power circuit and the low voltages used in the control circuits, the gate pulses are usually coupled to the thyristor by means of small pulse transformers. Most converters also include what is known as 'inverse cosine-weighted' firing circuitry: this means that the firing circuit is so arranged that the 'cosine' relationship is incorporated internally so that the mean output voltage of the converter becomes directly proportional to the input control voltage, which typically ranges from 0 to 10 V.

A.C. FROM D.C. SP – SP INVERSION

The business of getting a.c. from d.c. is known as inversion, and nine times out of ten we would ideally like to be able to produce sinusoidal output voltages of whatever frequency and amplitude we choose. Unfortunately the constraints imposed by the necessity to use a switching strategy means that we always have to settle for a voltage waveform which is composed of rectangular chunks, and is thus far from ideal. Nevertheless it turns out that a.c. motors are remarkably tolerant, and will operate satisfactorily despite the inferior waveforms produced by the inverter.

Single-phase inverter

We can illustrate the basis of inverter operation by considering the single-phase example shown in Figure 2.13. This inverter uses bipolar transistors as the switching elements, with diodes (not shown) to provide the freewheel paths needed when the load is inductive.

The input or d.c. side of the inverter (on the left in Figure 2.13) is usually referred to as the 'd.c. link', reflecting the fact that in the majority of cases the d.c. is obtained by rectifying the incoming constant-frequency mains. The output or a.c. side is taken from terminals A and B as shown in Figure 2.13.

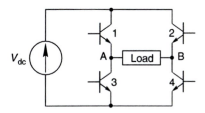

Figure 2.13 *Inverter circuit for single-phase output. (The four freewheel diodes have been omitted for the sake of clarity.)*

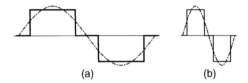

Figure 2.14   *Inverter output voltage waveforms – resistive load*

When transistors 1 and 4 are switched on, the load voltage is positive, and equal to the d.c. link voltage, while when 2 and 3 are on it is negative. If no devices are switched on, the output voltage is zero. Typical output voltage waveforms at low and high switching frequencies are shown in Figure 2.14(a) and (b), respectively.

Here each pair of devices is on for one-third of a cycle, and all the devices are off for two periods of one-sixth of a cycle. The output waveform is clearly not a sine wave, but at least it is alternating and symmetrical. The fundamental component is shown dotted in Figure 2.14.

Within each cycle the pattern of switching is regular, and easily programmed using appropriate logic circuitry. Frequency variation is obtained by altering the clock frequency controlling the four-step switching pattern. (The oscillator providing the clock signal can be controlled by an analogue voltage, or it can be generated using software.) The effect of varying the switching frequency is shown in Figure 2.14, from which we can see that the amplitude of the fundamental component of voltage remains constant, regardless of frequency. Unfortunately (as explained in Chapter 7), this is not what we want for supplying an induction motor: to prevent the air-gap flux in the motor from falling as the frequency is raised we need to be able to increase the voltage in proportion to the frequency. We will look at voltage control shortly, after a brief digression to discuss the problem of 'shoot-through'.

Inverters with the configurations shown in Figures 2.13 and 2.16 are subject to a potentially damaging condition which can arise if both transistors in one 'leg' of the inverter inadvertently turn on simultaneously. This should never happen if the devices are switched correctly, but if something goes wrong and both devices are on together – even for a very short time – they form a short-circuit across the d.c. link. This fault condition is referred to as 'shoot-through' because a high current is established very rapidly, destroying the devices. A good inverter therefore includes provision for protecting against the possibility of shoot-through, usually by imposing a minimum time-delay between one device in the leg going off and the other coming on.

Output voltage control

There are two ways in which the amplitude of the output voltage can be controlled. First, if the d.c. link is provided from a.c. mains via a controlled rectifier or from a battery via a chopper, the d.c. link voltage can be varied. We can then set the amplitude of the output voltage to any value within the range of the link. For a.c. motor drives (see Chapter 7) we can arrange for the link voltage to track the output frequency of the inverter, so that at high output frequency we obtain a high output voltage and vice-versa. This method of voltage control results in a simple inverter, but requires a controlled (and thus relatively expensive) rectifier for the d.c. link.

The second method, which predominates in small and medium sizes, achieves voltage control by pulse-width-modulation (PWM) within the inverter itself. A cheaper uncontrolled rectifier can then be used to provide a constant-voltage d.c. link.

The principle of voltage control by PWM is illustrated in Figure 2.15.

At low-output frequencies, a low-output voltage is usually required, so one of each pair of devices is used to chop the voltage, the mark–space ratio being varied to achieve the desired voltage at the output. The low fundamental voltage component at low frequency is shown dotted in Figure 2.15(a). At a higher frequency a higher voltage is needed, so the chopping device is allowed to conduct for a longer fraction of each

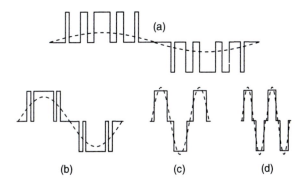

Figure 2.15 *Inverter output voltage and frequency control with pulse-width modulation*

cycle, giving the higher fundamental output as shown in Figure 2.15(b). As the frequency is raised still higher, the separate 'on' periods eventually merge, giving the waveform as shown in Figure 2.15(c). Any further increase in frequency takes place without further increase in the output voltage, as shown in Figure 2.15(d).

In drives applications, the range of frequencies over which the voltage/frequency ratio can be kept constant is known as the 'PWM' region, and the upper limit of the range is usually taken to define the 'base speed' of the motor. Above this frequency, the inverter can no longer match voltage to frequency, the inverter effectively having run out of steam as far as voltage is concerned. The maximum voltage is thus governed by the link voltage, which must therefore be sufficiently high to provide whatever fundamental voltage the motor needs at its base speed, which is usually 50 or 60 Hz.

Beyond the PWM region the voltage waveform is as shown in Figure 2.15(d): this waveform is usually referred to as 'quasi-square', though in the light of the overall object of the exercise (to approximate to a sine wave) a better description might be 'quasi-sine'.

When supplying an inductive motor load, fast recovery freewheel diodes are needed in parallel with each device. These may be discrete devices, or fitted in a common package with the transistor, or even integrated to form a single transistor/diode device.

Sinusoidal PWM

So far we have emphasised the importance of being able to control the amplitude of the fundamental output voltage by modulating the width of the pulses which make up the output waveform. If this was the only requirement, we would have an infinite range of modulation patterns which would be sufficient. But as well as the right fundamental amplitude, we want the harmonic content to be minimised, i.e. we want the output waveform to be as close as possible to a pure sine wave. It is particularly important to limit the amplitude of the low-order harmonics, since these are the ones which are most likely to provoke an unwanted response from the motor.

The number, width, and spacing of the pulses is therefore optimised to keep the harmonic content as low as possible. A host of sophisticated strategies have been developed, almost all using a microprocessor-based system to store and/or generate the modulation patterns. There is an obvious advantage in using a high switching frequency, since there are then more pulses to play with. Ultrasonic frequencies are now widely

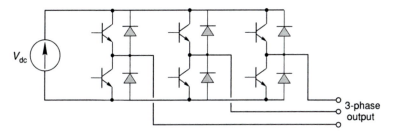

Figure 2.16 *Three-phase inverter power circuit*

used, and as devices improve the switching frequencies continue to rise. Most manufacturers claim their particular system is better than the competition, but it is not clear which will ultimately emerge as best for motor operation. Some early schemes used comparatively few pulses per cycle, and changed the number of pulses in discrete steps rather than smoothly, which earned them the nickname 'gear-changers'. These inverters were noisy and irritating.

3-phase inverter

A 3-phase output can be obtained by adding only two more switches to the four needed for a single-phase inverter, giving the typical power-circuit configuration shown in Figure 2.16. As usual, a freewheel diode is required in parallel with each transistor to protect against overvoltages caused by an inductive (motor) load.

We note that the circuit configuration in Figure 2.16 is the same as for the 3-phase controlled rectifier looked at earlier. We mentioned then that the controlled rectifier could be used to regenerate, i.e. to convert power from d.c. to a.c., and this is of course 'inversion' as we now understand it.

Forced and natural commutation – historical perspective

We have assumed in this discussion that the switching devices can turn-off or 'commutate' on demand, so that the output (load) current is either reduced to zero or directed to another leg of the inverter. Transistors and gate-turn-off thyristors (see Section 2.4.6) can operate like this, but, as explained earlier, conventional thyristors cannot turn off on command. Nevertheless thyristors are widely used to invert power from d.c. to a.c. as we will see when we look at the d.c. motor drive in Chapter 4, so we should look briefly at how this is possible.

There are two distinct ways in which thyristors are used in inverters. In the first, where the inverter is used to supply an essentially passive a.c. load, such as an induction motor, each thyristor has to be equipped with its own auxiliary 'forced commutating' circuit, whose task is to force the current through the thyristor to zero when the time comes for it to turn off. The commutating circuits are quite complex, and require substantial capacitors to store the energy required for commutation. Force-commutated thyristor converters therefore tend to be bulky and expensive. At one time they were the only alternative, but they are now more or less obsolete, having been superseded by MOSFET, IGBT or GTO versions (see Section 2.5).

The other way in which thyristors can be used to invert power from d.c. to a.c. is when the a.c. side of the bridge is connected to a three-phase stiff (i.e. low source impedance) mains supply. This is the normal 'controlled rectifier' arrangement introduced earlier. In this case it turns out that the currents in the thyristors are naturally forced to zero by the active mains voltages, thereby allowing the thyristors to commutate or turn off naturally. This mode of operation continues to be important, as we will see when we look at d.c.motor drives.

Matrix converters

Each of the converters we have looked at so far has been developed with a specific function in mind (e.g. rectifying a.c. to obtain d.c.), but they share an important common feature in that they all involve the use of sequentially operated switching devices to connect the output terminals to the supply terminals. In all of these circuits the switching devices (transistors, thyristors, and diodes) can only conduct current in one direction, and this results in limitations on the ability of the circuit to operate with reverse power flow: for some circuits, e.g. the simple diode rectifier, reverse power flow is not possible at all.

The ideal power-electronic converter would allow power conversion in either direction between two systems of any voltage and frequency, including d.c., and would not involve any intermediate stage, such as a d.c. link. In principle this can be achieved by means of an array of switches that allow any one of a set of input terminals to be connected to any one of a set of output terminals, at any desired instant. Not surprisingly, the generic name for such converters is 'matrix converter'.

By way of example, Figure 2.17 shows nine switches arranged to permit power conversion between two three-phase systems.

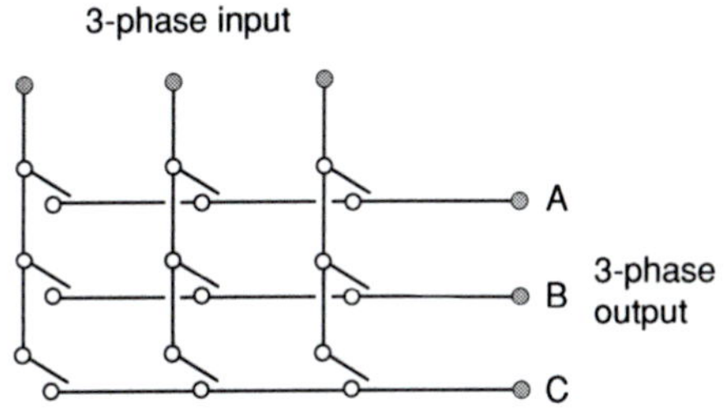

3-phase input

Figure 2.17 *Principle of matrix converter*

The operation can best be understood by imagining that the input voltage is that of the sinusoidal three-phase mains system, so at every instant we know exactly what each incoming line-to-line voltage will be. Let us assume that we wish to synthesise a three-phase sinusoidal output of a known voltage and frequency. At every instant we know what voltage we want, say between the lines A and B, and we know what the voltages are between the three incoming lines. So we switch on whichever pair of the A and B switches connects us to the two incoming lines whose voltage at the time is closest to the desired output line-to-line voltage, and we stay with it while it offers the best approximation to what we want. When a different combination of switches would allow us to hook onto a more appropriate pair of input lines, the switching pattern changes.

A little thought will make clear that since at any instant there are only three different incoming line-to-line voltages to choose from, we cannot expect to synthesise a decent sinusoidal waveform with this sort of discrete switching, although this is the approach taken to derive a low-frequency output in the cycloconverter drive (see Chapter 7). To obtain a better approximation to a sinusoidal waveform we must use chopping, where we switch rapidly on and off to modulate the output voltage amplitude (see also Chapter 7). This means that the switches have to be capable of operating at much higher frequencies than the fundamental output frequency, so that switching loss becomes an important consideration.

In order for such converters to be able to transmit power in either direction (so that the terms input and output cease to have specific meanings), the switches (shown in Figure 2.17) must be capable of carrying current in either direction. The lack of single affordable devices with this bi-directional current capability explains why matrix converters are not yet significant in the drives market, but their development is attracting attention and they look likely to emerge as improved control strategies and bi-directional switching devices become more practicable.

INVERTER SWITCHING DEVICES

As far as the user is concerned, it does not really matter what type of switching device is used inside the inverter, but it is probably helpful to mention the four most important families of devices in current use so that the terminology is familiar and the symbols used for each device can be recognised. The feature which unites all four devices is that they can be switched on and off by means of a low-power control signal, i.e. they are self-commutating. We have seen earlier that this ability to be turned on or off on demand is essential in any inverter which feeds a passive load, such as an induction motor.

Each device is discussed briefly below, with a broad indication of its most likely range of application. Because there is considerable overlap between competing devices, it is not possible to be dogmatic and specify which device is best, and the reader should not be surprised to find that one manufacturer may offer a 5 kW inverter which uses MOSFETs while another chooses to use IGBTs. The whole business of power electronics is still developing: there are other devices (such as those based on silicon carbide) that are yet to emerge onto the drives scene. One trend which continues is the integration of the drive and protection circuitry in the same package as the switching device (or devices). This obviously leads to considerable simplification and economy in the construction of the complete converter.

Bipolar junction transistor (BJT)

Historically the bipolar junction transistor was the first to be used for power switching. Of the two versions (npn and pnp) only the npn has been widely used in inverters for drives, mainly in applications ranging up to a few kilowatts and several hundred volts.

The npn version is shown in Figure 2.18: the main (load) current flows into the collector (C) and out of the emitter (E), as shown by the arrow on the device symbol. To switch the device on (i.e. to make the resistance of the collector–emitter circuit low, so that load current can flow), a small current must be caused to flow from the base (B) to the emitter. When the base–emitter current is zero, the resistance of the collector–emitter circuit is very high, and the device is switched off.

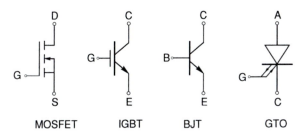

Figure 2.18   *Circuit symbols for self-commutating devices*

The advantage of the bipolar transistor is that when it is turned on, the collector–emitter voltage is low (see Figure 2.3) and hence the power dissipation is small in comparison with the load power, i.e. the device is an efficient power switch. The disadvantage is that although the power required in the base–emitter circuit is tiny in comparison with the load power, it is not insignificant and in the largest power transistors can amount to several tens of watts. This means that the complexity and cost of the base-drive circuitry can be considerable.

Metal oxide semiconductor field effect transistor (MOSFET)

Since the 1980s the power MOSFET has superseded the BJT in inverters for drives. Like the BJT, the MOSFET is a three-terminal device and is available in two versions, the n-channel and the p-channel. The n-channel is the most widely used, and is shown in Figure 2.18. The main (load) current flows into the drain (D) and out of the source (S). (Confusingly, the load current in this case flows in the *opposite* direction to the arrow on the symbol.) Unlike the BJT, which is controlled by the base current, the MOSFET is controlled by the gate-source voltage.

To turn the device on, the gate-source voltage must be comfortably above a threshold of a few volts. When the voltage is first applied to the gate, currents flow in the parasitic gate-source and gate-drain capacitances, but once these capacitances have been charged the input current to the gate is negligible, so the steady-state gate drive power is minimal. To turn the device off, the parasitic capacitances must be discharged and the gate-source voltage must be held below the threshold level.

The principal advantage of the MOSFET is that it is a voltage-controlled device which requires negligible power to hold it in the on state. The gate drive circuitry is thus less complex and costly than the base-drive circuitry of an equivalent bipolar device. The disadvantage of the MOSFET is that in the 'on' state the effective resistance of the drain source is higher than an equivalent bipolar device, so the power dissipation is higher and the device is rather less efficient as a power switch. MOSFETs are used in low and medium power inverters up to a few kilowatts, with voltages generally not exceeding 700 V.

Insulated gate bipolar transistor (IGBT)

The IGBT (Figure 2.18) is a hybrid device which combines the best features of the MOSFET (i.e. ease of gate turn on and turn off from low-power logic circuits) and the BJT (relatively low power dissipation in the main collector–emitter circuit). These obvious advantages give the IGBT the edge over the MOSFET and BJT, and account for their dominance in all but small drives. They are particularly well suited to the medium power, medium voltage range (up to several hundred kilowatts).

The path for the main (load) current is from collector to emitter, as in the npn bipolar device.

Gate turn-off thyristor (GTO)

The GTO (Figure 2.18) is turned on by a pulse of current in the gate-cathode circuit in much the same way as a conventional thyristor. But unlike an ordinary thyristor, which cannot be turned off by gate action, the GTO can be turned off by a negative gate-cathode current. The main (load) current flows from anode to cathode, as in a conventional thyristor. The twin arrowed paths on the gate lead (Figure 2.18) indicate that control action is achieved by both forward and reverse gate currents. (In US literature, a single gate lead with a short crossbar is used instead of the two arrows.)

The gate drive requirements are more demanding than for a conventional thyristor, and the on-state performance is worse, with a forward volt-drop of perhaps 3 V compared with 1.5 V, but these are the penalties to be paid in return for the added flexibility. The GTO has considerably higher voltage and current ratings (up to 3 kV and 2 kA) than the other three devices and is therefore used in high-power inverters.

CONVERTER WAVEFORMS AND ACOUSTIC NOISE

In common with most textbooks, the waveforms shown in this chapter (and later in the book) are what we would hope to see under ideal conditions. It makes sense to concentrate on these ideal waveforms from the point of view of gaining a basic understanding, but we ought to be warned that what we see on an oscilloscope may well look rather different!

We have seen that the essence of power electronics is the switching process, so it should not come as much of a surprise to learn that in practice the switching is seldom achieved in such a clear-cut fashion as we have assumed. Usually, there will be some sort of high-frequency oscillation or 'ringing' evident, particularly on the voltage waveforms following each transition due to switching. This is due to the effects of stray capacitance and inductance: it should be anticipated at the design stage, and steps should be taken to minimise it by fitting 'snubbing' circuits at the appropriate places in the converter. However complete suppression of all these transient phenomena is seldom economically worthwhile so the user should not be too alarmed to see remnants of the transient phenomena in the output waveforms.

Acoustic noise is also a matter which can worry newcomers. Most power electronic converters emit whining or humming sounds at frequencies corresponding to the fundamental and harmonics of the switching frequency, though when the converter is used to feed a motor, the sound from the motor is usually a good deal louder than the sound from the converter itself. These sounds are very difficult to describe in words, but typically range from a high-pitched hum through a whine to a piercing whistle. They vary in intensity with the size of converter and the load, and to the trained ear can give a good indication of the health of the motor and converter.

COOLING OF POWER SWITCHING DEVICES

Thermal resistance

We have seen that by adopting a switching strategy the power loss in the switching devices is small in comparison with the power throughput, so the converter has a high efficiency. Nevertheless almost all the heat which is produced in the switching devices is released in the active region of the semiconductor, which is itself very small and will overheat and breakdown unless it is adequately cooled. It is therefore essential to ensure that even under the most onerous operating conditions, the temperature of the active junction inside the device does not exceed the safe value.

Consider what happens to the temperature of the junction region of the device when we start from cold (i.e. ambient) temperature and operate the device so that its average power dissipation remains constant. At first, the junction temperature begins to rise, so some of the heat generated is conducted to the metal case, which stores some heat as its temperature rises. Heat then flows into the heatsink (if fitted), which begins to warm up, and heat begins to flow to the surrounding air, at ambient temperature. The temperatures of the junction, case and heatsink continue to rise until eventually an equilibrium is reached when the total rate of loss of heat to ambient temperature is equal to the power dissipation inside the device.

The final steady-state junction temperature thus depends on how difficult it is for the power loss to escape down the temperature gradient to ambient, or in other words on the total 'thermal resistance' between the junction inside the device and the surrounding medium (usually air). Thermal resistance is usually expressed in °C/W, which directly indicates how much temperature rise will occur in the steady state for every watt of dissipated power. It follows that for a given power dissipation, the higher the thermal resistance, the higher the temperature rise, so in order to minimise the temperature rise of the device, the total thermal resistance between it and the surrounding air must be made as small as possible.

The device designer aims to minimise the thermal resistance between the semiconductor junction and the case of the device, and provides a large and flat metal mounting surface to minimise the thermal resistance between the case and the heatsink. The converter designer must ensure good thermal contact between the device and the heatsink, usually by a bolted joint liberally smeared with heat-conducting compound to fill any microscopic voids, and must design the heatsink to minimise the thermal resistance to air (or in some cases oil or water). Heatsink design offers the only real scope for appreciably reducing the total resistance, and involves careful selection of the material, size, shape and orientation of the heatsink, and the associated air-moving system (see below).

One drawback of the good thermal path between the junction and case of the device is that the metal mounting surface (or surfaces in the case of the popular hockeypuck package) can be electrically 'live'. This poses a difficulty for the converter designer, because mounting the device directly on the heatsink causes the latter to be dangerous. In addition, several separate isolated heatsinks may be required in order to avoid short-circuits. The alternative is for the devices to be electrically isolated from the heatsink using thin mica spacers, but then the thermal resistance is appreciably increased.

Increasingly devices come in packaged 'modules' with an electrically isolated metal base to get round the 'live' problem. The packages contain combinations of transistors, diodes or thyristors, from which various converter circuits can be built up. Several modules can be mounted on a single heatsink, which does not have to be isolated from the enclosure or cabinet. They are available in ratings suitable for converters up to hundreds of kilowatts, and the range is expanding. This development, coupled with a move to fan-assisted cooling of heatsinks has resulted in a dramatic reduction in the overall size of complete converters, so that a modern 20 kW drive converter is perhaps only the size of a small briefcase.

Arrangement of heatsinks and forced air cooling

The principal factors which govern the thermal resistance of a heatsink are the total surface area, the condition of the surface and the air flow. Most converters use extruded aluminium heatsinks, with multiple fins to increase the effective cooling surface area and lower the resistance, and with a machined face or faces for mounting the devices. Heatsinks are usually mounted vertically to improve natural air convection. Surface finish is important, with black anodised aluminium being typically 30% better than bright.

A typical layout for a medium-power (say 200 kW) converter is shown in Figure 2.19.

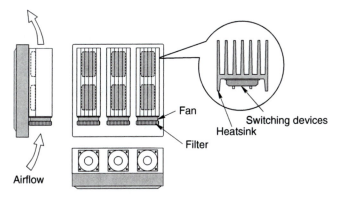

Figure 2.19 *Layout of converter showing heatsink and cooling fans*

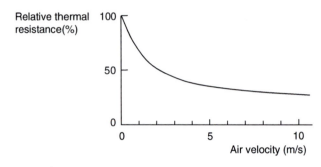

Figure 2.20   *Sketch showing the influence of air velocity on effective thermal resistance. (The thermal resistance in still air is taken as 100%.)*

The fans are positioned either at the top or bottom of the heatsink, and draw external air upwards, assisting natural convection. The value of even a modest air-flow is shown by the sketch in Figure 2.20. With an air velocity of only 2 m/s, for example, the thermal resistance is halved as compared with the naturally cooled setup, which means that for a given temperature rise the heatsink can be half the size of the naturally cooled one. Only a little of this saving in space is taken up by the fans, as shown in Figure 2.19. Large increases in the air velocity bring diminishing returns, as shown in Figure 2.20, and also introduce additional noise which is generally undesirable.

Cooling fans

Cooling fans have integral hub-mounted inside-out motors, i.e. the rotor is outside the stator and carries the blades of the fan. The rotor diameter/length ratio is much higher than for most conventional motors in order to give a slimline shape to the fan assembly, which is well-suited for mounting at the end of an extruded heatsink (Figure 2.19). The rotor inertia is thus relatively high, but this is unimportant because the total inertia is dominated by the impeller, and there is no need for high accelerations.

Mains voltage 50 or 60 Hz fans have external rotor single-phase shaded-pole motors, which normally run at a fixed speed of around 2700 rev/min, and have input powers typically between 10 and 50 W (see Chapter 6). The torque required in a fan is roughly proportional to the cube of the speed, so the starting torque requirement is low and the motor can be designed to have a high running efficiency (see Chapter 6). Slower-speed (but less efficient) versions are used where acoustic noise is a problem.

Low-voltage (5, 12 or 24 V) d.c. fans employ brushless motors with Hall effect rotor position detection (see Chapter 10). The absence of sparking from a conventional commutator is important to limit interference with adjacent sensitive equipment. These fans are generally of lower power than their a.c. counterparts, typically from as little as 1 W up to about 10 W, and with running speeds of typically between 3000 and 5000 rev/min. They are mainly used for cooling circuit boards directly, and have the advantage that the speed can be controlled by voltage variation, thereby permitting a trade-off between noise and volume flow.

REVIEW QUESTIONS

1) In the circuit of Figure Q.1, both voltage sources and the diodes can be treated as ideal, and the load is a resistor. (Note: this question is specifically aimed at reinforcing the understanding of how diodes behave: it is not representative of any practical circuit.)

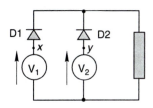

Figure Q.1

Sketch the voltage across the load under the following conditions:

(a) V_1 is a sinusoid of amplitude 20 V, and V_2 is a constant voltage of $+10$ V
(b) V_1 is a sinusoid of amplitude 20 V, and V_2 is a constant voltage of -10 V
(c) The same as (a), except that the diode D1 is placed below, rather above, V_1.

2) What is the maximum d.c. voltage available from a fully controlled bridge converter supplying a motor and operating from low-impedance 230 V mains?

3) Estimate the firing angle required to produce a mean output voltage of 300 V from a fully controlled 3-phase converter supplied from stiff 415 V, 50 Hz mains. Assume that the load current is continuous. How would the firing angle have to change if the supply frequency was 60 Hz rather than 50 Hz?

4) Sketch the waveform of voltage across one of the thyristors in a fully controlled single-phase converter with a firing angle delay of 60°. Assume that the d.c. load current is continuous. Figure 2.9 may prove helpful.

5) Sketch the current waveform in the a.c. supply when a single-phase fully controlled converter with $\alpha = 45°$ is supplying a highly inductive load which draws a smooth current of 25 A. If the a.c. supply is 240 V, 50 Hz, and losses in the devices are neglected, calculate the peak and average supply power per cycle.

6) A d.c. chopper circuit is often said to be like an a.c. transformer. Explain what this means by considering the input and output power relationships for a chopper-fed inductive motor load supplied with an average voltage of 20 V from a 100 V battery. Assume that the motor current remains constant throughout at 5 A.

7) A 5 kHz step-down transistor chopper operating from a 150 V battery supplies an R/L load which draws an almost-constant current of 5 A. The resistance of the load is 8Ω.
Treating all devices as ideal, estimate:
(a) the mark:space ratio of the chopper; (b) the average power in the load; and (c) the average power from the source.

8) This question relates to the switching circuit of Figure Q.8, and in particular to the function of the diode.

Some possible answers to the question 'what is the purpose of the diode' are given below:
(a)   to prevent reverse current in the switch;
(b)   to protect the inductor from high voltages;
(c)   to limit the rate of change of current in the supply;
(d)   to limit the voltage across the MOSFET;
(e)   to dissipate the stored energy in the inductance;
Discuss these answers and identify which one(s) are correct.

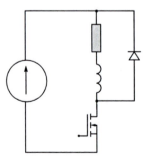

Figure Q.8

9) In the circuit of Figure Q.8, assume that the supply voltage is 100 V, and that the forward volt-drop across the diode is 0.7 V. Some common answers to the question 'when the current is freewheeling, what is the voltage across the MOSFET' are given below:

(a) 99.3 V;

(b) 0.7 V;

(c) 0 V;

(d) depends on the inductance;

(e) 100.7 V.

Discuss these answers and identify which one is correct.

3

CONVENTIONAL D.C. MOTORS

INTRODUCTION

Until the 1980s the conventional (brushed) d.c. machine was the automatic choice where speed or torque control is called for, and large numbers remain in service despite a declining market share that reflects the move to inverter-fed induction motors. Applications range from steel rolling mills, railway traction, to a very wide range of industrial drives, robotics, printers, and precision servos. The range of power outputs is correspondingly wide, from several megawatts at the top end down to only a few watts, but except for a few of the small low-performance ones, such as those used in toys, all have the same basic structure, as shown in Figure 3.1.

The motor has two separate circuits. The smaller pair of terminals connect to the field windings, which surround each pole and are normally in series: in the steady state all the input power to the field windings is dissipated as heat – none of it is converted to mechanical output power. The main terminals convey the 'power' or 'work' current to the brushes which make sliding contact to the armature winding on the rotor. The supply to the field is separate from that for the armature, hence the description 'separately excited'.

As in any electrical machine it is possible to design a d.c. motor for any desired supply voltage, but for several reasons it is unusual to find rated voltages lower than about 6 V or much higher than 700 V. The lower limit arises because the brushes (see below) give rise to an unavoidable volt-drop of perhaps 0.5–1 V, and it is clearly not good practice to let this 'wasted' voltage became a large fraction of the supply voltage. At the other end of the scale, it becomes prohibitively expensive to insulate the commutator segments to withstand higher voltages. The function and operation of the commutator is discussed later, but it is appropriate to mention here that brushes and commutators are

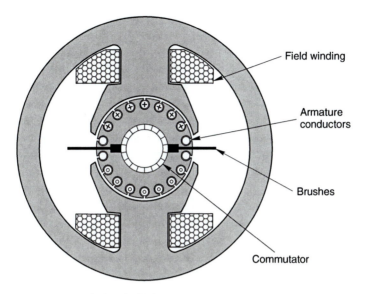

Field winding

Armature
conductors

Brushes

Commutator

Figure 3.1   *Conventional (brushed) d.c. motor*

troublesome at very high speeds. Small d.c. motors, say up to hundreds of watts output, can run at perhaps 12 000 rev/min, but the majority of medium and large motors are usually designed for speeds below 3000 rev/min.

Increasingly, motors are being supplied with power-electronic drives, which draw power from the a.c. mains and convert it to d.c. for the motor. Since the mains voltages tend to be standardised (e.g. 110 V, 220–240 V, or 380–440 V, 50 or 60 Hz), motors are made with rated voltages which match the range of d.c. outputs from the converter (see Chapter 2).

As mentioned above, it is quite normal for a motor of a given power, speed and size to be available in a range of different voltages. In principle, all that has to be done is to alter the number of turns and the size of wire making up the coils in the machine. A 12 V, 4 A motor, for example, could easily be made to operate from 24 V instead, by winding its coils with twice as many turns of wire having only half the cross-sectional area of the original. The full speed would be the same at 24 V as the original was at 12 V, and the rated current would be 2 A, rather than 4 A. The input power and output power would be un-changed, and externally there would be no change in appearance, except that the terminals might be a bit smaller.

Traditionally d.c. motors were classified as shunt, series or separately excited. In addition it was common to see motors referred to as 'compound-wound'. These descriptions date from the period before the advent of power electronics, and a strong association built up, linking one or other 'type' of d.c. machine with a particular application. There is really no fundamental difference between shunt, series or separately excited machines, and the names simply reflect the way in which the field and armature circuits are interconnected. The terms still persist, however, and we will refer to them again later. But first we must gain an understanding of how the basic machine operates, so that we are equipped to understand what the various historic terms mean, and hence see how modern practice is deployed to achieve the same ends.

We should make clear at this point that whereas in an a.c. machine the number of poles is of prime importance in determining the speed, the pole number in a d.c. machine is of little consequence as far as the user is concerned. It turns out to be more economical to use two or four poles in small or medium size d.c. motors, and more (e.g. ten or twelve or even more) in large ones, but the only difference to the user is that the 2-pole type will have two brushes at 180°, the 4-pole will have four brushes at 90°, and so on. Most of our discussion centres on the 2-pole version in the interests of simplicity, but there is no essential difference as far as operating characteristics are concerned.

TORQUE PRODUCTION

Torque is produced by the interaction between the axial current-carrying conductors on the rotor and the radial magnetic flux produced by the stator. The flux or 'excitation' can be furnished by permanent magnets (Figure 3.2(a)) or by means of field windings (Figures 3.1 and 3.2(b)).

Permanent magnet versions are available in motors with outputs from a few watts up to a few kilowatts, while wound-field machines begin at about 100 W and extend to the largest (MW) outputs. The advantages of the permanent magnet type are that no electrical supply is required for the field, and the overall size of the motor can be smaller. On the other hand, the strength of the field cannot be varied, so one possible option for control is ruled out.

Ferrite magnets have been used for many years. They are relatively cheap and easy to manufacture but their energy product (a measure of their effectiveness as a source of excitation) is poor. Rare earth magnets (e.g. neodymium–iron–boron or samarium–cobalt) provide much higher energy products, and yield high torque/volume ratios: they are used in high-performance servo motors, but are relatively expensive and

Plate 3.1   *Cutaway view of 4-pole d.c. motor. The skewed armature windings on the rotor are connected at the right-hand-end via risers to the commutator segments. There are four sets of brushes, and each brush arm holds four brush-boxes. The sectioned coils surrounding two of the four poles are visible close to the left-hand armature end-windings. (Photo courtesy of ABB)*

difficult to manufacture and handle. Nd–Fe–B magnets have the highest energy product but can only be operated at temperatures below about 150°C, which is not sufficient for some high-performance motors.

Although the magnetic field is essential to the operation of the motor, we should recall that in Chapter 1 we saw that none of the mechanical output power actually comes from the field system. The excitation acts like a catalyst in a chemical reaction, making the energy conversion possible but not contributing to the output.

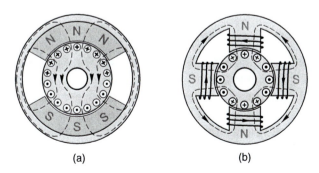

Figure 3.2   *Excitation (field) systems for d.c. motors* (a) *2-pole permanent magnet;* (b) *4-pole wound field*

The main (power) circuit consists of a set of identical coils wound in slots on the rotor, and known as the armature. Current is fed into and out of the rotor via carbon 'brushes' which make sliding contact with the 'commutator', which consists of insulated copper segments mounted on a cylindrical former. (The term 'brush' stems from the early attempts to make sliding contacts using bundles of wires bound together in much the same way as the willow twigs in a witch's broomstick. Not surprisingly these primitive brushes soon wore grooves in the commutator.)

The function of the commutator is discussed below, but it is worth stressing here that all the electrical energy which is to be converted into mechanical output has to be fed into the motor through the brushes and commutator. Given that a high-speed sliding electrical contact is involved, it is not surprising that to ensure trouble-free operation the commutator needs to be kept clean, and the brushes and their associated springs need to be regularly serviced. Brushes wear away, of course, though if properly set they can last for thousands of hours. All being well, the brush debris (in the form of graphite particles) will be carried out of harm's way by the ventilating air: any build up of dust on the insulation of the windings of a high-voltage motor risks the danger of short circuits, while debris on the commutator itself is dangerous and can lead to disastrous 'flashover' faults.

The axial length of the commutator depends on the current it has to handle. Small motors usually have one brush on each side of the commutator, so the commutator is quite short, but larger heavy-current motors may well have many brushes mounted on a common arm, each with its own brushbox (in which it is free to slide) and with all the brushes on one arm connected in parallel via their flexible copper leads or 'pigtails'. The length of the commutator can then be comparable with the 'active' length of the armature (i.e. the part carrying the conductors exposed to the radial flux).

Function of the commutator

Many different winding arrangements are used for d.c. armatures, and it is neither helpful nor necessary for us to delve into the nitty-gritty of winding and commutator design. These issues are best left to motor designers and repairers. What we need to do is to focus on what a well-designed commutator-winding actually achieves, and despite the apparent complexity, this can be stated quite simply.

The purpose of the commutator is to ensure that regardless of the position of the rotor, the pattern of current flow in the rotor is always as shown in Figure 3.3.

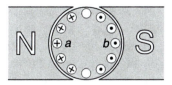

Figure 3.3   *Pattern of rotor (armature) currents in 2-pole d.c. motor*

Current enters the rotor via one brush, flows through all the rotor coils in the directions shown in Figure 3.3, and leaves via the other brush. The first point of contact with the armature is via the commutator segment or segments on which the brush is pressing at the time (the brush is usually wider than a single segment), but since the interconnections between the individual coils are made at each commutator segment, the current actually passes through all the coils via all the commutator segments in its path through the armature.

We can see from Figure 3.3 that all the conductors lying under the N pole carry current in one direction, while all those under the S pole carry current in the opposite direction. All the conductors under the N pole will therefore experience a downward force (which is proportional to the radial flux density B and the armature current I) while all the conductors under the S pole will experience an equal upward force. A torque is thus produced on the rotor, the magnitude of the torque being proportional to the product of the flux density and the current. In practice the flux density will not be completely uniform under the pole, so the force on some of the armature conductors will be greater than on others. However, it is straightforward to show that the total torque developed is given by

$$T = K_T \Phi I \qquad (3.1)$$

where Φ is the total flux produced by the field, and K_T is constant for a given motor. In the majority of motors the flux remains constant, so we see that the motor torque is directly proportional to the armature current. This extremely simple result means that if a motor is required to produce constant torque at all speeds, we simply have to keep the armature current constant. Standard drive packages usually include provision for doing this, as will be seen later. We can also see from equation (3.1) that the direction of the torque can be reversed by reversing either the armature current (I) or the flux (Φ). We obviously make use of this when we want the motor to run in reverse, and sometimes when we want regenerative braking.

The alert reader might rightly challenge the claim made above – that the torque will be constant regardless of rotor position. Looking at Figure 3.3, it should be clear that if the rotor turned just a few degrees, one of the five conductors shown as being under the pole will move out into the region where there is no radial flux, before the next one moves under the pole. Instead of five conductors producing force, there will then be only four, so won't the torque be reduced accordingly?

The answer to this question is yes, and it is to limit this unwelcome variation of torque that most motors have many more coils than shown in Figure 3.3. Smooth torque is of course desirable in most applications in order to avoid vibrations and resonances in the transmission and load, and is essential in machine tool drives where the quality of finish can be marred by uneven cutting if the torque and speed are not steady.

Broadly speaking, the higher the number of coils (and commutator segments) the better, because the ideal armature would be one in which the pattern of current on the rotor corresponded to a 'current sheet', rather than a series of discrete packets of current. If the number of coils was infinite, the rotor would look identical at every position, and the torque would therefore be absolutely smooth. Obviously this is not practicable, but it is closely approximated in most d.c. motors. For practical and economic reasons the number of slots is higher in large motors, which may well have a hundred or more coils, and hence very little ripple in their output torque.

Operation of the commutator – interpoles

Returning now to the operation of the commutator, and focusing on a particular coil (e.g. the one shown as *ab* in Figure 3.3) we note that for half a revolution – while side *a* is under the N pole and side *b* is under the S pole, the current needs to be positive in side *a* and negative in side *b* in order to produce a positive torque. For the other half revolution, while side *a* is under the S pole and side *b* is under the N pole, the current must flow in the opposite direction through the coil for it to continue to produce positive torque. This reversal of current takes place in each coil as it passes through the interpolar axis, the coil being 'switched-round' by the action of the commutator sliding under the brush. Each time a coil reaches this position it is said to be undergoing commutation, and the relevant coil in Figure 3.3 has therefore been shown as having no current to indicate that its current is in the process of changing from positive to negative.

The essence of the current-reversal mechanism is revealed by the simplified sketch shown in Figure 3.4. This diagram shows a single coil

 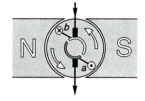

Figure 3.4 *Simplified diagram of single-coil motor to illustrate the current-reversing function of the commutator*

fed via the commutator, and brushes with current that always flows in at the top brush.

In the left-hand sketch, coil-side *a* is under the N pole and carries positive current because it is connected to the shaded commutator segment which in turn is fed from the top brush. Side *a* is therefore exposed to a flux density directed from left (N) to right (S) in the sketch, and will therefore experience a downward force. This force will remain constant while the coil-side remains under the N pole. Conversely, side *b* has negative current but it also lies in a flux density directed from right to left, so it experiences an upward force. There is thus an anti-clockwise torque on the rotor.

When the rotor turns to the position shown in the sketch on the right, the current in both sides is reversed, because side *b* is now fed with positive current via the unshaded commutator segment. The direction of force on each coil side is reversed, which is exactly what we want in order for the torque to remain clockwise. Apart from the short period when the coil is outside the influence of the flux, and undergoing commutation (current reversal), the torque is constant.

It should be stressed that the discussion above is intended to illustrate the principle involved, and the sketch should not be taken too literally. In a real multi-coil armature, the commutator arc is much smaller than that shown in Figure 3.4 and only one of the many coils is reversed at a time, so the torque remains very nearly constant regardless of the position of the rotor.

The main difficulty in achieving good commutation arises because of the self inductance of the armature coils and the associated stored energy. As we have seen earlier, inductive circuits tend to resist change in current, and if the current reversal has not been fully completed by the time the brush slides off the commutator segment in question, there will be a spark at the trailing edge of the brush.

In small motors some sparking is considered tolerable, but in medium and large wound-field motors small additional stator poles known as

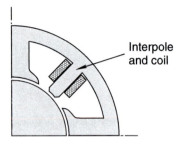

Figure 3.5   *Sketch showing location of interpole and interpole winding. (The main field windings have been omitted for the sake of clarity.)*

interpoles (or compoles) are provided to improve commutation and hence minimise sparking. These extra poles are located midway between the main field poles, as shown in Figure 3.5. Interpoles are not normally required in permanent magnet motors because the absence of stator iron close to the rotor coils results in much lower armature coil inductance.

The purpose of the interpoles is to induce a motional e.m.f. in the coil undergoing commutation, in such a direction as to speed-up the desired reversal of current, and thereby prevent sparking. The e.m.f. needed is proportional to the current (armature current) which has to be commutated, and to the speed of rotation. The correct e.m.f. is therefore achieved by passing the armature current through the coils on the interpoles, thereby making the flux from the interpoles proportional to the armature current. The interpole coils therefore consist of a few turns of thick conductor, connected permanently in series with the armature.

MOTIONAL E.M.F.

Readers who have skipped Chapter 1 are advised to check that they are familiar with the material covered in Section 1.7 before reading the rest of this chapter, as not all of the lessons discussed in Chapter 1 are repeated explicitly here.

When the armature is stationary, no motional e.m.f. is induced in it. But when the rotor turns, the armature conductors cut the radial magnetic flux and an e.m.f. is induced in them.

As far as each individual coil on the armature is concerned, an alternating e.m.f. will be induced in it when the rotor rotates. For the coil *ab* in Figure 3.3, for example, side *a* will be moving upward through the flux if the rotation is clockwise, and an e.m.f. directed out-of-the-plane of the paper will be generated. At the same time the 'return' side of the coil *b* will be moving downwards, so the same magnitude of e.m.f.

will be generated, but directed into the paper. The resultant e.m.f. in the coil will therefore be twice that in the coil-side, and this e.m.f. will remain constant for almost half a revolution, during which time the coil sides are cutting a constant flux density. For the comparatively short time when the coil is not cutting any flux, the e.m.f. will be zero and then the coil will begin to cut through the flux again, but now each side is under the other pole, so the e.m.f. is in the opposite direction. The resultant e.m.f. waveform in each coil is therefore a rectangular alternating wave, with magnitude and frequency proportional to the speed of rotation.

The coils on the rotor are connected in series, so if we were to look at the e.m.f. across any given pair of diametrically opposite commutator segments, we would see a large alternating e.m.f. (We would have to station ourselves on the rotor to do this, or else make sliding contacts using slip-rings.)

The fact that the induced voltage in the rotor is alternating may come as a surprise, since we are talking about a d.c. motor rather than an a.c. motor. But any worries we may have should be dispelled when we ask what we will see by way of induced e.m.f. when we 'look in' at the brushes. We will see that the brushes and commutator effect a remarkable transformation, bringing us back into the reassuring world of d.c.

The first point to note is that the brushes are stationary. This means that although a particular segment under each brush is continually being replaced by its neighbour, the circuit lying between the two brushes always consists of the same number of coils, with the same orientation with respect to the poles. As a result the e.m.f. at the brushes is direct (i.e. constant), rather than alternating.

The magnitude of the e.m.f. depends on the position of the brushes around the commutator, but they are invariably placed at the point where they continually 'see' the peak value of the alternating e.m.f. induced in the armature. In effect, the commutator and brushes can be regarded as mechanical rectifier, which converts the alternating e.m.f. in the rotating reference frame to a direct e.m.f. in the stationary reference frame. It is a remarkably clever and effective device, its only real drawback being that it is a mechanical system, and therefore subject to wear and tear.

We saw earlier that to obtain smooth torque it was necessary to have a large number of coils and commutator segments, and we find that much the same considerations apply to the smoothness of the generated e.m.f. If there are only a few armature coils, the e.m.f. will have a noticeable ripple superimposed on the mean d.c. level. The higher we make the number of coils, the smaller the ripple, and the better the d.c. we produce. The small ripple we inevitably get with a finite number of segments is seldom any problem with motors used in drives, but can sometimes give rise to difficulties when a d.c. machine is used to provide a speed feedback signal in a closed-loop system (see Chapter 4).

In Chapter 1 we saw that when a conductor of length l moves at velocity v through a flux density B, the motional e.m.f. induced is given by $e = Blv$. In the complete machine we have many series-connected conductors; the linear velocity (v) of the primitive machine examined in Chapter 1 is replaced by the tangential velocity of the rotor conductors, which is clearly proportional to the speed of rotation (n); and the average flux density cut by each conductor (B) is directly related to the total flux (Φ). If we roll together the other influential factors (number of conductors, radius, active length of rotor) into a single constant (K_E), it follows that the magnitude of the resultant e.m.f. (E) which is generated at the brushes is given by

$$E = K_E \Phi n \tag{3.2}$$

This equation reminds us of the key role of the flux, in that until we switch on the field no voltage will be generated, no matter how fast the rotor turns. Once the field is energised, the generated voltage is directly proportional to the speed of rotation, so if we reverse the direction of rotation, we will also reverse the polarity of the generated e.m.f. We should also remember that the e.m.f. depends only on the flux and the speed, and is the same regardless of whether the rotation is provided by some external source (i.e. when the machine is being driven as a generator) or when the rotation is produced by the machine itself (i.e. when it is acting as a motor).

It has already been mentioned that the flux is usually constant at its full value, in which case equations (3.1) and (3.2) can be written in the form

$$T = k_t I \tag{3.3}$$

$$E = k_e \omega \tag{3.4}$$

where k_t is the motor torque constant, k_e is the e.m.f. constant, and ω is the angular speed in rad/s.

In this book, the international standard (SI) system of units is used throughout. In the SI system (which succeeded the MKS (meter, kilogram, second) system, the units for k_t are the units of torque (newton metre) divided by the unit of current (ampere), i.e. Nm/A; and the units of k_e units are volts/rad/s. (Note, however, that k_e is more often given in volts/1000 rev/min.)

It is not at all clear that the units for the torque constant (Nm/A) and the e.m.f. constant (V/rad/s), which on the face of it measure very different physical phenomena, are in fact the same, i.e. 1 Nm/A = 1 V/rad/s. Some readers will be content simply to accept it, others may be puzzled, a few may even find it obvious. Those who are surprised and puzzled may feel more comfortable by progressively replacing one set of units by their equivalent, to lead us in the direction of the other, e.g.

$$\frac{(newton)\ (metre)}{ampere} = \frac{joule}{ampere} = \frac{(watt)\ (second)}{ampere}$$

$$= \frac{(volt)\ (ampere)\ (second)}{ampere}$$

$$= (volt)(second)$$

This still leaves us to ponder what happened to the 'radians' in k_e, but at least the underlying unity is demonstrated, and after all a radian is a dimensionless quantity. Delving deeper, we note that 1 volt × 1 second = 1 weber, the unit of magnetic flux. This is hardly surprising because the production of torque and the generation of motional e.m.f. are both brought about by the catalytic action of the magnetic flux.

Returning to more pragmatic issues, we have now discovered the extremely convenient fact that in SI units, the torque and e.m.f. constants are equal, i.e. $k_t = k_e = k$. The torque and e.m.f. equations can thus be further simplified as

$$T = kI \tag{3.5}$$

$$E = k\omega \tag{3.6}$$

We will make use of these two delightfully simple equations time and again in the subsequent discussion. Together with the armature voltage equation (see below), they allow us to predict all aspects of behaviour of a d.c. motor. There can be few such versatile machines for which the fundamentals can be expressed so simply.

Though attention has been focused on the motional e.m.f. in the conductors, we must not overlook the fact that motional e.m.f.s are also induced in the body of the rotor. If we consider a rotor tooth, for example, it should be clear that it will have an alternating e.m.f. induced in it as the rotor turns, in just the same way as the e.m.f. induced in the adjacent conductor. In the machine shown in Figure 3.1, for example, when the e.m.f. in a tooth under a N pole is positive, the e.m.f in the diametrically opposite tooth (under a S pole) will be negative. Given that the rotor steel conducts electricity, these e.m.f.s will tend to set up circulating currents in the body of the rotor, so to prevent this happening, the rotor is made not from a solid piece but from thin steel laminations (typically less than 1 mm thick), which have an insulated coating to prevent the flow of unwanted currents. If the rotor was not laminated the induced current would not only produce large quantities of waste heat, but also exert a substantial braking torque.

Equivalent circuit

The equivalent circuit can now be drawn on the same basis as we used for the primitive machine in Chapter 1, and is shown in Figure 3.6.

The voltage V is the voltage applied to the armature terminals (i.e. across the brushes), and E is the internally developed motional e.m.f. The resistance and inductance of the complete armature are represented by R and L in Figure 3.6. The sign convention adopted is the usual one when the machine is operating as a motor. Under motoring conditions, the motional e.m.f. E always opposes the applied voltage V, and for this reason it is referred to as 'back e.m.f.' For current to be forced into the motor, V must be greater than E, the armature circuit voltage equation being given by

$$V = E + IR + L\frac{\mathrm{d}I}{\mathrm{d}t} \tag{3.7}$$

The last term in equation (3.7) represents the inductive volt-drop due to the armature self-inductance. This voltage is proportional to the rate of change of current, so under steady-state conditions (when the current is

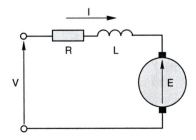

Figure 3.6   *Equivalent circuit of a d.c. motor*

constant), the term will be zero and can be ignored. We will see later that the armature inductance has an unwelcome effect under transient conditions, but is also very beneficial in smoothing the current waveform when the motor is supplied by a controlled rectifier.

D.C. MOTOR – STEADY-STATE CHARACTERISTICS

From the user's viewpoint the extent to which speed falls when load is applied, and the variation in speed with applied voltage are usually the first questions that need to be answered in order to assess the suitability of the motor for the job in hand. The information is usually conveyed in the form of the steady-state characteristics, which indicate how the motor behaves when any transient effects (caused for example by a sudden change in the load) have died away and conditions have once again become steady. Steady-state characteristics are usually much easier to predict than transient characteristics, and for the d.c. machine they can all be deduced from the simple equivalent circuit in Figure 3.6.

Under steady-state conditions, the armature current I is constant and equation (3.7) simplifies to

$$V = E + IR \quad \text{or} \quad I = \frac{(V - E)}{R} \tag{3.8}$$

This equation allows us to find the current if we know the applied voltage, the speed (from which we get E via equation (3.6)) and the armature resistance, and we can then obtain the torque from equation (3.5). Alternatively, we may begin with torque and speed, and work out what voltage will be needed.

We will derive the steady-state torque–speed characteristics for any given armature voltage V in Section 3.4.3, but first we begin by establishing the relationship between the no-load speed and the armature voltage, since this is the foundation on which the speed control philosophy is based.

No-load speed

By 'no-load' we mean that the motor is running light, so that the only mechanical resistance is due to its own friction. In any sensible motor the frictional torque will be small, and only a small driving torque will therefore be needed to keep the motor running. Since motor torque is proportional to current (equation (3.5)), the no-load current will also be small. If we assume that the no-load current is in fact zero, the calcula-tion of no-load speed becomes very simple. We note from equation (3.8) that zero current implies that the back e.m.f. is equal to the applied voltage, while equation (3.2) shows that the back e.m.f. is proportional to speed. Hence under true no-load (zero torque) conditions, we obtain

$$V = E = K_E \Phi n \quad \text{or} \quad n = \frac{V}{K_E \Phi} \tag{3.9}$$

where n is the speed. (We have used equation (3.8) for the e.m.f., rather than the simpler equation (3.4) because the latter only applies when the flux is at its full value, and in the present context it is important for us to see what happens when the flux is reduced.)

At this stage we are concentrating on the steady-state running speeds, but we are bound to wonder how it is that the motor attains speed from rest. We will return to this when we look at transient behaviour, so for the moment it is sufficient to recall that we came across an equation identical to equation (3.9) when we looked at the primitive linear motor in Chapter 1. We saw that if there was no resisting force opposing the motion, the speed would rise until the back e.m.f. equalled the supply voltage. The same result clearly applies to the frictionless and unloaded d.c. motor here.

We see from equation (3.9) that the no-load speed is directly propor-tional to armature voltage, and inversely proportional to field flux. For the moment we will continue to consider the case where the flux is constant, and demonstrate by means of an example that the approxima-tions used in arriving at equation (3.9) are justified in practice. Later, we can use the same example to study the torque–speed characteristic.

Performance calculation – example

Consider a 500 V, 9.1 kW, 20 A, permanent-magnet motor with an armature resistance of 1 Ω. (These values tell us that the normal oper-ating voltage is 500 V, the current when the motor is fully loaded is 20 A, and the mechanical output power under these full-load conditions is 9.1 kW.) When supplied at 500 V, the unloaded motor is found to run at 1040 rev/min., drawing a current of 0.8 A.

Whenever the motor is running at a steady speed, the torque it produces must be equal (and opposite) to the total opposing or load torque: if the motor torque was less than the load torque, it would decelerate, and if the motor torque was higher than the load torque it would accelerate. From equation (3.3), we see that the motor torque is determined by its current, so we can make the important statement that, in the steady state, the motor current will be determined by the mechanical load torque. When we make use of the equivalent circuit (Figure 3.6) under steady-state conditions, we will need to get used to the idea that the current is determined by the load torque – i.e. one of the principal 'inputs' which will allow us to solve the circuit equations is the mechanical load torque, which is not shown on the diagram. For those who are not used to electromechanical interactions this can be a source of difficulty.

Returning to our example, we note that because it is a real motor, it draws a small current (and therefore produces some torque) even when unloaded. The fact that it needs to produce torque, even though no load torque has been applied and it is not accelerating, is attributable to the inevitable friction in the cooling fan, bearings, and brushgear.

If we want to estimate the no-load speed at a different armature voltage (say 250 V), we would ignore the small no-load current and use equation (3.9), giving

$$\text{no-load speed at } 250\,V = (250/500) \times 1040 = 520\,\text{rev/min}$$

Since equation (3.9) is based on the assumption that the no-load current is zero, this result is only approximate.

If we insist on being more precise, we must first calculate the original value of the back e.m.f., using equation (3.8), which gives

$$E = 500 - (0.8 \times 1) = 499.2\,V$$

As expected, the back e.m.f. is almost equal to the applied voltage. The corresponding speed is 1040 rev/min, so the e.m.f. constant must be 499.2/1040 or 480 V/1000 rev/min. To calculate the no-load speed for $V = 250\,V$, we first need to know the current. We are not told anything about how the friction torque varies with speed so all we can do is to assume that the friction torque is constant, in which case the no-load current will be 0.8 A regardless of speed. With this assumption, the back e.m.f. will be given by

$$E = 250 - (0.8 \times 1) = 249.2V$$

And hence the speed will be given by

$$\text{no-load speed at } 250\,V = \frac{249.2}{480} \times 1000 = 519.2\,\text{rev/min}$$

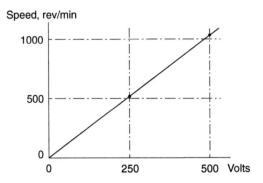

Figure 3.7   *No-load speed of d.c. motor as a function of armature voltage*

The difference between the approximate and true no-load speeds is very small and is unlikely to be significant. Hence we can safely use equation (3.9) to predict the no-load speed at any armature voltage and obtain the set of no-load speeds shown in Figure 3.7. This diagram illustrates the very simple linear relationship between the speed of an unloaded d.c. motor and the armature voltage.

Behaviour when loaded

Having seen that the no-load speed of the motor is directly proportional to the armature voltage, we need to explore how the speed will vary when we change the load on the shaft.

The usual way we quantify 'load' is to specify the torque needed to drive the load at a particular speed. Some loads, such as a simple drum-type hoist with a constant weight on the hook, require the same torque regardless of speed, but for most loads the torque needed varies with the speed. For a fan, for example, the torque needed varies roughly with the square of the speed. If we know the torque/speed characteristic of the load, and the torque/speed characteristic of the motor, we can find the steady-state speed simply by finding the intersection of the two curves in the torque–speed plane. An example (not specific to a d.c. motor) is shown in Figure 3.8.

At point X the torque produced by the motor is exactly equal to the torque needed to keep the load turning, so the motor and load are in equilibrium and the speed remains steady. At all lower speeds, the motor torque will be higher than the load torque, so the nett torque will be positive, leading to an acceleration of the motor. As the speed rises towards X, the acceleration reduces until the speed stabilises at X. Conversely, at speeds above X the motor's driving torque is less than

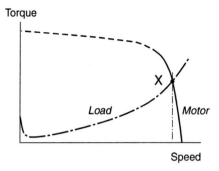

Figure 3.8   *Steady-state torque–speed curves for motor and load showing location (X) of steady-state operating condition*

the braking torque exerted by the load, so the nett torque is negative and the system will decelerate until it reaches equilibrium at X. This example is one which is inherently stable, so that if the speed is disturbed for some reason from the point X, it will always return there when the disturbance is removed.

Turning now to the derivation of the torque/speed characteristics of the d.c. motor, we can profitably use the previous example to illustrate matters. We can obtain the full-load speed for $V = 500$ V by first calculating the back e.m.f. at full load (i.e. when the current is 20 A). From equation (3.8) we obtain

$$E = 500 - (20 \times 1) = 480 \, V$$

We have already seen that the e.m.f. constant is 480 V/1000 rev/min, so the full load speed is clearly 1000 rev/min. From no-load to full-load, the speed falls linearly, giving the torque–speed curve for $V = 500$ V shown in Figure 3.9. Note that from no-load to full-load, the speed falls from 1040 rev/min to 1000 rev/min, a drop of only 4%. Over the same range the back e.m.f. falls from very nearly 500 V to 480 V, which of course also represents a drop of 4%.

We can check the power balance using the same approach as in Section 1.7 of Chapter 1. At full load the electrical input power is given by VI, i.e. 500 V $\times$ 20 A = 10 kW. The power loss in the armature resistance is $I^2R = 400 \times 1 = 400$ W. The power converted from electrical to mechanical form is given by EI, i.e. $480 \, V \times 20 \, A = 9600 \, W$. We can see from the no-load data that the power required to overcome friction and iron losses (eddy currents and hysteresis, mainly in the rotor) at no-load is approximately $500 \, V \times 0.8 \, A = 400 \, W$, so this leaves about 9.2 kW. The rated output power of 9.1 kW indicates that 100 W of additional losses (which we will not attempt to explore here) can be expected under full-load conditions.

Two important observations follow from these calculations. Firstly, the speed drop with load is very small. This is very desirable for most applications, since all we have to do to maintain almost constant speed is to set the appropriate armature voltage and keep it constant. Secondly, a delicate balance between V and E is revealed. The current is in fact proportional to the difference between V and E (equation (3.8)), so that quite small changes in either V or E give rise to disproportionately large changes in the current. In the example, a 4% reduction in E causes the current to rise to its rated value. Hence to avoid excessive currents (which cannot be tolerated in a thyristor supply, for example), the difference between V and E must be limited. This point will be taken up again when transient performance is explored.

A representative family of torque–speed characteristics for the motor discussed above is shown in Figure 3.9. As already explained, the no-load speeds are directly proportional to the applied voltage, while the slope of each curve is the same, being determined by the armature resistance: the smaller the resistance the less the speed falls with load. These operating characteristics are very attractive because the speed can be set simply by applying the correct voltage.

The upper region of each characteristic in Figure 3.9 is shown dotted because in this region the armature current is above its rated value, and the motor cannot therefore be operated continuously without overheating. Motors can and do operate for short periods above rated current, and the fact that the d.c. machine can continue to provide torque in proportion to current well into the overload region makes it particularly well-suited to applications requiring the occasional boost of excess torque.

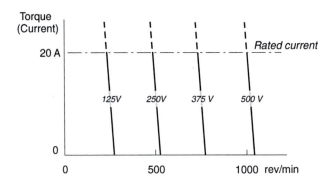

Figure 3.9   *Family of steady-state torque–speed curves for a range of armature voltages*

A cooling problem might be expected when motors are run continuously at full current (i.e. full torque) even at very low speed, where the natural ventilation is poor. This operating condition is considered quite normal in converter-fed motor drive systems, and motors are accordingly fitted with a small air-blower motor as standard.

This book is about motors, which convert electrical power into mechanical power. But, in common with all electrical machines, the d.c. motor is inherently capable of operating as a generator, converting mechanical power into electrical power. And although the overwhelming majority of motors will spend most of their working lives in motoring mode, there are applications such as rolling mills where frequent reversal is called for, and others where rapid braking is required. In the former, the motor is controlled so that it returns the stored kinetic energy to the supply system each time the rolls have to be reversed, while in the latter case the energy may also be returned to the supply, or dumped as heat in a resistor. These transient modes of operation may better be described as 'regeneration' since they only involve recovery of energy originally provided by the motor.

Continuous generation is of course possible using a d.c. machine provided we have a source of mechanical power, such as an internal combustion (IC) engine. In the example discussed above we saw that when connected to a 500 V supply, the unloaded machine ran at 1040 rev/min, at which point the back e.m.f. was very nearly 500 V and only a tiny positive current was flowing. As we applied mechanical load to the shaft the steady-state speed fell, thereby reducing the back e.m.f. and increasing the armature current until the motor toque was equal to the opposing load torque and equilibrium is attained. We saw that the smaller the armature resistance, the less the drop in speed with load.

Conversely, if instead of applying an opposing (load) torque, we use the IC engine to supply torque in the opposite direction, i.e. trying to increase the speed of the motor, the increase in speed will cause the motional e.m.f. to be greater than the supply voltage (500 V). This means that the current will flow from the d.c. machine to the supply, resulting in a reversal of power flow. Stable generating conditions will be achieved when the motor torque (current) is equal and opposite to the torque provided by the IC engine. In the example, the full-load current is 20 A, so in order to drive this current through its own resistance and overcome the supply voltage, the e.m.f. must be given by

$$E = IR + V = (20 \times 1) + 500 = 520\,\text{V}$$

The corresponding speed can be calculated by reference to the no-load e.m.f. (499.2 V at 1040 rev/min) from which the steady generating speed is given by

$$\frac{N_{\text{gen}}}{1040} = \frac{520}{499.2} \text{ i.e. } N_{\text{gen}} = 1083 \text{ rev/min}.$$

On the torque–speed plot (Figure 3.9) this condition lies on the downward projection of the 500 V characteristic at a current of −20 A. We note that the full range of operation, from full-load motoring to full-load generating is accomplished with only a modest change in speed from 1000 to 1083 rev/min.

It is worth emphasising that in order to make the unloaded motor move into the generating mode, all that we had to do was to start supplying mechanical power to the motor shaft. No physical changes had to be made to the motor to make it into a generator – the hardware is equally at home functioning as a motor or as a generator – which is why it is best referred to as a 'machine'. (How nice it would be if the IC engine could do the same; whenever we slowed down, we could watch the rising gauge as its kinetic energy is converted back into hydrocarbon fuel in the tank!)

To complete this section we will derive the analytical expression for the steady-state speed as a function of the two variables that we can control, i.e. the applied voltage (V), and the load torque (T_L). Under steady-state conditions the armature current is constant and we can therefore ignore the armature inductance term in equation (3.7); and because there is no acceleration, the motor torque is equal to the load torque. Hence by eliminating the current I between equations (3.5) and (3.7), and substituting for E from equation (3.6) the speed is given by

$$\omega = \frac{V}{k} - \frac{R}{k^2} T_L \qquad (3.10)$$

This equation represents a straight line in the speed/torque plane, as we saw with our previous worked example. The first term shows that the no-load speed is directly proportional to the armature voltage, while the second term gives the drop in speed for a given load torque. The gradient or slope of the torque–speed curve is $-R/k^2$, showing again that the smaller the armature resistance, the smaller the drop in speed when load is applied.

Base speed and field weakening

Returning to our consideration of motor operating characteristics, when the field flux is at its full value the speed corresponding to full armature voltage and full current (i.e. the rated full-load condition) is known as base speed (see Figure 3.10).

The motor can operate at any speed up to base speed, and at any torque (current) up to the rated value by appropriate choice of armature voltage. This full flux region of operation is indicated by the shaded area *Oabc* in Figure 3.10, and is often referred to as the 'constant torque' region of the torque–speed plane. In this context 'constant torque' signifies that at any speed below base speed the motor is capable of producing its full rated torque. Note that the term constant torque does not mean that the motor *will* produce constant torque, but rather it signifies that the motor *can* produce constant torque if required: as we have already seen, it is the mechanical load we apply to the shaft that determines the steady-state torque produced by the motor.

When the current is at maximum (i.e. along the line *ab* in Figure 3.10), the torque is at its maximum (rated) value. Since mechanical power is given by torque times speed, the power output along *ab* is proportional to the speed, and the maximum power thus corresponds to the point *b* in Figure 3.10. At point *b*, both the voltage and current have their full rated values.

To run faster than base speed the field flux must be reduced, as indicated by equation (3.9). Operation with reduced flux is known as 'field weakening', and we have already discussed this perhaps surprising mode in connection with the primitive linear motor in Chapter 1. For example, by halving the flux (and keeping the armature voltage at its full value), the no-load speed is doubled (point *d* in Figure 3.10). The

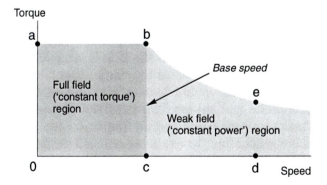

Figure 3.10   *Region of continuous operation in the torque–speed plane*

increase in speed is however obtained at the expense of available torque, which is proportional to flux times current (see equation (3.1)). The current is limited to rated value, so if the flux is halved, the speed will double but the maximum torque which can be developed is only half the rated value (point *e* in Figure 3.10). Note that at the point *e* both the armature voltage and the armature current again have their full rated values, so the power is at maximum, as it was at point *b*. The power is constant along the curve through *b* and *e*, and for this reason the shaded area to the right of the line *bc* is referred to as the 'constant power' region. Obviously, field weakening is only satisfactory for applications which do not demand full torque at high speeds, such as electric traction.

The maximum allowable speed under weak field conditions must be limited (to avoid excessive sparking at the commutator), and is usually indicated on the motor rating plate. A marking of 1200/1750 rev/min, for example, would indicate a base speed of 1200 rev/min, and a maximum speed with field weakening of 1750 rev/min. The field weakening range varies widely depending on the motor design, but maximum speed rarely exceeds three or four times the base speed.

To sum up, the speed is controlled as follows:

- Below base speed, the flux is maximum, and the speed is set by the armature voltage. Full torque is available at any speed.
- Above base speed, the armature voltage is at (or close to) maximum and the flux is reduced in order to raise the speed. The maximum torque available reduces in proportion to the flux.

To judge the suitability of a motor for a particular application we need to compare the torque–speed characteristic of the prospective load with the operating diagram for the motor: if the load torque calls for operation outside the shaded areas of Figure 3.10, a larger motor is clearly called for.

Finally, we should note that according to equation (3.9), the no-load speed will become infinite if the flux is reduced to zero. This seems unlikely; after all, we have seen that the field is essential for the motor to operate, so it seems unreasonable to imagine that if we removed the field altogether, the speed would rise to infinity. In fact, the explanation lies in the assumption that 'no-load' for a real motor means zero torque. If we could make a motor which had no friction torque whatsoever, the speed would indeed continue to rise as we reduced the field flux towards zero. But as we reduced the flux, the torque per ampere of armature current would become smaller and smaller, so in a real machine with friction, there will come a point where the torque being produced by the motor is equal to the friction torque, and the speed will therefore be limited. Nevertheless, it is quite dangerous to open circuit the field winding, especially in a large unloaded motor. There may be sufficient 'residual' magnetism left in the poles to produce significant accelerating torque to lead to a run-away situation. Usually, field and armature circuits are interlocked so that if the field is lost, the armature circuit is switched off automatically.

Armature reaction

In addition to deliberate field-weakening, as discussed above, the flux in a d.c. machine can be weakened by an effect known as 'armature reaction'. As its name implies, armature reaction relates to the influence that the armature MMF has on the flux in the machine: in small machines it is negligible, but in large machines the unwelcome field weakening caused by armature reaction is sufficient to warrant extra design features to combat it. A full discussion would be well beyond the needs of most users, but a brief explanation is included for the sake of completeness.

The way armature reaction occurs can best be appreciated by looking at Figure 3.1 and noting that the MMF of the armature conductors acts along the axis defined by the brushes, i.e. the armature MMF acts in quadrature to the main flux axis which lies along the stator poles. The reluctance in the quadrature direction is high because of the large air spaces that the flux has to cross, so despite the fact that the rotor MMF at full current can be very large, the quadrature flux is relatively small; and because it is perpendicular to the main flux, the average value of the latter would not be expected to be affected by the quadrature flux, even though part of the path of the reaction flux is shared with the main flux as it passes (horizontally in Figure 3.1) through the main pole-pieces.

A similar matter was addressed in relation to the primitive machine in Chapter 1. There it was explained that it was not necessary to take account of the flux produced by the conductor itself when calculating the electromagnetic force on it. And if it were not for the nonlinear phenomenon of magnetic saturation, the armature reaction flux would have no effect on the average value of the main flux in the machine shown in Figure 3.1: the flux density on one edge of the pole-pieces would be increased by the presence of the reaction flux, but decreased by the same amount on the other edge, leaving the average of the main flux unchanged. However if the iron in the main magnetic circuit is already some way into saturation, the presence of the rotor MMF will cause less of an increase on one edge than it causes by way of decrease on the other, and there will be a nett reduction in main flux.

We know that reducing the flux leads to an increase in speed, so we can now see that in a machine with pronounced armature reaction, when the load on the shaft is increased and the armature current increases to produce more torque, the field is simultaneously reduced and the motor speeds up. Though this behaviour is not a true case of instability, it is not generally regarded as desirable!

Large motors often carry additional windings fitted into slots in the pole-faces and connected in series with the armature. These 'compensating' windings produce an MMF in opposition to the armature MMF, thereby reducing or eliminating the armature reaction effect.

Maximum output power

We have seen that if the mechanical load on the shaft of the motor increases, the speed falls and the armature current automatically increases until equilibrium of torque is reached and the speed again becomes steady. If the armature voltage is at its maximum (rated) value, and we increase the mechanical load until the current reaches its rated value, we are clearly at full-load, i.e. we are operating at the full speed (determined by voltage) and the full torque (determined by current). The maximum current is set at the design stage, and reflects the tolerable level of heating of the armature conductors.

Clearly if we increase the load on the shaft still more, the current will exceed the safe value, and the motor will begin to overheat. But the question which this prompts is 'if it were not for the problem of overheating, could the motor deliver more and more power output, or is there a limit'?

We can see straightaway that there will be a maximum by looking at the torque–speed curves in Figure 3.9. The mechanical output power is the product of torque and speed, and we see that the power will be zero when either the load torque is zero (i.e. the motor is running light) or the speed is zero (i.e. the motor is stationary). There must be maximum between these two zeroes, and it is easy to show that the peak mechanical power occurs when the speed is half of the no-load speed. However, this operating condition is only practicable in very small motors: in the majority of motors, the supply would simply not be able to provide the very high current required.

Turning to the question of what determines the theoretical maximum power, we can apply the maximum power transfer theorem (from circuit theory) to the equivalent circuit in Figure 3.6. The inductance can be ignored because we assume d.c. conditions. If we regard the armature resistance R as if it were the resistance of the source V, the theorem tells us that in order to transfer maximum power to the load (represented by the motional e.m.f. on the right-hand side of Figure 3.6) we must make the load 'look like' a resistance equal to the source resistance, R. This condition is obtained when the applied voltage V divides equally so that half of it is dropped across R and the other half is equal to the e.m.f. E. (We note that the condition $E = V/2$ corresponds to the motor running at half the no-load speed, as stated above.) At maximum power, the current is $V/2R$, and the mechanical output power (EI) is given by $V^2/4R$.

The expression for the maximum output power is delightfully simple. We might have expected the maximum power to depend on other motor parameters, but in fact it is determined solely by the armature voltage and the armature resistance. For example, we can say immediately that a 12 V motor with an armature resistance of $1\,\Omega$ cannot possibly produce more than 36 W of mechanical output power.

We should of course observe that under maximum power conditions the overall efficiency is only 50% (because an equal power is burned off as heat in the armature resistance); and emphasise again that only very small motors can ever be operated continuously in this condition. For the vast majority of motors, it is of academic interest only because the current ($V/2R$) will be far too high for the supply.

TRANSIENT BEHAVIOUR – CURRENT SURGES

It has already been pointed out that the steady-state armature current depends on the small difference between the back e.m.f. E and the applied voltage V. In a converter-fed drive it is vital that the current is kept within safe bounds, otherwise the thyristors or transistors (which have very limited overcurrent capacity) will be destroyed, and it follows from equation (3.8) that in order to prevent the current from exceeding its rated value we cannot afford to let V and E differ by more than IR, where I is the rated current.

It would be unacceptable, for example, to attempt to bring all but the smallest of d.c. motors up to speed simply by switching on rated voltage. In the example studied earlier, rated voltage is 500 V, and the armature resistance is $1\,\Omega$. At standstill, the back e.m.f. is zero, and hence the initial current would be $500/1 = 500\,\text{A}$, or 25 times rated current! This would destroy the thyristors in the supply converter (and/or blow the fuses). Clearly the initial voltage we must apply is much less than 500 V;

and if we want to limit the current to rated value (20 A in the example) the voltage needed will be 20 × 1, i.e. only 20 V. As the speed picks up, the back e.m.f. rises, and to maintain the full current V must also be ramped up so that the difference between V and E remains constant at 20 V. Of course, the motor will not accelerate nearly so rapidly when the current is kept in check as it would if we had switched on full voltage, and allowed the current to do as it pleased. But this is the price we must pay in order to protect the converter.

Similar current-surge difficulties occur if the load on the motor is suddenly increased, because this will result in the motor slowing down, with a consequent fall in E. In a sense we welcome the fall in E because this brings about the increase in current needed to supply the extra load, but of course we only want the current to rise to its rated value; beyond that point we must be ready to reduce V, to prevent an excessive current.

The solution to the problem of overcurrents lies in providing closed-loop current-limiting as an integral feature of the motor/drive package. The motor current is sensed, and the voltage V is automatically adjusted so that rated current is either never exceeded or is allowed to reach perhaps twice rated value for a few seconds. We will discuss the current control loop in Chapter 4.

Dynamic behaviour and time-constants

The use of the terms 'surge' and 'sudden' in the discussion above would have doubtless created the impression that changes in the motor current or speed can take place instantaneously, whereas in fact a finite time is always necessary to effect changes in both. (If the current changes, then so does the stored energy in the armature inductance; and if speed changes, so does the rotary kinetic energy stored in the inertia. For either of these changes to take place in zero time it would be necessary for there to be a pulse of infinite power, which is clearly impossible.)

The theoretical treatment of the transient dynamics of the d.c. machine is easier than for any other type of electric motor but is nevertheless beyond our scope. However it is worth summarising the principal features of the dynamic behaviour, and highlighting the fact that all the transient changes that occur are determined by only two time-constants. The first (and most important from the user's viewpoint) is the electro-mechanical time-constant, which governs the way the speed settles to a new level following a disturbance such as a change in armature voltage or load torque. The second is the electrical (or armature) time-constant, which is usually much shorter and governs the rate of change of armature current immediately following a change in armature voltage.

When the motor is running, there are two 'inputs' that we can change suddenly, namely the applied voltage and the load torque. When either of these is changed, the motor enters a transient period before settling to its new steady state. It turns out that if we ignore the armature induc- tance (i.e. we take the armature time-constant to be zero), the transient period is characterised by first-order exponential responses in the speed and current. This assumption is valid for all but the very largest motors. We obtained a similar result when we looked at the primitive linear motor in Chapter 1 (see Figure 1.16).

For example, if we suddenly increased the armature voltage of a frictionless and unloaded motor from V_1 to V_2, its speed and current would vary as shown in Figure 3.11.

There is an immediate increase in the current (note that we have ignored the inductance), reflecting the fact that the applied voltage is suddenly more than the back e.m.f.; the increased current produces more torque and hence the motor accelerates; the rising speed is accom- panied by an increase in back e.m.f., so the current begins to fall; and the process continues until a new steady speed is reached corresponding to the new voltage. In this particular case, the steady-state current is zero because we have assumed that there is no friction or load torque, but the shape of the dynamic response would be the same if there had been an initial load, or if we had suddenly changed the load.

The expression describing the current as a function of time (t) is:

$$i = \left(\frac{V_2 - V_1}{R}\right) e^{-t/\tau} \tag{3.11}$$

The expression for the change in speed is similar, the time dependence again featuring the exponential transient term $e^{-t/\tau}$. The significance of

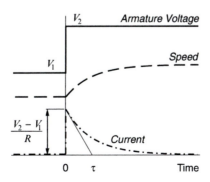

Figure 3.11   *Response of d.c. motor to step increase in armature voltage*

the time-constant (τ) is shown in Figure 3.11. If the initial gradient of the current–time graph is projected, it intersects the final value after one time-constant. In theory, it takes an infinite time for the response to settle, but in practice the transient is usually regarded as over after about four or five time-constants. We note that the transient response is very satisfactory: as soon as the voltage is increased the current immediately increases to provide more torque and begin the acceleration, but the accelerating torque is reduced progressively to ensure that the new target speed is approached smoothly. Happily, because the system is first-order, there is no suggestion of an oscillatory response with overshoots.

Analysis yields the relationship between the time-constant and the motor/system parameters as

$$\tau = \frac{RJ}{k^2} \tag{3.12}$$

where R is the armature resistance, J is the total rotary inertia of motor plus load, and k is the motor constant (equations 3.5 and 3.6). The appropriateness of the term 'electromechanical time-constant' should be clear from equation (3.12), because τ depends on the electrical parameters (R and k) and the mechanical parameter, J. The fact that if the inertia was doubled, the time-constant would double and transients would take twice as long is perhaps to be expected, but the influence of the motor parameters R and k is probably not so obvious.

The electrical or armature time-constant is defined in the usual way for a series L, R circuit, i.e.

$$\tau_a = \frac{L}{R} \tag{3.13}$$

If we were to hold the rotor of a d.c. motor stationary and apply a step voltage V to the armature, the current would climb exponentially to a final value of V/R with a time-constant τ_a.

If we always applied pure d.c. voltage to the motor, we would probably want τ_a to be as short as possible, so that there was no delay in the build-up of current when the voltage is changed. But given that most motors are fed with voltage waveforms which are far from smooth (see Chapter 2), we are actually rather pleased to find that because of the inductance and associated time-constant, the current waveforms (and hence the torque) are smoother than the voltage waveform. So the unavoidable presence of armature inductance turns out (in most cases) to be a blessing in disguise.

So far we have looked at the two time-constants as if they were unrelated in the influence they have on the current. We began with the electromechanical time-constant, assuming that the armature time-constant was zero, and saw that the dominant influence on the current during the transient was the motional e.m.f. We then examined the current when the rotor was stationary (so that the motional e.m.f. is zero), and saw that the growth or decay of current is governed by the armature inductance, manifested via the armature time-constant.

In reality, both time-constants influence the current simultaneously, and the picture is more complicated than we have implied, as the system is in fact a second-order one. However, the good news is that for most motors, and for most purposes, we can take advantage of the fact that the armature time-constant is much shorter than the electromechanical time-constant. This allows us to approximate the behaviour by decoupling the relatively fast 'electrical transients' in the armature circuit from the much slower 'electromechanical transients' which are apparent to the user. From the latter's point of view, only the electromechanical transient is likely to be of interest.

SHUNT, SERIES AND COMPOUND MOTORS

Before variable-voltage supplies became readily available, most d.c. motors were obliged to operate from a single d.c. supply, usually of constant voltage. The armature and field circuits were therefore designed either for connection in parallel (shunt), or in series. As we will see shortly, the operating characteristics of shunt and series machines differ widely, and hence each type tends to claim its particular niche: shunt motors were judged to be good for constant-speed applications, while series motors were (and still are) widely used for traction applications.

In a way it is unfortunate that these historical patterns of association have become so deep-rooted. The fact is that a converter-fed separately excited motor, free of any constraint between field and armature, can do everything that a shunt or series motor can, and more; and it is doubtful if shunt and series motors would ever have become widespread if variable-voltage supplies had always been around. Both shunt and series motors are handicapped in comparison with the separately excited motor, and we will therefore be well advised to view their oft-proclaimed merits with this in mind.

The operating characteristics of shunt, series and compound (a mixture of both) motors are explored below, but first we should say something of the physical differences. At a fundamental level these amount

Plate 3.2 *Totally-enclosed fan-ventilated wound-field d.c. motors. The smaller motor is rated at 500 W (0.67 h.p.) at 1500 rev/min, while the larger is rated at 10 kW (13.4 h.p.) at 2000 rev/min. Both motors have finned aluminium frames. (Photograph by courtesy of Brook Crompton)*

to very little, but in the detail of the winding arrangement, they are considerable.

For a given continuous output power rating at a given speed, we find that shunt and series motors are the same size, with the same rotor diameter, the same poles, and the same quantities of copper in the armature and field windings. This is to be expected when we recall that the power output depends on the specific magnetic and electric loadings, so we anticipate that to do a given job, we will need the same amounts of active material.

The differences emerge when we look at the details of the windings, especially the field system, and they can best be illustrated by means of an example which contrasts shunt and series motors for the same output power.

Suppose that for the shunt version the supply voltage is 500 V, the rated armature (work) current is 50 A, and the field coils are required to provide an MMF of 500 ampere-turns (AT). The field might typically consist of say 200 turns of wire with a total resistance of $200\,\Omega$. When connected across the supply (500 V), the field current will be 2.5 A, and the MMF will be 500 AT, as required. The power dissipated as heat in the field will be $500\,\text{V} \times 2.5\,\text{A} = 1.25\,\text{kW}$, and the total power input at rated load will be $500\,\text{V} \times 52.5\,\text{A} = 26.25\,\text{kW}$.

To convert the machine into the equivalent series version, the field coils need to be made from much thicker conductor, since they have to carry the armature current of 50 A, rather than the 2.5 A of the shunt motor. So, working at the same current density, the cross section of each turn of the series field winding needs to be 20 times that of the shunt field wires, but conversely only one-twentieth of the turns (i.e. 10) are required for the same ampere-turns. The resistance of a wire of length l and cross-sectional area A, made from material of resistivity ρ is given by $R = \rho l / A$, so we can use this formula twice to show that the resistance of the new field winding will be much lower, at $0.5\,\Omega$.

We can now calculate the power dissipated as heat in the series field. The current is 50 A, the resistance is $0.5\,\Omega$, so the volt-drop across the series field is 25 V, and the power wasted as heat is 1.25 kW. This is the same as for the shunt machine, which is to be expected since both sets of field coils are intended to do the same job.

In order to allow for the 25 V dropped across the series field, and still meet the requirement for 500 V at the armature, the supply voltage must now be 525 V. The rated current is 50 A, so the total power input is $525\,\mathrm{V} \times 50\,\mathrm{A} = 26.25\,\mathrm{kW}$, the same as for the shunt machine.

This example illustrates that in terms of their energy-converting capabilities, shunt and series motors are fundamentally no different. Shunt machines usually have field windings with a large number of turns of fine wire, while series machines have a few turns of thick conductor. But the total amount and disposition of copper is the same, so the energy-converting abilities of both types are identical. In terms of their operating characteristics, however, the two types differ widely, as we will now see.

Shunt motor – steady-state operating characteristics

A basic shunt-connected motor has its armature and field in parallel across a single d.c. supply, as shown in Figure 3.12(a). Normally, the voltage will be constant and at the rated value for the motor, in which case the steady-state torque/speed curve will be similar to that of a separately excited motor at rated field flux, i.e. the speed will drop slightly with load, as shown by the line *ab* in Figure 3.12(b). Over the normal operating region the torque–speed characteristic is similar to that of the induction motor (see Chapter 6), so shunt motors are suited to the same duties, i.e. what are usually referred to as 'constant speed' applications.

Except for small motors (say less than about 1 kW), it will be necessary to provide an external 'starting resistance' (R_s in Figure 3.12) in series with the armature, to limit the heavy current which would flow if

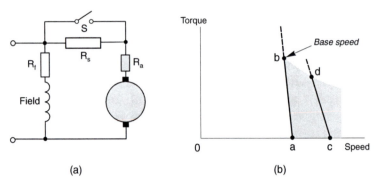

(a)   (b)

Figure 3.12   *Shunt-connected d.c. motor and steady-state torque–speed curve*

the motor was simply switched directly onto the supply. This starting resistance is progressively reduced as the motor picks up speed, the current falling as the back e.m.f. rises from its initial value of zero. In a manual starter the resistance is controlled by the operator, while in an automatic starter the motor armature voltage or current are sensed and the resistance is shorted out in predetermined stages.

We should ask what happens if the supply voltage varies for any reason, and as usual the easiest thing to look at is the case where the motor is running light, in which case the back e.m.f. will almost equal the supply voltage. If we reduce the supply voltage, intuition might lead us to anticipate a fall in speed, but in fact two contrary effects occur which leave the speed almost unchanged.

If the voltage is halved, for example, both the field current and the armature voltage will be halved, and if the magnetic circuit is not saturated the flux will also halve. The new steady value of back e.m.f. will have to be half its original value, but since we now have only half as much flux, the speed will be the same. The maximum output power will of course be reduced, since at full load (i.e. full current) the power available is proportional to the armature voltage. Of course if the magnetic circuit is saturated, a modest reduction in applied voltage may cause very little drop in flux, in which case the speed will fall in proportion to the drop in voltage. We can see from this discussion why, broadly speaking, the shunt motor is not suitable for operation below base speed.

Some measure of speed control is possible by weakening the field (by means of the resistance (R_f) in series with the field winding), and this allows the speed to be raised above base value, but only at the expense of torque. A typical torque–speed characteristic in the field-weakening region is shown by the line *cd* in Figure 3.12(b).

Reverse rotation is achieved by reversing the connections to either the field or the armature. The field is usually preferred since the current rating of the switch or contactor will be lower than for the armature.

Series motor – steady-state operating characteristics

The series connection of armature and field windings (Figure 3.13(a)) means that the field flux is directly proportional to the armature current, and the torque is therefore proportional to the square of the current. Reversing the direction of the applied voltage (and hence current) therefore leaves the direction of torque unchanged. This unusual property is put to good use in the universal motor, but is a handicap when negative (braking) torque is required, since either the field or armature connections must then be reversed.

If the armature and field resistance volt-drops are neglected, and the applied voltage (V) is constant, the current varies inversely with the speed, hence the torque (T) and speed (n) are related by

$$T \propto \left(\frac{V}{n}\right)^2 \tag{3.14}$$

A typical torque–speed characteristic is shown in Figure 3.13(b). The torque at zero speed is not infinite of course, because of the effects of saturation and resistance, both of which are ignored in equation (3.14).

It is important to note that under normal running conditions the volt-drop across the series field is only a small part of the applied voltage, most of the voltage being across the armature, in opposition to the back e.m.f. This is of course what we need to obtain an efficient energy-conversion. Under starting conditions, however, the back e.m.f. is

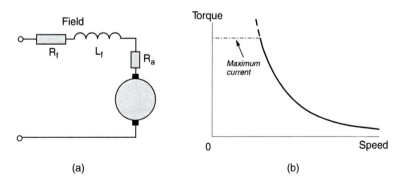

(a)   (b)

Figure 3.13   *Series-connected d.c. motor and steady-state torque–speed curve*

zero, and if the full voltage was applied the current would be excessive, being limited only by the armature and field resistances. Hence for all but small motors a starting resistance is required to limit the current to a safe value.

Returning to Figure 3.13(b), we note that the series motor differs from most other motors in having no clearly defined no-load speed, i.e. no speed (other than infinity) at which the torque produced by the motor falls to zero. This means that when running light, the speed of the motor depends on the windage and friction torques, equilibrium being reached when the motor torque equals the total mechanical resisting torque. In large motors, the windage and friction torque is often relatively small, and the no-load speed is then too high for mechanical safety. Large series motors should therefore never be run uncoupled from their loads. As with shunt motors, the connections to either the field or armature must be reversed in order to reverse the direction of rotation.

Large series motors have traditionally been used for traction. Often, books say this is because the series motor has a high starting torque, which is necessary to provide acceleration to the vehicle from rest. In fact any d.c. motor of the same frame size will give the same starting torque, there being nothing special about the series motor in this respect. The real reason for its widespread use is that under the simplest possible supply arrangement (i.e. constant voltage) the overall shape of the torque–speed curve fits well with what is needed in traction applications. This was particularly important in the days when it was simply not feasible to provide any control of the armature voltage.

The inherent suitability of the series motor for traction is illustrated by the curves in Figure 3.14, which relate to a railway application. The solid line represents the motor characteristic, while the dotted line is the

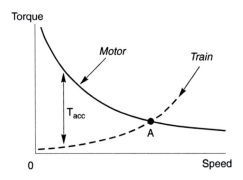

Figure 3.14   *Torque–speed curves illustrating the application of a series-connected d.c. motor to traction*

steady-state torque–speed curve for the train, i.e. the torque which the motor must provide to overcome the rolling resistance and keep the train running at each speed.

At low speeds the rolling resistance is low, the motor torque is much higher, and therefore the nett torque (T_{acc}) is large and the train accelerates at a high rate. As the speed rises, the nett torque diminishes and the acceleration tapers off until the steady speed is reached at point A in Figure 3.14.

Some form of speed control is obviously necessary in the example above if the speed of the train is not to vary when it encounters a gradient, which will result in the rolling resistance curve shifting up or down. There are basically three methods which can be used to vary the torque–speed characteristics, and they can be combined in various ways.

Firstly, resistors can be placed in parallel with the field or armature, so that a specified fraction of the current bypasses one or the other. Field 'divert' resistors are usually preferred since their power rating is lower than armature divert resistors. For example, if a resistor with the same resistance as the field winding is switched in parallel with it, half of the armature current will now flow through the resistor and half will flow through the field. At a given speed and applied voltage, the armature current will increase substantially, so the flux will not fall as much as might be expected, and the torque will rise, as shown in Figure 3.15(a). This method is inefficient because power is wasted in the resistors, but is simple and cheap to implement. A more efficient method is to provide 'tappings' on the field winding, which allow the number of turns to be varied, but of course this can only be done if the motor has the tappings brought out.

Secondly, if a multicell battery is used to supply the motor, the cells may be switched progressively from parallel to series to give a range of

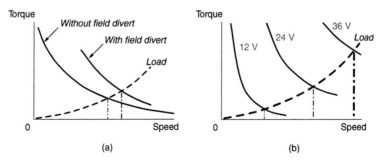

Figure 3.15   *Series motor characteristics with (a) field divert control and (b) voltage control*

discrete steps of motor voltage, and hence a series of torque–speed curves. Road vehicles with 12 V lead–acid batteries often use this approach to provide say 12, 24, and 36 V for the motor, thereby giving three discrete 'speed' settings, as shown in Figure 3.15(b).

Finally, where several motors are used (e.g. in a multiple-unit railway train) and the supply voltage is fixed, the motors themselves can be switched in various series/parallel groupings to vary the voltage applied to each.

Universal motors

In terms of numbers the main application area for the series commutator motor is in portable power tools, foodmixers, vacuum cleaners etc., where paradoxically the supply is a.c. rather than d.c. Such motors are often referred to as 'universal' motors because they can run from either a d.c. or an a.c. supply.

At first sight the fact that a d.c. machine will work on a.c. is hard to believe. But when we recall that in a series motor the field flux is set up by the current which also flows in the armature, it can be seen that reversal of the current will be accompanied by reversal of the direction of the magnetic flux, thereby ensuring that the torque remains positive. When the motor is connected to a 50 Hz supply for example, the (sinusoidal) current will change direction every 10 msec, and there will be a peak in the torque 100 times per second. But the torque will always remain unidirectional, and the speed fluctuations will not be noticeable because of the smoothing effect of the armature inertia.

Series motors for use on a.c. supplies are always designed with fully laminated construction (to limit eddy current losses produced by the pulsating flux in the magnetic circuit), and are intended to run at high speeds, say 8–12 000 rev/min. at rated voltage. Commutation and sparking are worse than when operating from d.c., and output powers are seldom greater than 1 kW. The advantage of high speed in terms of power output per unit volume was emphasised in Chapter 1, and the universal motor is perhaps the best everyday example which demonstrates how a high power can be obtained with small size by designing for a high speed.

Until recently the universal motor offered the only relatively cheap way of reaping the benefit of high speed from single-phase a.c. supplies. Other small a.c. machines, such as induction motors and synchronous motors, were limited to maximum speeds of 3000 rev/min at 50 Hz (or 3600 rev/min at 60 Hz), and therefore could not compete in terms of power per unit volume. The availability of high-frequency inverters (see

Chapter 8) has opened up the prospect of higher specific outputs from induction motors, but currently the universal motor remains the dominant force in small low-cost applications, because of the huge investment that has been made over many years to produce them in vast numbers.

Speed control of small universal motors is straightforward using a triac (in effect a pair of thyristors connected back to back) in series with the a.c. supply. By varying the firing angle, and hence the proportion of each cycle for which the triac conducts, the voltage applied to the motor can be varied to provide speed control. This approach is widely used for electric drills, fans etc. If torque control is required (as in hand power tools, for example), the current is controlled rather than the voltage, and the speed is determined by the load.

Compound motors

By arranging for some of the field MMF to be provided by a series winding and some to be provided by a shunt winding, it is possible to obtain motors with a wide variety of inherent torque–speed characteristics. In practice most compound motors have the bulk of the field MMF provided by a shunt field winding, so that they behave more or less like a shunt connected motor. The series winding MMF is relatively small, and is used to allow the torque–speed curve to be trimmed to meet a particular load requirement.

When the series field is connected so that its MMF reinforces the shunt field MMF, the motor is said to be 'cumulatively compounded'. As the load on the motor increases, the increased armature current in the series field causes the flux to rise, thereby increasing the torque per ampere but at the same time, resulting in a bigger drop in speed as compared with a simple shunt motor. On the other hand, if the series field winding opposes the shunt winding, the motor is said to be 'differentially compounded'. In this case an increase in current results in a weakening of the flux, a reduction in the torque per ampere, but a smaller drop in speed than in a simple shunt motor. Differential compounding can therefore be used where it is important to maintain as near constant-speed as possible.

FOUR-QUADRANT OPERATION AND REGENERATIVE BRAKING

As we saw in Section 3.4, the beauty of the separately excited d.c. motor is the ease with which it can be controlled. Firstly, the steady-state speed is determined by the applied voltage, so we can make the motor run at any desired speed in either direction simply by applying the appropriate magnitude and polarity of the armature voltage. Secondly, the torque is directly proportional to the armature current, which in turn depends on the difference between the applied voltage V and the back e.m.f. E. We can therefore make the machine develop positive (motoring) or negative (generating) torque simply by controlling the extent to which the applied voltage is greater or less than the back e.m.f. An armature voltage controlled d.c. machine is therefore inherently capable of what is known as 'four-quadrant' operation, with reference to the numbered quadrants of the torque–speed plane shown in Figure 3.16.

Figure 3.16 looks straightforward but experience shows that to draw the diagram correctly calls for a clear head, so it is worth spelling out the key points in detail. A proper understanding of this diagram is invaluable as an aid to seeing how controlled-speed drives operate.

Firstly, one of the motor terminals is shown with a dot, and in all four quadrants the dot is uppermost. The purpose of this convention is to indicate the sign of the torque: if current flows into the dot, the machine produces positive torque, and if current flows out of the dot, the torque is negative.

Secondly, the supply voltage is shown by the old-fashioned battery symbol, as use of the more modern circle symbol for a voltage source would make it more difficult to differentiate between the source and the circle representing the machine armature. The relative magnitudes of applied voltage and motional e.m.f. are emphasised by the use of two battery cells when $V > E$ and one when $V < E$.

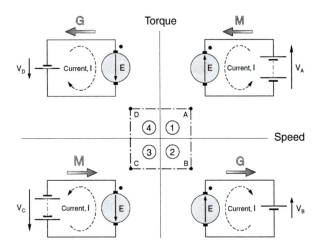

Figure 3.16   *Operation of d.c. motor in the four quadrants of the torque–speed plane*

We have seen that in a d.c. machine speed is determined by applied voltage and torque is determined by current. Hence on the right-hand side of the diagram the supply voltage is positive (upwards), while on the left-hand side the supply voltage is negative (downwards). And in the upper half of the diagram current is positive (into the dot), while in the lower half it is negative (out of the dot). For the sake of convenience, each of the four operating conditions (A, B, C, D) have the same magnitude of speed and the same magnitude of torque: these translate to equal magnitudes of motional e.m.f. and current for each condition.

When the machine is operating as a motor and running in the forward direction, it is operating in quadrant 1. The applied voltage V_A is positive and greater than the back e.m.f. E, and positive current therefore flows into the motor: in Figure 3.16, the arrow representing V_A has accordingly been drawn larger than E. The power drawn from the supply $(V_A I)$ is positive in this quadrant, as shown by the shaded arrow labelled M to represent motoring. The power converted to mechanical form is given by EI, and an amount $I^2 R$ is lost as heat in the armature. If E is much greater than IR (which is true in all but small motors), most of the input power is converted to mechanical power, i.e. the conversion process is efficient.

If, with the motor running at position A, we suddenly reduce the supply voltage to a value V_B which is less than the back e.m.f., the current (and hence torque) will reverse direction, shifting the operating point to B in Figure 3.16. There can be no sudden change in speed, so the e.m.f. will remain the same. If the new voltage is chosen so that $E - V_B = V_A - E$, the new current will have the same amplitude as at position A, so the new (negative) torque will be the same as the original positive torque, as shown in Figure 3.16. But now power is supplied from the machine to the supply, i.e. the machine is acting as a generator, as shown by the shaded arrow.

We should be quite clear that all that was necessary to accomplish this remarkable reversal of power flow was a modest reduction of the voltage applied to the machine. At position A, the applied voltage was $E + IR$, while at position B it is $E - IR$. Since IR will be small compared with E, the change $(2IR)$ is also small.

Needless to say the motor will not remain at point B if left to its own devices. The combined effect of the load torque and the negative machine torque will cause the speed to fall, so that the back e.m.f. again falls below the applied voltage V_B, the current and torque become positive again, and the motor settles back into quadrant 1, at a lower speed corresponding to the new (lower) supply voltage. During the deceleration phase, kinetic energy from the motor and load inertias is returned to the supply. This is therefore an example of regenerative braking, and it occurs naturally every time we reduce the voltage in order to lower the speed.

If we want to operate continuously at position B, the machine will have to be driven by a mechanical source. We have seen above that the natural tendency of the machine is to run at a lower speed than that corresponding to point B, so we must force it to run faster, and create an e.m.f greater than V_B, if we wish it to generate continuously.

It should be obvious that similar arguments to those set out above apply when the motor is running in reverse (i.e. V is negative). Motoring then takes place in quadrant 3 (point C), with brief excursions into quadrant 4 (point D, accompanied by regenerative braking), whenever the voltage is reduced in order to lower the speed.

Full speed regenerative reversal

To illustrate more fully how the voltage has to be varied during sustained regenerative braking, we can consider how to change the speed of an unloaded motor from full speed in one direction to full speed in the other, in the shortest possible time.

At full forward speed the applied armature voltage is taken to be $+V$ (shown as 100% in Figure 3.17), and since the motor is unloaded the no-load current will be very small and the back e.m.f. will be almost equal to V. Ultimately, we will clearly need an armature voltage of $-V$ to make the motor run at full speed in reverse. But we cannot simply reverse the applied voltage: if we did, the armature current immediately afterwards would be given by $(-V - E)/R$, which would be disastrously high. (The motor might tolerate it for the short period for which it would last, but the supply certainly would not!).

What we need to do is adjust the voltage so that the current is always limited to rated value, and in the right direction. Since we want to decelerate as fast as possible, we must aim to keep the current negative, and at rated value (i.e. -100%) throughout the period of deceleration and for the run up to full speed in reverse. This will give us constant torque throughout, so the deceleration (and subsequent acceleration) will be constant, and the speed will change at a uniform rate, as shown in Figure 3.17.

We note that to begin with, the applied voltage has to be reduced to less than the back e.m.f., and then ramped down linearly with time so that the difference between V and E is kept constant, thereby keeping the current constant at its rated value. During the reverse run-up, V has to be numerically greater than E, as shown in Figure 3.17. (The differ-

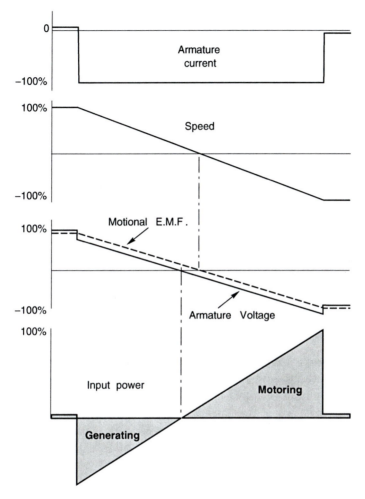

Figure 3.17 *Regenerative reversal of d.c. motor from full-speed forward to full-speed reverse, at maximum allowable torque (current)*

ence between V and E has been exaggerated in Figure 3.17 for clarity: in a large motor, the difference may only be one or two percent at full speed.)

The power to and from the supply is shown in the bottom plot in Figure 3.17, the energy being represented by the shaded areas. During the deceleration period most of the kinetic energy of the motor (lower shaded area) is progressively returned to the supply, the motor acting as a generator for the whole of this time. The total energy recovered in this way can be appreciable in the case of a large drive such as a steel rolling mill. A similar quantity of energy (upper shaded area) is supplied and stored as kinetic energy as the motor picks up speed in the reverse sense.

Three final points need to be emphasised. Firstly, we have assumed throughout the discussion that the supply can provide positive or negative voltages, and can accept positive or negative currents. A note of caution is therefore appropriate, because many simple power electronic converters do not have this flexibility. Users need to be aware that if full four-quadrant operation (or even two-quadrant regeneration) is called for, a basic converter will probably not be adequate. This point is taken up again in Chapter 4. Secondly, we should not run away with the idea that in order to carry out the reversal in Figure 3.17 we would have to work out in advance how to profile the applied voltage as a function of time. Our drive system will normally have the facility for automatically operating the motor in constant-current mode, and all we will have to do is to set the new target speed. This is also taken up in Chapter 4. And finally, we must remember that the discussion above relates to separately excited motors. If regenerative braking is required for a series motor, the connections to either the field or armature must be reversed in order to reverse the direction of torque.

Dynamic braking

A simpler and cheaper but less effective method of braking can be achieved by dissipating the kinetic energy of the motor and load in a resistor, rather than returning it to the supply. A version of this technique is employed in the cheaper power electronic converter drives, which have no facility for returning power to the mains.

When the motor is to be stopped, the supply to the armature is removed and a resistor is switched across the armature brushes. The motor e.m.f. drives a (negative) current through the resistor, and the negative torque results in deceleration. As the speed falls, so does the e.m.f., the current, and the braking torque. At low speeds the braking torque is therefore very small. Ultimately, all the kinetic energy is converted to heat in the motor's own armature resistance and external resistance. Very rapid initial braking is obtained by using a low resistance (or even simply short-circuiting the armature). Dynamic braking is still widely used in traction because of its simplicity, though most new rapid transit schemes employ the more energy-efficient regenerative braking process.

TOY MOTORS

The motors used in model cars, trains etc. are rather different in construction from those discussed so far, primarily because they are designed to be cheap to make. They also run at high speeds, so it is not important for the torque to be smooth. A typical arrangement used for rotor diameters from 1 cm to perhaps 3 cm is shown in Figure 3.18.

The rotor, made from laminations with a small number (typically three or five) of multi-turn coils in very large 'slots,' is simple to manufacture, and because the commutator has few segments, it too is cheap to make. The field system (stator) consists of radially magnetised ceramic magnets with a steel backplate to complete the magnetic circuit.

The rotor clearly has very pronounced saliency, with three very large projections which are in marked contrast to the rotors we looked at earlier where the surface was basically cylindrical. It is easy to imagine that even when there is no current in the rotor coils, there is a strong tendency for the stator magnets to pull one or other of the rotor saliencies into alignment with a stator pole, so that the rotor would tend to lock in any one of six positions. This cyclic 'detent' torque is due to the variation of reluctance with rotor position, an effect which is exploited in a.c. reluctance motors (see Chapter 9), but is unwanted here. To combat the problem the rotor laminations are skewed before the coils are fitted, as shown in the lower sketch on the left.

Each of the three rotor poles carries a multi-turn coil, the start of which is connected to a commutator segment, as shown in the cross section on the right of Figure 3.18. The three ends are joined together. The brushes are wider than the inter-segment space, so in some rotor positions the current from say the positive brush will divide and flow first through two coils in parallel until it reaches the common point, then

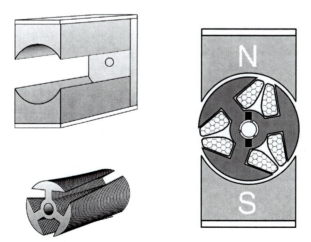

Figure 3.18   *Miniature d.c. motor for use in model toys*

through the third coil to the negative brush, while in other positions the current flows through only two coils.

A real stretch of imagination is required to picture the mechanism of torque production using the '*BIl*' approach we have followed previously, as the geometry is so different. But we can take a more intuitive approach by considering, for each position, the polarity and strength of the magnetisation of each of the three rotor poles, which depend of course on the direction and magnitude of the respective currents. Following the discussion in the previous paragraph, we can see that in some positions there will be say a strong N and two relatively weak S poles, while at others there will be one strong N pole, one strong S pole, and one unexcited pole.

Thus although the rotor appears to be a 3-pole device, the magnetisation pattern is always 2-pole (because two adjacent weak S poles function as a single stronger S pole). When the rotor is rotating, the stator N pole will attract the nearest rotor S pole, pulling it round towards alignment. As the force of attraction diminishes to zero, the commutator reverses the rotor current in that pole so that it now becomes a N pole and is pushed away towards the S stator pole.

There is a variation of current with angular position as a result of the different resistances seen by the brushes as they alternately make contact with either one or two segments, and the torque is far from uniform, but operating speeds are typically several thousand rev/min and the torque pulsations are smoothed out by the rotor and load inertia.

REVIEW QUESTIONS

The first nine questions test general understanding; questions 10 to 18 are numerical problems based mainly on the equivalent circuit; questions 19 to 26 are discursive questions related to d.c. machines; and the remaining questions are more challenging, with an applications bias.

1) What is the primary (external) parameter that determines the speed of an unloaded d.c. motor?

2) What is the primary external factor that determines the steady-state running current of a d.c. motor, for any given armature voltage?

3) What determines the small current drawn by a d.c. motor when running without any applied mechanical load?

4) What determines how much the speed of a d.c. motor reduces when the load on its shaft is increased? Why do little motors slow down more than large ones?

5) What has to be done to reverse the direction of rotation of:
   (a) a separately excited motor;
   (b) a shunt motor;
   (c) a series motor?

6) Most d.c. motors can produce much more than their continuous-lyrated torque. Why is it necessary to limit continuous torque?

7) What is the basic difference between a d.c. motor and a d.c. generator?

8) From the point of view of supply, an unloaded d.c. motor running light looks like a high resistance, but when running at full load it looks like a much lower resistance. Why is this?

9) Why do d.c. motors run faster when their field flux is reduced?

10) A separately excited d.c. motor runs from a 220 V supply and draws an armature current of 15 A. The armature resistance is 0.8 Ω. Calculate the generated voltage (back e.m.f).

   If the field current is suddenly reduced by 10%, calculate
   (a) the value to which the armature current rises momentarily, and
   (b) the percentage increase in torque when the current reaches the value in (a).

   Ignore armature inductance, neglect saturation, and assume that the field flux is directly proportional to the field current.

11) A shunt-connected d.c. machine driven by a diesel engine supplies a current of 25 A to a 110 V battery. The armature and field resistances are 0.5 Ω and 110 Ω respectively, and the friction, windage and other losses total 220 W. Calculate
   (a) the generated e.m.f, (b) the efficiency.

12) A 250 V d.c. motor with an armature resistance of 1 Ω is running unloaded at 1800 rev/min and drawing a current of 2 A. Estimate the friction torque.

13) What voltage would you expect across the armature terminals of the motor in question 12 immediately after it had suddenly been disconnected from the 250 V supply?

14) The full-load current of the motor in question 12 is 32 A. Estimate the full-load speed and the rated torque.

15) (a) When driven at 1500 rev/min the open-circuit armature voltage of a d.c. machine is 110 V. Calculate the e.m.f. constant in

volts per radian/s. Calculate also the machine torque when the armature current is 10 A.

(b) Suppose the machine was at rest, and a weight of 5 kg was suspended from a horizontal bar of length 80 cm attached to the shaft, as shown in Figure Q.15.

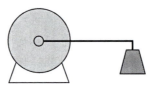

Figure Q.15

What current must be applied to the armature to keep the arm horizontal? Will the equilibrium be a stable one? (Neglect the mass of the bar; $g = 9.81\,\text{m/s}^2$.)

(c) When the machine runs as a motor drawing 25 A from a 110 V supply, the speed is 1430 rev/min. Calculate the armature resistance. Hence find the voltage needed to keep the bar horizontal.

(d) At what speed must the machine be driven to supply 3.5 kW to a 110 V system? Calculate the corresponding torque. If the field circuit consumes 100 W and the friction losses are 200 W, calculate the efficiency of the generator.

16) The manufacturer of a 12 V toy motor with an armature resistance of $8\,\Omega$ claims that it can produce a torque of 20 mNm at 5000 rev/min. Show that this claim is not justified.

17) The following test results were obtained for a small permanent-magnet ironless-rotor d.c. motor for use in a print-head drive:

Armature resistance $= 2.9\,\Omega$
Speed when running unloaded with 6 V applied $= 8000\,\text{rev/min}$
Current when running unloaded with 6 V applied $= 70\,\text{mA}$
Rotor inertia $= 0.138 \times 10^{-6}\,\text{kg m}^2$.

Calculate the induced e.m.f. and the friction torque when running light with 6 V applied, and estimate the initial acceleration from rest when the motor is switched directly onto a 6 V supply.

18) The equations expressing torque in terms of current ($T = kI$) and motional e.m.f. in terms of speed ($E = k\omega$) are central in under-

standing the operation of a d.c. machine. Using only these equations, show that the mechanical output power is given by $W = EI$.

19) A customer finds a 24 W, 5000 rev/min motor with an armature resistance of $0.8\,\Omega$, which suits his application, except that it is a 12 V motor and the only supply he has is 24 V. The motor vendor says that he can supply a 24 V version instead. How would the parameters of the 24 V version differ from the 12 V one?

20) Explain briefly why:
   (a) large d.c. motors cannot normally be started by applying full voltage;
   (b) the no-load speed of a permanent-magnet motor is almost proportional to the armature voltage;
   (c) a d.c. motor draws more current from the supply when the load on the shaft is increased;
   (d) the field windings of a d.c. motor consume energy continuously even though they do not contribute to the mechanical output power;
   (e) the field poles of a d.c. machine are not always laminated.

21) A separately-excited d.c. motor is running light with a constant armature voltage. When the field current is suddenly reduced a little, the armature current increases substantially, and the speed rises to settle at a higher level. Explain these events with the aid of an equivalent circuit and discussion of the relevant equations.

22) A separately-excited motor used in traction has a field control circuit that ensures that the field flux is directly proportional to the armature current. Sketch a family of torque–speed curves for the motor when operating with a range of armature voltages.

23) A small permanent-magnet d.c. motor has an armature resistance of $1\,\Omega$. What is the maximum possible mechanical output power when the armature is supplied at 12 V? Why is this maximum power condition only of theoretical interest for large motors?

24) Explain briefly why a universal motor is able to operate on either a.c. or d.c.

25) It is claimed in this book that a motor of a given size and power can be made available for operation at any voltage. But it is clear that when it comes to battery-powered hand tools of a given size and speed, the higher-voltage versions are more powerful. What accounts for this contradiction?

26) Series motors will work on a.c. because field and armature currents reverse simultaneously, so the torque remains unidirectional. If a shunt-connected motor was supplied with a.c., the voltage on its armature and field would reverse together, so would it not also work satisfactorily?

27) This question relates to a permanent-magnet d.c. machine with an armature resistance of $0.5\,\Omega$.

   When the rotor is driven at 1500 rev/min by an external mechanical source, its open-circuit armature voltage is 220 V. All parts of the question relate to steady-state conditions, i.e. after all transients have died away.

   The machine is to be used as a generator and act as a dynamic brake to restrain a lowering load, as shown in Figure Q.27(a). The hanging mass of 14.27 kg is suspended by a rope from a 20 cm diameter winding drum on the motor shaft.

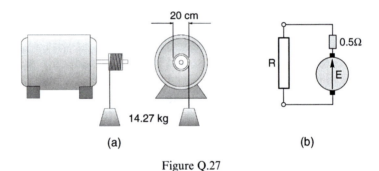

Figure Q.27

The majority of the generated power is to be dissipated in an external resistor (R) connected across the armature, as shown in Figure Q.27(b).

(a) Calculate the value of resistance required so that the mass descends at a steady speed of 15 m/s. (Take $g = 9.81\,\text{m/s}^2$.)

(b) What is the power dissipated in (i) the external resistor, (ii) the armature? Where does the energy dissipated in the resistors come from?

28) This question is about dynamic (transient) behaviour during dynamic braking in which the machine acts as a generator to convert its kinetic energy to heat in a resistor. Familiarity with first-order differential equations is needed to answer the question fully.

A permanent-magnet d.c. motor with armature resistance of 1 Ω is running at 3000 rev/min and drawing a current of 10 A from a 200 V supply. Calculate the motor constant.

If the motor runs unloaded from a 100 V supply and the supply is then suddenly disconnected and the armature terminals are connected to a 4 Ω resistor, show that the speed decays exponentially with a time-constant of 2.73 s. Hence calculate the time taken for the speed to reduce to 100 rev/min. Take the effective inertia as 0.2 kg m^2 and ignore friction.

29) When the armature of an unloaded permanent-magnet d.c. motor is supplied with a sinusoidally varying current at a frequency of 0.5 Hz, the speed varies from +2000 rev/min to −2000 rev/min. Estimate the speed range when the frequency is increased to 5 Hz, the peak current remaining the same.

30) Two new and identical 200 kW, 520 V, 420 A, 1000 rev/min d.c. machines are to be tested at close to full load by coupling their shafts together, so that one will act as a generator to supply power to the other, which will act as a motor and drive the generator (no, this is not perpetual motion.....read on). In this way it is hoped that there will be no need for the manufacturer to provide high-power electrical supplies, nor heavy-duty mechanical loading rigs.

The two armatures are connected in parallel, as shown in Figure Q.30; care being taken to ensure that when the duo rotate, the polarities of the motional e.m.f.s. in the two machines are such that there is no tendency for current to circulate between them (i.e. both are say upwards in Fig Q.30).

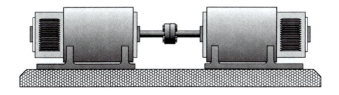

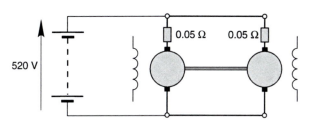

Figure Q.30

Both machines are separately excited, and their field currents can be adjusted manually. The field currents of both machines are first set at the normal (rated) level, and the output of the d.c. supply is gradually raised (to avoid excessive current surges) until the armature voltage is 520 V and the motors are running at a no-load speed of 1040.2 rev/min. In this condition, assuming that there is perfect balance, both motors draw 18.5 A, making a total of 37 A from the d.c. source. This current reflects mainly the torque required to overcome windage, friction, and iron losses.

If the armature circuit resistance of each machine is 0.05 Ω, and brush volt-drop is ignored, estimate for each machine:

(a)   the motional e.m.f. at no load;
(b)   the no-load armature circuit power loss;
(c)   the mechanical (friction and windage) power loss;
(d)   the no-load torque.

Now suppose that the field current (and hence the flux) of machine 1 is gradually reduced until the current in machine 1 reaches its rated value of 420 A.

Assuming that the d.c. supply acts as an ideal voltage source, set up the two armature equations and the torque balance equation and solve them to find:

(a)   the speed of the machines;
(b)   the current in machine 2;
(c)   the nett torque;
(d)   the power supplied by the d.c. source, as a percentage of the rated power of each machine.

In the light of the answers above, comment on the extent to which it can be claimed that the machines have been tested at full-load, and on the economy in terms of the power supply that is required to carry out the test.

4

D.C. MOTOR DRIVES

INTRODUCTION

The thyristor d.c. drive remains an important speed-controlled industrial drive, especially where the higher maintenance cost associated with
the d.c. motor brushes (c.f. induction motor) is tolerable. The controlled
(thyristor) rectifier provides a low-impedance adjustable 'd.c.' voltage
for the motor armature, thereby providing speed control.

Until the 1960s, the only really satisfactory way of obtaining the
variable-voltage d.c. supply needed for speed control of an industrial
d.c. motor was to generate it with a d.c. generator. The generator was
driven at fixed speed by an induction motor, and the field of the generator
was varied in order to vary the generated voltage. The motor/generator
(MG) set could be sited remote from the d.c. motor, and multi-drive sites
(e.g. steelworks) would have large rooms full of MG sets, one for each
variable-speed motor on the plant. Three machines (all of the same power
rating) were required for each of these 'Ward Leonard' drives, which was
good business for the motor manufacturer. For a brief period in the 1950s
they were superseded by grid-controlled mercury arc rectifiers, but these
were soon replaced by thyristor converters which offered cheaper first
cost, higher efficiency (typically over 95%), smaller size, reduced maintenance, and faster response to changes in set speed. The disadvantages of
rectified supplies are that the waveforms are not pure d.c., that the
overload capacity of the converter is very limited, and that a single
converter is not capable of regeneration.

Though no longer pre-eminent, study of the d.c. drive is valuable for
several reasons:

- The structure and operation of the d.c. drive are reflected in almost all
  other drives, and lessons learned from the study of the d.c. drive
  therefore have close parallels to other types.

- The d.c. drive tends to remain the yardstick by which other drives are judged.
- Under constant-flux conditions the behaviour is governed by a relatively simple set of linear equations, so predicting both steady-state and transient behaviour is not difficult. When we turn to the successors of the d.c. drive, notably the induction motor drive, we will find that things are much more complex, and that in order to overcome the poor transient behaviour, the strategies adopted are based on emulating the d.c. drive.

The first and major part of this chapter is devoted to thyristor-fed drives, after which we will look briefly at chopper-fed drives that are used mainly in medium and small sizes, and finally turn attention to small servo-type drives.

THYRISTOR D.C. DRIVES – GENERAL

For motors up to a few kilowatts the armature converter can be supplied from either single-phase or three-phase mains, but for larger motors three-phase is always used. A separate thyristor or diode rectifier is used to supply the field of the motor: the power is much less than the armature power, so the supply is often single-phase, as shown in Figure 4.1.

The arrangement shown in Figure 4.1 is typical of the majority of d.c. drives and provides for closed-loop speed control. The function of the two control loops will be explored later, but readers who are not familiar

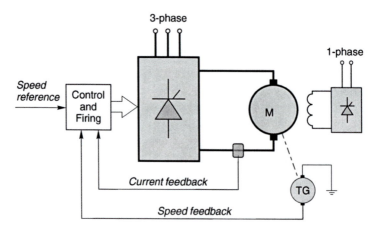

Figure 4.1   *Schematic diagram of speed-controlled d.c. motor drive*

Plate 4.1 *High performance force-ventilated d.c. motor. The motor is of all-laminated construction and designed for use with a thyristor converter. The small blower motor is an induction machine that runs continuously, thereby allowing the main motor to maintain full torque at low speed without overheating. (Photo courtesy of Control Techniques)*

with the basics of feedback and closed-loop systems may find it helpful to read through the Appendix at this point.

The main power circuit consists of a six-thyristor bridge circuit (as discussed in Chapter 2), which rectifies the incoming a.c. supply to produce a d.c. supply to the motor armature. The assembly of thyristors, mounted on a heatsink, is usually referred to as the 'stack'. By altering the firing angle of the thyristors the mean value of the rectified voltage can be varied, thereby allowing the motor speed to be controlled.

We saw in Chapter 2 that the controlled rectifier produces a crude form of d.c. with a pronounced ripple in the output voltage. This ripple component gives rise to pulsating currents and fluxes in the motor, and in order to avoid excessive eddy-current losses and commutation problems, the poles and frame should be of laminated construction. It is accepted practice for motors supplied for use with thyristor drives to have laminated construction, but older motors often have solid poles and/or frames, and these will not always work satisfactorily with a rectifier supply. It is also the norm for drive motors to be supplied with an attached 'blower' motor as standard. This provides continuous through ventilation and allows the motor to operate continuously at full torque even down to the lowest speeds without overheating.

Low power control circuits are used to monitor the principal variables of interest (usually motor current and speed), and to generate appropriate firing pulses so that the motor maintains constant speed despite variations in the load. The 'speed reference' (Figure 4.1) is typically an analogue voltage varying from 0 to 10 V, and obtained from a manual speed-setting potentiometer or from elsewhere in the plant.

The combination of power, control, and protective circuits constitutes the converter. Standard modular converters are available as off-the-shelf items in sizes from 0.5 kW up to several hundred kW, while larger drives will be tailored to individual requirements. Individual converters may be mounted in enclosures with isolators, fuses etc., or groups of converters may be mounted together to form a multi-motor drive.

Motor operation with converter supply

The basic operation of the rectifying bridge has been discussed in Chapter 2, and we now turn to the matter of how the d.c. motor behaves when supplied with 'd.c.' from a controlled rectifier.

By no stretch of imagination could the waveforms of armature voltage looked at in Chapter 2 (e.g. Figure 2.12) be thought of as good d.c., and it would not be unreasonable to question the wisdom of feeding such an unpleasant looking waveform to a d.c. motor. In fact it turns out that the motor works almost as well as it would if fed with pure d.c., for two main reasons. Firstly, the armature inductance of the motor causes the waveform of armature current to be much smoother than the waveform of armature voltage, which in turn means that the torque ripple is much less than might have been feared. And secondly, the inertia of the armature is sufficiently large for the speed to remain almost steady despite the torque ripple. It is indeed fortunate that such a simple arrangement works so well, because any attempt to smooth-out the voltage waveform (perhaps by adding smoothing capacitors) would prove to be prohibitively expensive in the power ranges of interest.

Motor current waveforms

For the sake of simplicity we will look at operation from a single-phase (2-pulse) converter, but similar conclusions apply to the 6-pulse converter. The voltage (V_a) applied to the motor armature is typically as shown in Figure 4.2(a): as we saw in Chapter 2, it consists of rectified 'chunks' of the incoming mains voltage, the precise shape and average value depending on the firing angle.

The voltage waveform can be considered to consist of a mean d.c. level (V_{dc}), and a superimposed pulsating or ripple component which we can denote loosely as V_{ac}. The mean voltage V_{dc} can be altered by varying the firing angle, which also incidentally alters the ripple (i.e. V_{ac}).

The ripple voltage causes a ripple current to flow in the armature, but because of the armature inductance, the amplitude of the ripple current is small. In other words, the armature presents a high impedance to a.c. voltages. This smoothing effect of the armature inductance is shown in Figure 4.2(b), from which it can be seen that the current ripple is relatively small in comparison with the corresponding voltage ripple. The average value of the ripple current is of course zero, so it has no effect on the average torque of the motor. There is nevertheless a variation in torque every half-cycle of the mains, but because it is of small amplitude and high frequency the variation in speed (and hence back e.m.f., E) will not usually be noticeable.

The current at the end of each pulse is the same as at the beginning, so it follows that the average voltage across the armature inductance (L) is zero. We can therefore equate the average applied voltage to the sum of the back e.m.f. (assumed pure d.c. because we are ignoring speed fluctuations) and the average voltage across the armature resistance, to yield

$$V_{dc} = E + I_{dc}R \tag{4.1}$$

which is exactly the same as for operation from a pure d.c. supply. This is very important, as it underlines the fact that we can control the mean motor voltage, and hence the speed, simply by varying the converter delay angle.

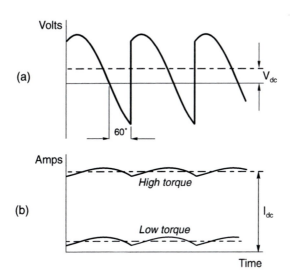

Figure 4.2 *Armature voltage* (a) *and armature current* (b) *waveforms for continuous-current operation of a d.c. motor supplied from a single-phase fully-controlled thyristor converter, with firing angle of 60°*

The smoothing effect of the armature inductance is important in achieving successful motor operation: the armature acts as a low-pass filter, blocking most of the ripple, and leading to a more or less constant armature current. For the smoothing to be effective, the armature time-constant needs to be long compared with the pulse duration (half a cycle with a 2-pulse drive, but only one sixth of a cycle in a 6-pulse drive). This condition is met in all 6-pulse drives, and in many 2-pulse ones. Overall, the motor then behaves much as it would if it was supplied from an ideal d.c. source (though the I^2R loss is higher than it would be if the current was perfectly smooth).

The no-load speed is determined by the applied voltage (which depends on the firing angle of the converter); there is a small drop in speed with load and as we have previously noted, the average current is determined by the load. In Figure 4.2, for example, the voltage waveform in (a) applies equally for the two load conditions represented in (b), where the upper current waveform corresponds to a high value of load torque while the lower is for a much lighter load; the speed being almost the same in both cases. (The small difference in speed is due to IR as explained in Chapter 3). We should note that the current ripple remains the same – only the average current changes with load. Broadly speaking, therefore, we can say that the speed is determined by the converter firing angle, which represents a very satisfactory state because we can control the firing angle by low-power control circuits and thereby regulate the speed of the drive.

The current waveforms in Figure 4.2(b) are referred to as 'continuous', because there is never any time during which the current is not flowing. This 'continuous current' condition is the norm in most drives, and it is highly desirable because it is only under continuous current conditions that the average voltage from the converter is determined solely by the firing angle, and is independent of the load current. We can see why this is so with the aid of Figure 2.7, imagining that the motor is connected to the output terminals and that it is drawing a continuous current. For half of a complete cycle, the current will flow into the motor from T1 and return to the mains via T4, so the armature is effectively switched across the supply and the armature voltage is equal to the supply voltage, which is assumed to be ideal, i.e. it is independent of the current drawn. For the other half of the time, the motor current flows from T2 and returns to the supply via T3, so the motor is again hooked-up to the supply, but this time the connections are reversed. Hence the average armature voltage – and hence to a first approximation the speed – are defined once the firing angle is set.

Discontinuous current

We can see from Figure 4.2 that as the load torque is reduced, there will come a point where the minima of the current ripple touches the zero-current line, i.e. the current reaches the boundary between continuous and discontinuous current. The load at which this occurs will also depend on the armature inductance, because the higher the inductance the smoother the current (i.e. the less the ripple). Discontinuous current mode is therefore most likely to be encountered in small machines with low inductance (particularly when fed from two-pulse converters) and under light-load or no-load conditions.

Typical armature voltage and current waveforms in the discontinuous mode are shown in Figure 4.3, the armature current consisting of discrete pulses of current that occur only while the armature is connected to the supply, with zero current for the period (represented by θ in Figure 4.3) when none of the thyristors are conducting and the motor is coasting free from the supply.

The shape of the current waveform can be understood by noting that with resistance neglected, equation (3.7) can be rearranged as

$$\frac{di}{dt} = \frac{1}{L}(V - E) \tag{4.2}$$

which shows that the rate of change of current (i.e. the gradient of the lower graph in Figure 4.3) is determined by the instantaneous difference between the applied voltage V and the motional e.m.f. E. Values of $(V - E)$ are shown by the vertical hatchings in Figure 4.3, from which it can be seen that if $V > E$, the current is increasing, while if $V < E$, the current is falling. The peak current is thus determined by the area of the upper or lower shaded areas of the upper graph.

The firing angle in Figures 4.2 and 4.3 is the same, at 60°, but the load is less in Figure 4.3 and hence the average current is lower (though, for the sake of the explanation offered below the current axis in Figure 4.3 is expanded as compared with that in Figure 4.2). It should be clear by comparing these figures that the armature voltage waveforms (solid lines) differ because, in Figure 4.3, the current falls to zero before the next firing pulse arrives and during the period shown as θ the motor floats free, its terminal voltage during this time being simply the motional e.m.f. (E). To simplify Figure 4.3 it has been assumed that the armature resistance is small and that the corresponding volt-drop ($I_a R_a$) can be ignored. In this case, the average armature voltage (V_{dc}) must be equal to the motional e.m.f., because there can be no average voltage across the armature inductance when there is no nett change in the

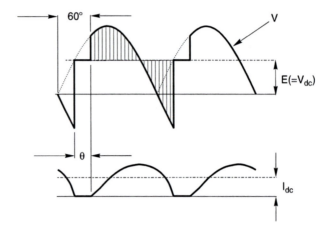

Figure 4.3   *Armature voltage current waveforms for discontinuous-current operation of a d.c. motor supplied from a single-phase fully-controlled thyristor converter, with firing angle of 60°*

current over one pulse: the hatched areas – representing the volt-seconds in the inductor – are therefore equal.

The most important difference between Figures 4.2 and 4.3 is that the average voltage is higher when the current is discontinuous, and hence the speed corresponding to the conditions in Figure 4.3 is higher than in 4.2 despite both having the same firing angle. And whereas in continuous mode a load increase can be met by an increased armature current without affecting the voltage (and hence speed), the situation is very different when the current is discontinuous. In the latter case, the only way that the average current can increase is when speed (and hence E) falls so that the shaded areas in Figure 4.3 become larger.

This means that from the user's viewpoint the behaviour of the motor in discontinuous mode is much worse than in the continuous current mode, because as the load torque is increased, there is a serious drop in speed. The resulting torque–speed curve therefore has a very unwelcome 'droopy' characteristic in the discontinuous current region, as shown in Figure 4.4, and in addition the I^2R loss is much higher than it would be with pure d.c.

Under very light or no-load conditions, the pulses of current become virtually non-existent, the shaded areas in Figure 4.3 become very small and the motor speed reaches a point at which the back e.m.f. is equal to the peak of the supply voltage.

It is easy to see that inherent torque–speed curves with sudden discontinuities of the form shown in Figure 4.4 are very undesirable. If for example the firing angle is set to zero and the motor is fully loaded, its

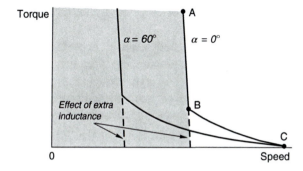

Figure 4.4 *Torque-speed curves illustrating the undesirable 'droopy' characteristic associated with discontinuous current. The improved characteristic (shown dotted) corresponds to operation with continuous current*

speed will settle at point A, its average armature voltage and current having their full (rated) values. As the load is reduced, current remaining continuous, there is the expected slight rise in speed, until point B is reached. This is the point at which the current is about to enter the discontinuous phase. Any further reduction in the load torque then produces a wholly disproportionate – not to say frightening – increase in speed, especially if the load is reduced to zero when the speed reaches point C.

There are two ways by which we can improve these inherently poor characteristics. Firstly, we can add extra inductance in series with the armature to further smooth the current waveform and lessen the likelihood of discontinuous current. The effect of adding inductance is shown by the dotted lines in Figure 4.4. And secondly, we can switch from a single-phase converter to a 3-phase converter which produces smoother voltage and current waveforms, as discussed in Chapter 2.

When the converter and motor are incorporated in a closed-loop control the user should be unaware of any shortcomings in the inherent motor/converter characteristics because the control system automatically alters the firing angle to achieve the target speed at all loads. In relation to Figure 4.4, for example, as far as the user is concerned the control system will confine operation to the shaded region, and the fact that the motor is theoretically capable of running unloaded at the high speed corresponding to point C is only of academic interest.

Converter output impedance: overlap

So far we have tacitly assumed that the output voltage from the converter was independent of the current drawn by the motor, and depended only on the delay angle α. In other words we have treated the converter as an ideal voltage source.

In practice the a.c. supply has a finite impedance, and we must therefore expect a volt-drop which depends on the current being drawn by the motor. Perhaps surprisingly, the supply impedance (which is mainly due to inductive leakage reactances in transformers) manifests itself at the output stage of the converter as a supply resistance, so the supply volt-drop (or regulation) is directly proportional to the motor armature current.

It is not appropriate to go into more detail here, but we should note that the effect of the inductive reactance of the supply is to delay the transfer (or commutation) of the current between thyristors; a phenomenon known as overlap. The consequence of overlap is that instead of the output voltage making an abrupt jump at the start of each pulse, there is a short period when two thyristors are conducting simultaneously. During this interval the output voltage is the mean of the voltages of the incoming and outgoing voltages, as shown typically in Figure 4.5. It is important for users to be aware that overlap is to be expected, as otherwise they may be alarmed the first time they connect an oscilloscope to the motor terminals. When the drive is connected to a 'stiff' (i.e. low impedance) industrial supply the overlap will only last for perhaps a few microseconds, so the 'notch' shown in Figure 4.5 would be barely visible on an oscilloscope. Books always exaggerate the width of the overlap for the sake of clarity, as in Figure 4.5: with a 50 or 60 Hz supply, if the overlap lasts for more than say 1 ms, the implication is that the supply system impedance is too high for the size of converter in question, or conversely, the converter is too big for the supply.

Returning to the practical consequences of supply impedance, we simply have to allow for the presence of an extra 'source resistance' in series with the output voltage of the converter. This source resistance is

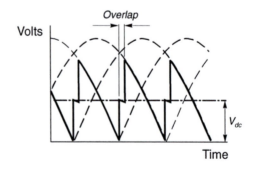

Figure 4.5   *Distortion of converter output voltage waveform caused by rectifier overlap*

in series with the motor armature resistance, and hence the motor torque–speed curves for each value of α have a somewhat steeper droop than they would if the supply impedance was zero.

Four-quadrant operation and inversion

So far we have looked at the converter as a rectifier, supplying power from the a.c. mains to a d.c. machine running in the positive direction and acting as a motor. As explained in Chapter 3, this is known as one-quadrant operation, by reference to quadrant 1 of the complete torque–speed plane shown in Figure 3.16.

But suppose we want to run the machine as a motor in the opposite direction, with negative speed and torque, i.e. in quadrant 3; how do we do it? And what about operating the machine as a generator, so that power is returned to the a.c. supply, the converter then 'inverting' power rather than rectifying, and the system operating in quadrant 2 or quadrant 4. We need to do this if we want to achieve regenerative braking. Is it possible, and if so how?

The good news is that as we saw in Chapter 3 the d.c. machine is inherently a bidirectional energy converter. If we apply a positive voltage V greater than E, a current flows into the armature and the machine runs as a motor. If we reduce V so that it is less than E, the current, torque and power automatically reverse direction, and the machine acts as a generator, converting mechanical energy (its own kinetic energy in the case of regenerative braking) into electrical energy. And if we want to motor or generate with the reverse direction of rotation, all we have to do is to reverse the polarity of the armature supply. The d.c. machine is inherently a four-quadrant device, but needs a supply which can provide positive or negative voltage, and simultaneously handle either positive or negative current.

This is where we meet a snag: a single thyristor converter can only handle current in one direction, because the thyristors are unidirectional devices. This does not mean that the converter is incapable of returning power to the supply however. The d.c. current can only be positive, but (provided it is a fully controlled converter) the d.c. output voltage can be either positive or negative (see Chapter 2). The power flow can therefore be positive (rectification) or negative (inversion).

For normal motoring where the output voltage is positive (and assuming a fully controlled converter), the delay angle (α) will be up to 90°. (It is common practice for the firing angle corresponding to rated d.c. voltage to be around 20° when the incoming a.c. voltage is normal: if the a.c. voltage falls for any reason, the firing angle can then

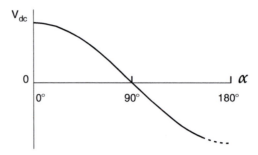

Figure 4.6   *Average d.c. output voltage from a fully-controlled thyristor converter as a function of the firing angle delay* α

be further reduced to compensate and allow full d.c. voltage to be maintained.)

When α is greater than 90°, however, the output voltage is negative, as indicated by equation (2.5), and is shown in Figure 4.6. A single fully controlled converter therefore has the potential for two-quadrant operation, though it has to be admitted that this capability is not easily exploited unless we are prepared to employ reversing switches in the armature or field circuits. This is discussed next.

Single-converter reversing drives

We will consider a fully controlled converter supplying a permanent-magnet motor, and see how the motor can be regeneratively braked from full speed in one direction, and then accelerated up to full speed in reverse. We looked at this procedure in principle at the end of Chapter 3, but here we explore the practicalities of achieving it with a converter-fed drive. We should be clear from the outset that in practice, all the user has to do is to change the speed reference signal from full forward to full reverse: the control system in the drive converter takes care of matters from then on. What it does, and how, is discussed below.

When the motor is running at full speed forward, the converter delay angle will be small, and the converter output voltage V and current I will both be positive. This condition is shown in Figure 4.7(a), and corresponds to operation in quadrant 1.

In order to brake the motor, the torque has to be reversed. The only way this can be done is by reversing the direction of armature current. The converter can only supply positive current, so to reverse the motor torque we have to reverse the armature connections, using a mechanical

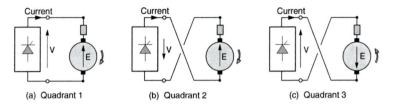

Figure 4.7   *Stages in motor reversal using a single-converter drive and mechanical reversing switch*

switch or contactor, as shown in Figure 4.7(b). (Before operating the contactor, the armature current would be reduced to zero by lowering the converter voltage, so that the contactor is not required to interrupt current.) Note that because the motor is still rotating in the positive direction, the back e.m.f. remains in its original sense; but now the motional e.m.f. is seen to be assisting the current and so to keep the current within bounds the converter must produce a negative voltage V which is just a little less than E. This is achieved by setting the delay angle at the appropriate point between 90° and 180°. (The dotted line in Figure 4.6 indicates that the maximum acceptable negative voltage will generally be somewhat less than the maximum positive voltage: this restriction arises because of the need to preserve a margin for commutation of current between thyristors.) Note that the converter current is still positive (i.e. upwards in Figure 4.7(b)), but the converter voltage is negative, and power is thus flowing back to the mains. In this condition the system is operating in quadrant 2, and the motor is decelerating because of the negative torque. As the speed falls, E reduces, and so V must be reduced progressively to keep the current at full value. This is achieved automatically by the action of the current-control loop, which is discussed later.

The current (i.e. torque) needs to be kept negative in order to run up to speed in the reverse direction, but after the back e.m.f. changes sign (as the motor reverses), the converter voltage again becomes positive and greater than E, as shown in Figure 4.7(c). The converter is then rectifying, with power being fed into the motor, and the system is operating in quadrant 3.

Schemes using reversing contactors are not suitable where the reversing time is critical, because of the delay caused by the mechanical reversing switch, which may easily amount to 200–400 msec. Field reversal schemes operate in a similar way, but reverse the field current instead of the armature current. They are even slower, because of the relatively long time-constant of the field winding.

Double-converter reversing drives

Where full four-quadrant operation and rapid reversal is called for, two converters connected in anti-parallel are used, as shown in Figure 4.8. One converter supplies positive current to the motor, while the other supplies negative current.

The bridges are operated so that their d.c. voltages are almost equal thereby ensuring that any d.c. circulating current is small, and a reactor is placed between the bridges to limit the flow of ripple currents which result from the unequal ripple voltages of the two converters. Alternatively, the reactor can be dispensed with by only operating one converter at a time. The changeover from one converter to the other can only take place after the firing pulses have been removed from one converter, and the armature current has decayed to zero. Appropriate zero-current detection circuitry is provided as an integral part of the drive, so that as far as the user is concerned, the two converters behave as if they were a single ideal bidirectional d.c. source.

Prospective users need to be aware of the fact that a basic single converter can only provide for operation in one quadrant. If regenerative braking is required, either field or armature reversing contactors will be needed; and if rapid reversal is essential, a double converter has to be used. All these extras naturally push up the purchase price.

Power factor and supply effects

One of the drawbacks of a converter-fed d.c. drive is that the supply power factor is very low when the motor is operating at high torque (i.e. high current) and low speed (i.e. low armature voltage), and is less than unity even at base speed and full load. This is because the supply current waveform lags the supply voltage waveform by the delay angle α, as shown (for a 3-phase converter) in Figure 4.9, and also the supply current is approximately rectangular (rather than sinusoidal).

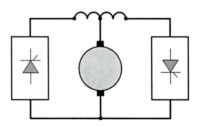

Figure 4.8   *Double-converter reversing drive*

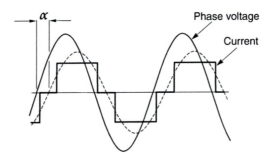

Figure 4.9 *Supply voltage and current waveforms for single-phase converter-fed d.c. motor drive*

It is important to emphasise that the supply power factor is always lagging, even when the converter is inverting. There is no way of avoiding the low power factor, so users of large drives need to be prepared to augment their existing power factor correcting equipment if necessary.

The harmonics in the mains current waveform can give rise to a variety of interference problems, and supply authorities generally impose statutory limits. For large drives (say hundreds of kilowatts), filters may have to be provided to prevent these limits from being exceeded.

Since the supply impedance is never zero, there is also inevitably some distortion of the mains voltage waveform, as shown in Figure 4.10 which indicates the effect of a 6-pulse converter on the supply line-to-line voltage waveform. The spikes and notches arise because the mains is momentarily short-circuited each time the current commutates from one thyristor to the next, i.e. during the overlap period discussed earlier. For the majority of small and medium drives, connected to stiff industrial supplies, these notches are too small to be noticed (they are greatly exaggerated for the sake of clarity in Figure 4.10); but they can pose a

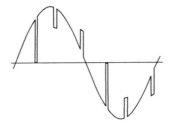

Figure 4.10 *Distortion of line voltage waveform caused by overlap in three-phase fully-controlled converter. (The width of the notches has been exaggerated for the sake of clarity.)*

serious interference problem for other consumers when a large drive is connected to a weak supply.

CONTROL ARRANGEMENTS FOR D.C. DRIVES

The most common arrangement, which is used with only minor variations from small drives of say 0.5 kW up to the largest industrial drives of several megawatts, is the so-called two-loop control. This has an inner feedback loop to control the current (and hence torque) and an outer loop to control speed. When position control is called for, a further outer position loop is added. A two-loop scheme for a thyristor d.c. drive is discussed first, but the essential features are the same in a chopper-fed drive. Later the simpler arrangements used in low-cost small drives are discussed.

The discussion is based on analogue control, and as far as possible is limited to those aspects which the user needs to know about and understand. In practice, once a drive has been commissioned, there are only a few potentiometer adjustments (or presets in the case of a digital control) to which the user has access. Whilst most of them are self-explanatory (e.g. max. speed, min. speed, accel. and decel. rates), some are less obvious (e.g. 'current stability', 'speed stability', 'IR comp'.) so these are explained.

To appreciate the overall operation of a two-loop scheme we can consider what we would do if we were controlling the motor manually. For example, if we found by observing the tachogenerator that the speed was below target, we would want to provide more current (and hence torque) in order to produce acceleration, so we would raise the armature voltage. We would have to do this gingerly however, being mindful of the danger of creating an excessive current because of the delicate balance that exists between the back e.m.f., E and applied voltage, V. We would doubtless wish to keep our eye on the ammeter at all times to avoid blowing-up the thyristor stack, and as the speed approached the target, we would trim back the current (by lowering the applied voltage) so as to avoid overshooting the set speed. Actions of this sort are carried out automatically by the drive system, which we will now explore.

A standard d.c. drive system with speed and current control is shown in Figure 4.11. The primary purpose of the control system is to provide speed control, so the 'input' to the system is the speed reference signal on the left, and the output is the speed of the motor (as measured by the tachogenerator TG) on the right. As with any closed-loop system, the overall performance is heavily dependent on the quality of the feedback signal, in this case the speed-proportional voltage provided by the tachogenerator. It is therefore important to ensure that the tacho is of high quality (so that its output voltage does not vary with ambient temperature, and is ripple-free) and as a result the cost of the tacho often represents a significant fraction of the total cost.

We will take an overview of how the scheme operates first, and then examine the function of the two loops in more detail.

To get an idea of the operation of the system we will consider what will happen if, with the motor running light at a set speed, the speed reference signal is suddenly increased. Because the set (reference) speed is now greater than the actual speed there will be a speed error signal (see also Figure 4.12), represented by the output of the left-hand summing junction in Figure 4.11. A speed error indicates that acceleration is required, which in turn means torque, i.e. more current. The speed error is amplified by the speed controller (which is more accurately described as a speed-error amplifier) and the output serves as the reference or input signal to the inner control system. The inner feedback loop is a current-control loop, so when the current reference increases, so does the motor armature current, thereby providing extra torque and initiating acceleration. As the speed rises the speed error reduces, and the current and torque therefore reduce to obtain a smooth approach to the target speed.

We will now look in more detail at the inner (current -control) loop, as its correct operation is vital to ensure that the thyristors are protected against excessive overcurrents.

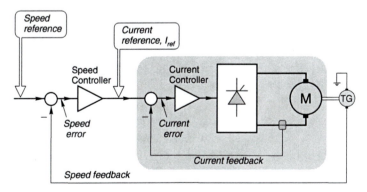

Figure 4.11   *Schematic diagram of analogue controlled-speed drive with current and speed feedback control loops*

Current control

The closed-loop current controller, or current loop, is at the heart of the drive system and is indicated by the shaded region in Figure 4.11. The purpose of the current loop is to make the actual motor current follow the current reference signal (I_{ref}) shown in Figure 4.11. It does this by comparing a feedback signal of actual motor current with the current reference signal, amplifying the difference (or current error), and using the resulting amplified current error signal (an analogue voltage) to control the firing angle α – and hence the output voltage – of the converter. The current feedback signal is obtained either from a d.c. current transformer (which gives an isolated analogue voltage output), or from a.c current transformer/rectifiers in the mains supply lines.

The job of comparing the reference (demand) and actual current signals and amplifying the error signal is carried out by the current-error amplifier. By giving the current error amplifier a high gain, the actual motor current will always correspond closely to the current reference signal, i.e. the current error will be small, regardless of motor speed. In other words, we can expect the actual motor current to follow the 'current reference' signal at all times, the armature voltage being automatically adjusted by the controller so that, regardless of the speed of the motor, the current has the correct value.

Of course no control system can be perfect, but it is usual for the current-error amplifier to be of the proportional plus integral (PI) type (see below), in which case the actual and demanded currents will be exactly equal under steady-state conditions.

The importance of preventing excessive converter currents from flowing has been emphasised previously, and the current control loop provides the means to this end. As long as the current control loop functions properly, the motor current can never exceed the reference value. Hence by limiting the magnitude of the current reference signal (by means of a clamping circuit), the motor current can never exceed the specified value. This is shown in Figure 4.12, which represents a small portion of Figure 4.11. The characteristics of the speed controller are shown in the shaded panel, from which we can see that for small errors in speed, the current reference increases in proportion to the speed, thereby ensuring 'linear system' behaviour with a smooth approach to the target speed. However, once the speed error exceeds a limit, the output of the speed-error amplifier saturates and there is thus no further increase in the current reference. By arranging for this maximum current reference to correspond to the full (rated) current of the system there is no possibility of the current in the motor and converter exceeding its

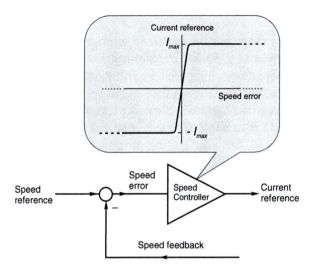

Figure 4.12 *Detail showing characteristic of speed error amplifier*

rated value, no matter how large the speed error becomes. This point is explored further in Section 4.3.3.

This 'electronic current limiting' is by far the most important protective feature of any drive. It means that if for example the motor suddenly stalls because the load seizes (so that the back e.m.f. falls dramatically), the armature voltage will automatically reduce to a very low value, thereby limiting the current to its maximum allowable level.

The first thing we should aim at when setting up a drive is a good current loop. In this context, 'good' means that the steady-state motor current should correspond exactly with the current reference, and the transient response to step changes in the current reference should be fast and well damped. The first of these requirements is satisfied by the integral term in the current-error amplifier, while the second is obtained by judicious choice of the amplifier proportional gain and time-constant. As far as the user is concerned, the 'current stability' adjustment is provided to allow him to optimise the transient response of the current loop.

On a point of jargon, it should perhaps be mentioned that the current-error amplifier is more often than not called either the 'current controller' (as in Figure 4.11) or the 'current amplifier'. The first of these terms is quite sensible, but the second can be very misleading: there is after all no question of the motor current itself being amplified.

Torque control

For applications requiring the motor to operate with a specified torque regardless of speed (e.g. in line tensioning), we can dispense with the outer (speed) loop, and simply feed a current reference signal directly to the current controller (usually via the 'torque ref' terminal on the control board). This is because torque is directly proportional to current, so the current controller is in effect also a torque controller. We may have to make an allowance for accelerating torque by means of a transient 'inertia compensating' signal, but this is usually provided for via a potentiometer adjustment or digital preset.

In the current-control mode, the current remains constant at the set value, and the steady running speed is determined by the load. If the torque reference signal was set at 50%, for example, and the motor was initially at rest, it would accelerate with a constant current of half rated value until the load torque was equal to the motor torque. Of course, if the motor was running without any load, it would accelerate quickly, the applied voltage ramping up so that it always remained higher than the back e.m.f. by the amount needed to drive the specified current into the armature. Eventually the motor would reach a speed (a little above normal 'full' speed) at which the converter output voltage had reached its upper limit, and it is therefore no longer possible to maintain the set current: thereafter, the motor speed would remain steady.

Speed control

The outer loop in Figure 4.11 provides speed control. Speed feedback is provided by a d.c. tachogenerator and the actual and required speeds are fed into the speed-error amplifier (often known simply as the speed amplifier or the speed controller).

Any difference between the actual and desired speed is amplified, and the output serves as the input to the current loop. Hence if for example the actual motor speed is less than the desired speed, the speed amplifier will demand current in proportion to the speed error, and the motor will therefore accelerate in an attempt to minimise the speed error.

When the load increases, there is an immediate deceleration and the speed-error signal increases, thereby calling on the inner loop for more current. The increased torque results in acceleration and a progressive reduction of the speed error until equilibrium is reached at the point where the current reference (I_{ref}) produces a motor current that gives a torque equal and opposite to the load torque. Looking at Figure 4.12, where the speed controller is shown as simple proportional amplifier (P control), it will be readily appreciated that in order for there to be a steady-state value of I_{ref}, there would have to be a finite speed error, i.e. a P controller would not allow us to reach exactly the target speed. (We could approach the ideal by increasing the gain of the amplifier, but that might lead us to instability.)

To eliminate the steady-state speed error we can easily arrange for the speed controller to have an integral (I) term as well as a proportional (P) term (see Appendix). A PI controller can have a finite output even when the input is zero, which means that we can achieve zero steady-state error if we employ PI control.

The speed will be held at the value set by the speed reference signal for all loads up to the point where full armature current is needed. If the load torque increases any more the speed will drop because the current-loop will not allow any more armature current to flow. Conversely, if the load attempted to force the speed above the set value, the motor current will be reversed automatically, so that the motor acts as a brake and regenerates power to the mains.

To emphasise further the vitally important protective role of the inner loop, we can see what happens when, with the motor at rest (and unloaded for the sake of simplicity), we suddenly increase the speed reference from zero to full value, i.e. we apply a step demand for full speed. The speed error will be 100%, so the output (I_{ref}) from the speed-error amplifier will immediately saturate at its maximum value, which has been deliberately clamped so as to correspond to a demand for the maximum (rated) current in the motor. The motor current will therefore be at rated value, and the motor will accelerate at full torque. Speed and back e.m.f (E) will therefore rise at a constant rate, the applied voltage (V) increasing steadily so that the difference ($V - E$) is sufficient to drive rated current (I) through the armature resistance. A very similar sequence of events was discussed in Chapter 3, and illustrated by the second half of Figure 3.17. (In some drives the current reference is allowed to reach 150% or even 200% of rated value for a few seconds, in order to provide a short torque boost. This is particularly valuable in starting loads with high static friction, and is known as 'two-stage current limit'.)

The output of the speed amplifier will remain saturated until the actual speed is quite close to the target speed, and for all this time the motor current will therefore be held at full value. Only when the speed is within a few percent of target will the speed-error amplifier come out of saturation. Thereafter, as the speed continues to rise, and the speed error falls, the output of the speed-error amplifier falls below the clamped level. Speed control then enters a linear regime, in which the correcting current (and hence the torque) is proportional to speed error, thus giving a smooth approach to final speed.

A 'good' speed controller will result in zero steady-state error, and have a well-damped response to step changes in the demanded speed. The integral term in the PI control caters for the requirement of zero steady-state error, while the transient response depends on the setting of the proportional gain and time-constant. The 'speed stability' potentiometer is provided to allow the user to optimise the transient speed response.

It should be noted that it is generally much easier to obtain a good transient response with a regenerative drive, which has the ability to supply negative current (i.e. braking torque) should the motor overshoot the desired speed. A non-regenerative drive cannot furnish negative current (unless fitted with reversing contactors), so if the speed overshoots the target the best that can be done is to reduce the armature current to zero and wait for the motor to decelerate naturally. This is not satisfactory, and every effort therefore has to be made to avoid controller settings which lead to an overshoot of the target speed.

As with any closed-loop scheme, problems occur if the feedback signal is lost when the system is in operation. If the tacho feedback became disconnected, the speed amplifier would immediately saturate, causing full torque to be applied. The speed would then rise until the converter output reached its maximum output voltage. To guard against this many drives incorporate tacho-loss detection circuitry, and in some cases armature voltage feedback (see later section) automatically takes over in the event of tacho failure.

Drives which use field-weakening to extend the speed range include automatic provision for controlling both armature voltage and field current when running above base speed. Typically, the field current is kept at full value until the armature voltage reaches about 95% of rated value. When a higher speed is demanded, the extra armature voltage applied is accompanied by a simultaneous reduction in the field current, in such a way that when the armature voltage reaches 100% the field current is at the minimum safe value. This process is known as 'spillover field weakening'.

Overall operating region

A standard drive with field-weakening provides armature voltage control of speed up to base speed, and field-weakening control of speed thereafter. Any torque up to the rated value can be obtained at any speed below base speed, and as explained in Chapter 3 this region is known as the 'constant torque' region. Above base speed, the maximum available torque reduces inversely with speed, so this is known as the 'constant power' region. For a converter-fed drive the operating region in quadrant 1 of the torque–speed plane is therefore shown in Figure 3.10. (If the drive is equipped for regenerative and reversing operation, the operating area is mirrored in the other three quadrants, of course.)

Armature voltage feedback and IR compensation

In low-power drives where precision speed-holding is not essential, and cost must be kept to a minimum, the tachogenerator is dispensed with and the armature voltage is used as a 'speed feedback' instead. Performance is clearly not as good as with tacho feedback, since whilst the steady-state no-load speed is proportional to armature voltage, the speed falls as the load (and hence armature current) increases.

We saw in Chapter 3 that the drop in speed with load was attributable to the armature resistance volt-drop (IR), and the drop in speed can therefore be compensated by boosting the applied voltage in proportion to the current. An adjustment labelled 'IR comp' or simply 'IR' is provided on the drive circuit for the user to adjust to suit the particular motor. The compensation is usually far from perfect, since it cannot cope with temperature variation of resistance, nor with the effects of armature reaction; but it is better than nothing.

Drives without current control

Cheaper drives often dispense with the full current control loop, and incorporate a crude but effective 'current-limit' which only operates when the maximum set current would otherwise be exceeded. These drives usually have an in-built ramp circuit which limits the rate of rise of the set speed signal so that under normal conditions the current limit is not activated.

CHOPPER-FED D.C. MOTOR DRIVES

If the source of supply is d.c. (for example in a battery vehicle or a rapid transit system) a chopper-type converter is usually employed. The basic operation of a single-switch chopper was discussed in Chapter 2, where it was shown that the average output voltage could be varied by periodically switching the battery voltage on and off for varying intervals. The principal difference between the thyristor-controlled rectifier and the chopper is that in the former the motor current always flows through the supply, whereas in the latter, the motor current only flows from the supply terminals for part of each cycle.

A single-switch chopper using a transistor, MOSFET or IGBT can only supply positive voltage and current to a d.c. motor, and is therefore restricted to quadrant 1 motoring operation. When regenerative and/or rapid speed reversal is called for, more complex circuitry is required, involving two or more power switches, and consequently leading to increased cost. Many different circuits are used and it is not possible to go into detail here, though it should be mentioned that the chopper circuit discussed in Chapter 2 only provides an output voltage in the range $0 < E$, where E is the battery voltage, so this type of chopper is only suitable if the motor voltage is less than the battery voltage. Where the motor voltage is greater than the battery voltage, a 'step-up' chopper using an additional inductance as an intermediate energy store is used.

Performance of chopper-fed d.c. motor drives

We saw earlier that the d.c. motor performed almost as well when fed from a phase-controlled rectifier as it does when supplied with pure d.c. The chopper-fed motor is, if anything, rather better than the phase-controlled, because the armature current ripple can be less if a high chopping frequency is used.

Typical waveforms of armature voltage and current are shown in Figure 4.13(c): these are drawn with the assumption that the switch is ideal. A chopping frequency of around 100 Hz, as shown in Figure 4.13, is typical of medium and large chopper drives, while small drives often use a much higher chopping frequency, and thus have lower ripple current. As usual, we have assumed that the speed remains constant despite the slightly pulsating torque, and that the armature current is continuous.

The shape of the armature voltage waveform reminds us that when the transistor is switched on, the battery voltage V is applied directly to the armature, and during this period the path of the armature current is indicated by the dotted line in Figure 4.13(a). For the remainder of the cycle the transistor is turned 'off' and the current freewheels through the diode, as shown by the dotted line in Figure 4.13(b). When the current is freewheeling through the diode, the armature voltage is clamped at (almost) zero.

The speed of the motor is determined by the average armature voltage, (V_{dc}), which in turn depends on the proportion of the total cycle

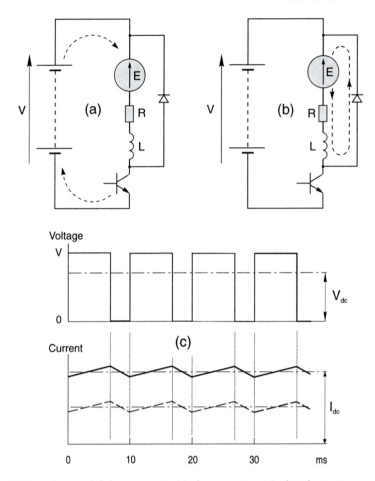

Figure 4.13   *Chopper-fed d.c. motor. In (a) the transistor is 'on' and armature current is flowing through the voltage source; in (b) the transistor is 'off' and the armature current freewheels through the diode. Typical armature voltage and current waveforms are shown at (c), with the dotted line representing the current waveform when the load torque is reduced by half*

time (T) for which the transistor is 'on'. If the on and off times are defined as $T_{on} = kT$ and $T_{off} = (1 - k)T$, where $0 < k < 1$, then the average voltage is simply given by

$$V_{dc} = kV \qquad (4.3)$$

from which we see that speed control is effected via the on time ratio, k.

Turning now to the current waveforms shown in Figure 4.13(c), the upper waveform corresponds to full load, i.e. the average current (I_{dc})

produces the full rated torque of the motor. If now the load torque on the motor shaft is reduced to half rated torque, and assuming that the resistance is negligible, the steady-state speed will remain the same but the new mean steady-state current will be halved, as shown by the lower dotted curve. We note however that although, as expected, the mean current is determined by the load, the ripple current is unchanged, and this is explained below.

If we ignore resistance, the equation governing the current during the 'on' period is

$$V = E + L\frac{di}{dt}, \quad \text{or} \quad \frac{di}{dt} = \frac{1}{L}(V - E) \tag{4.4}$$

Since V is greater than E, the gradient of the current (di/dt) is positive, as can be seen in Figure 4.13(c). During this 'on' period the battery is supplying power to the motor. Some of the energy is converted to mechanical output power, but some is also stored in the magnetic field associated with the inductance. The latter is given by $1/2Li^2$, and so as the current (i) rises, more energy is stored.

During the 'off' period, the equation governing the current is

$$0 = E + L\frac{di}{dt}, \quad \text{or} \quad \frac{di}{dt} = -\frac{E}{L} \tag{4.5}$$

We note that during the 'off' time the gradient of the current is negative (as shown in Figure 4.13(c)) and it is determined by the motional e.m.f. E. During this period, the motor is producing mechanical output power which is supplied from the energy stored in the inductance; not surprisingly the current falls as the energy previously stored in the 'on' period is now given up.

We note that the rise and fall of the current (i.e. the current ripple) is inversely proportional to the inductance, but is independent of the mean d.c. current, i.e. the ripple does not depend on the load.

To study the input/output power relationship, we note that the battery current only flows during the 'on' period, and its average value is therefore kI_{dc}. Since the battery voltage is constant, the power supplied is simply given by $V(kI_{dc}) = kVI_{dc}$. Looking at the motor side, the average voltage is given by $V_{dc} = kV$, and the average current (assumed constant) is I_{dc}, so the power input to the motor is again kVI_{dc}, i.e. there is no loss of power in the ideal chopper. Given that k is less than one, we see that the input (battery) voltage is higher than the output (motor) voltage, but conversely the input current is less than the output current, and in this respect we see that the chopper behaves in much the same way for d.c. as a conventional transformer does for a.c.

Torque–speed characteristics and control arrangements

Under open-loop conditions (i.e. where the mark–space ratio of the chopper is fixed at a particular value) the behaviour of the chopper-fed motor is similar to the converter-fed motor discussed earlier (see Figure 4.3). When the armature current is continuous the speed falls only slightly with load, because the mean armature voltage remains constant. But when the armature current is discontinuous (which is most likely at high speeds and light load) the speed falls off rapidly when the load increases, because the mean armature voltage falls as the load increases. Discontinuous current can be avoided by adding an inductor in series with the armature, or by raising the chopping frequency, but when closed-loop speed control is employed, the undesirable effects of discontinuous current are masked by the control loop.

The control philosophy and arrangements for a chopper-fed motor are the same as for the converter-fed motor, with the obvious exception that the mark–space ratio of the chopper is used to vary the output voltage, rather than the firing angle.

D.C. SERVO DRIVES

The precise meaning of the term 'servo' in the context of motors and drives is difficult to pin down. Broadly speaking, if a drive incorporates 'servo' in its description, the implication is that it is intended specifically for closed-loop or feedback control, usually of shaft torque, speed, or position. Early servomechanisms were developed primarily for military applications, and it quickly became apparent that standard d.c. motors were not always suited to precision control. In particular high torque to inertia ratios were needed, together with smooth ripple-free torque. Motors were therefore developed to meet these exacting requirements, and not surprisingly they were, and still are, much more expensive than their industrial counterparts. Whether the extra expense of a servo motor can be justified depends on the specification, but prospective users should always be on their guard to ensure they are not pressed into an expensive purchase when a conventional industrial drive could cope perfectly well.

The majority of servo drives are sold in modular form, consisting of a high-performance permanent magnet motor, often with an integral

Plate 4.2   *High-performance permanent-magnet brushed d.c. servo motors with integral tachno/encoders. (Photo courtesy of Control Techniques)*

tachogenerator, and a chopper-type power amplifier module. The drive amplifier normally requires a separate regulated d.c. power supply, if, as is normally the case, the power is to be drawn from the a.c. mains. Continuous output powers range from a few watts up to perhaps 2–3 kW, with voltages of 12, 24, 48, and multiples of 50 V being standard.

Servo motors

Although there is no sharp dividing line between servo motors and ordinary motors, the servo type will be intended for use in applications which require rapid acceleration and deceleration. The design of the motor will reflect this by catering for intermittent currents (and hence torques) of many times the continuously rated value. Because most servo motors are small, their armature resistances are relatively high: the short-circuit (locked-rotor) current at full armature voltage is therefore perhaps only five times the continuously rated current, and the drive amplifier will normally be selected so that it can cope with this condition without difficulty, giving the motor a very rapid acceleration from rest. The even more arduous condition in which the full armature voltage is suddenly reversed with the motor running at full speed is also quite normal. (Both of these modes of operation would of course be quite unthinkable with a large d.c. motor, because of the huge currents which would flow as a result of the much lower per-unit armature resistance.) Because the drive amplifier must have a high current capability to provide for the high accelerations demanded, it is not normally necessary to employ an inner current-loop of the type discussed earlier.

In order to maximise acceleration, the rotor inertia must be minimised, and one obvious way to achieve this is to construct a motor in which only the electric circuit (conductors) on the rotor move, the magnetic part (either iron or permanent magnet) remaining stationary. This principle is adopted in 'ironless rotor' and 'printed armature' motors.

In the ironless rotor or moving-coil type (Figure 4.14) the armature conductors are formed as a thin-walled cylinder consisting essentially of nothing more than varnished wires wound in skewed form together with the disc-type commutator (not shown). Inside the armature sits a 2-pole (upper N, lower S) permanent magnet, which provides the radial flux, and outside it is a steel cylindrical shell which completes the magnetic circuit.

Needless to say the absence of slots to support the armature winding results in a relatively fragile structure, which is therefore limited to diameters of not much over 1 cm. Because of their small size they are often known as micromotors, and are very widely used in cameras, video systems, card readers etc.

The printed armature type is altogether more robust, and is made in sizes up to a few kilowatts. They are generally made in disc or pancake form, with the direction of flux axial and the armature current radial. The armature conductors resemble spokes on a wheel; the conductors themselves being formed on a lightweight disc. Early versions were made by using printed-circuit techniques, but pressed fabrication is now more common. Since there are usually at least 100 armature conductors, the torque remains almost constant as the rotor turns, which allows them to produce very smooth rotation at low speed. Inertia and armature inductance are low, giving a good dynamic response, and the short and fat shape makes them suitable for applications such as machine tools and disc drives where axial space is at a premium.

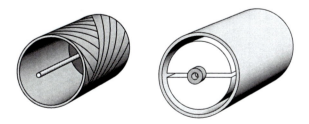

Figure 4.14 *Ironless rotor d.c. motor. The commutator (not shown) is usually of the disc type*

Position control

As mentioned earlier many servo motors are used in closed-loop position control applications, so it is appropriate to look briefly at how this is achieved. Later (in Chapter 8) we will see that the stepping motor provides an alternative open-loop method of position control, which can be cheaper for some applications.

In the example shown in Figure 4.15, the angular position of the output shaft is intended to follow the reference voltage (θ_{ref}), but it should be clear that if the motor drives a toothed belt linear outputs can also be obtained. The potentiometer mounted on the output shaft provides a feedback voltage proportional to the actual position of the output shaft. The voltage from this potentiometer must be a linear function of angle, and must not vary with temperature, otherwise the accuracy of the system will be in doubt.

The feedback voltage (representing the actual angle of the shaft) is subtracted from the reference voltage (representing the desired position) and the resulting position error signal is amplified and used to drive the motor so as to rotate the output shaft in the desired direction. When the output shaft reaches the target position, the position error becomes zero, no voltage is applied to the motor, and the output shaft remains at rest. Any attempt to physically move the output shaft from its target position immediately creates a position error and a restoring torque is applied by the motor.

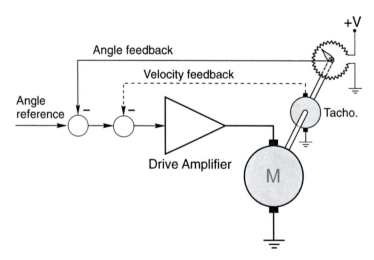

Figure 4.15   *Closed-loop angular position control using d.c. motor and angle feedback from a servo-type potentiometer*

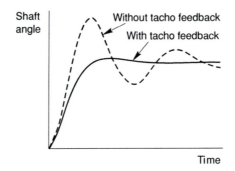

Figure 4.16 *Typical step responses for a closed-loop position control system, showing the improved damping obtained by the addition of tacho feedback*

The dynamic performance of the simple scheme described above is very unsatisfactory as it stands. In order to achieve a fast response and to minimise position errors caused by static friction, the gain of the amplifier needs to be high, but this in turn leads to a highly oscillatory response which is usually unacceptable. For some fixed-load applications, matters can be improved by adding a compensating network at the input to the amplifier, but the best solution is to use 'tacho' (speed) feedback (shown dotted in Figure 4.15) in addition to the main position feedback loop.

Tacho feedback clearly has no effect on the static behaviour (since the voltage from the tacho is proportional to the speed of the motor), but has the effect of increasing the damping of the transient response. The gain of the amplifier can therefore be made high in order to give a fast response, and the degree of tacho feedback can then be adjusted to provide the required damping (see Figure 4.16). Many servo motors have an integral tachogenerator for this purpose.

The example above dealt with an analogue scheme in the interests of simplicity, but digital position control schemes are now taking precedence, especially when brushless motors (see Chapter 9) are used. Complete 'controllers on a card' are available as off-the-shelf items, and these offer ease of interface to other systems as well as providing improved flexibility in shaping the dynamic response.

DIGITALLY CONTROLLED DRIVES

As in all forms of industrial and precision control, digital implementations have replaced analogue circuitry in many electric drive systems, but there are few instances where this has resulted in any real change to the structure of existing drives, and in most cases understanding how the drive functions is still best approached in the first instance by studying the analogue version. There are of course important systems which are predominantly digital, such as PWM inverter drives (see Chapter 7) and future drives that employ matrix converters may emerge and they are only possible using digital control. But as far as understanding d.c. drives is concerned, users who have developed a sound understanding of how the analogue version operates will find little to trouble them when considering the digital equivalent. Accordingly this section is limited to the consideration of a few of the advantages offered by digital implementations, and readers seeking more are recommended to consult a book such as that by Valentine (see page 400).

Many drives use digital speed feedback, in which a pulse train generated from a shaft-mounted encoder is compared (using a phase-locked loop) with a reference pulse train whose frequency corresponds to the desired speed. The reference frequency can easily be made accurate and drift-free; and noise in the encoder signal is easily rejected, so that very precise speed holding can be guaranteed. This is especially important when a number of independent motors must all be driven at identical speed. Phase-locked loops are also used in the firing-pulse synchronising circuits to overcome the problems caused by noise on the mains waveform.

Digital controllers offer freedom from drift, added flexibility (e.g. programmable ramp-up, ramp-down, maximum and minimum speeds etc.), ease of interfacing and linking to other drives and host computers and controllers, and self-tuning. User-friendly diagnostics represents another benefit, providing the local or remote user with current and historical data on the state of all the key drive variables. Many of these advantages are also offered with drives that continue to employ analogue control in the power electronic stages.

REVIEW QUESTIONS

1) A speed-controlled d.c. motor drive is running light at 50% of full speed. If the speed reference was raised to 100%, and the motor was allowed to settle, how would you expect the new steady-state values of armature voltage, tacho voltage and armature current to compare with the corresponding values when the motor was running at 50% speed?

2) A d.c. motor drive has a PI speed controller. The drive is initially running at 50% speed with the motor unloaded. A load torque of 100% is then applied to the shaft. How would you expect the new

steady-state values of armature voltage, tacho voltage and armature current to compare with the corresponding values before the load was applied?

3) An unloaded d.c. motor drive is started from rest by applying a sudden 100% speed demand. How would you expect the armature voltage and current to vary as the motor runs up to speed?

4) What would you expect to happen to a d.c. drive running with 50% torque at 50% speed if:
   a) the mains voltage fell by 10%;
   b) the tacho wires were inadvertently pulled off;
   c) the motor seized solid;
   d) a short-circuit was placed across the armature terminals;
   e) the current feedback signal was removed.

5) Why is discontinuous operation generally undesirable in a d.c. motor?

6) What is the difference between dynamic braking and regenerative braking?

7) Explain why, in the drives context, it is often said that the higher the armature circuit inductance of d.c. machine, the better. In what sense is high armature inductance not desirable?

8) The torque–speed characteristics shown in Figure Q.8 relate to a d.c. motor supplied from a fully controlled thyristor converter.

   Identify the axes. Indicate which parts of the characteristics display 'good' performance and which parts indicate 'bad' performance, and explain briefly what accounts for the abrupt change in behaviour. If curve A corresponds to a firing angle of 5°, estimate the firing angle for curve B. How might the shape of the curves change if

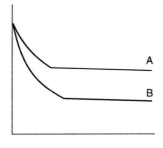

Figure Q.8

a substantial additional inductance was added in series with the armature of the motor?

9) The 250 kW drive for a tube-mill drawbench had to be engineered using two motors rated at 150 kW, 1200 rev/min and 100 kW, 1200 rev/min respectively, and coupled to a common shaft. Each motor was provided with its own speed-controlled drive. The specification called for the motors to share the load in proportion to their rating, so the controls were arranged as shown in Figure Q. 9 (the load is not shown).

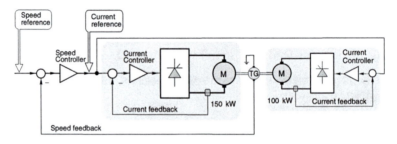

Figure Q.9

a) Explain briefly why this scheme is referred to as a master/slave arrangement.

b) How is load sharing achieved?

c) Discuss why this arrangement is preferable to one in which both drives have active outer speed loops.

d) Would there be any advantage in feeding the current reference for the smaller drive from the current feedback signal of the larger drive?

5

INDUCTION MOTORS – ROTATING FIELD, SLIP AND TORQUE

INTRODUCTION

Judged in terms of fitness for purpose coupled with simplicity, the induction motor must rank alongside the screwthread as one of mankind's best inventions. It is not only supremely elegant as an electro-mechanical energy converter, but is also by far the most important, with something like one-third of all the electricity generated being converted back to mechanical energy in induction motors. Despite playing a key role in industrial society, it remains largely unnoticed because of its workaday role in unglamorous surroundings driving machinery, pumps, fans, compressors, conveyors, hoists and a host of other routine but vital tasks. It will doubtless continue to dominate these fixed-speed applications, but thanks to the availability of reliable variable-frequency inverters, it is now also the leader in the controlled-speed arena.

Like the d.c. motor, the induction motor develops torque by the interaction of axial currents on the rotor and a radial magnetic field produced by the stator. But, whereas, in the d.c. motor the 'work' current has to be fed into the rotor by means of brushes and a commutator, the torque-producing currents in the rotor of the induction motor are induced by electromagnetic action, hence the name 'induction' motor. The stator winding therefore not only produces the magnetic field (the 'excitation'), but also supplies the energy that is converted to mechanical output. The absence of any sliding mechanical contacts and the consequent saving in terms of maintenance is a major advantage of the induction motor over its d.c. rival.

Other differences between the induction motor and the d.c. motor are: firstly that the supply to the induction motor is a.c. (usually 3-phase, but in smaller sizes single-phase); secondly that the magnetic field in the induction motor rotates relative to the stator, whereas in the d.c. motor it is stationary and thirdly that both stator and rotor in the induction motor are non-salient (i.e. effectively smooth) whereas the d.c. motor stator has projecting poles or saliencies which define the position of the field windings.

Given these differences we might expect to find major contrasts between the performance of the two types of motor, and it is true that their inherent characteristics exhibit distinctive features. But there are also many aspects of behaviour which are similar, as we shall see. Perhaps most important from the user's point of view is that there is no dramatic difference in size or weight between an induction motor and a d.c. motor giving the same power at the same base speed, though the induction motor will almost always be much cheaper. The similarity in size is a reflection of the fact that both types employ similar amounts of copper and iron, while the difference in price stems from the simpler construction of the induction motor.

Outline of approach

To understand how an induction motor operates, we must first unravel the mysteries of the rotating magnetic field. We shall see later that the rotor is effectively dragged along by the rotating field, but that it can never run quite as fast as the field. When we want to control the speed of the rotor, the best way is to control the speed of the field.

Our look at the mechanism of the rotating field will focus on the stator windings because they act as the source of the flux. In this part of the discussion we will ignore the presence of the rotor conductors. This makes it much easier to understand what governs the speed of rotation and the magnitude of the field, which are the two factors that mostly influence the motor behaviour.

Having established how the rotating field is set up, and what its speed and strength depend on, we move onto examine the rotor, concentrating on how it behaves when exposed to the rotating field, and discovering how the induced rotor currents and torque vary with rotor speed. In this section, we assume – again for the sake of simplicity – that the rotating flux set up by the stator is not influenced by the rotor.

Finally, we turn attention to the interaction between the rotor and stator, verifying that our earlier assumptions are well justified. Having done this we are in a position to examine the 'external characteristics' of the motor, i.e. the variation of motor torque and stator current with speed. These are the most important characteristics from the point of view of the user.

In discussing how the motor operates the approach leans heavily on first building up a picture of the main or air-gap flux. All the main characteristics which are of interest to the user can be explained and understood once a clear idea has been formed of what the flux wave is, what determines its amplitude and speed and how it interacts with the rotor to produce torque.

The use of mathematics has been kept to a minimum, and all but the simplest equivalent circuit have been avoided in favour of a physical explanation. This is because the aim throughout this book is to attempt to promote understanding not only of what happens, but also why.

The alternative approach, which is favoured in most textbooks on electrical machinery, is to move quickly to a position where the machine is represented by an (fairly complicated) equivalent circuit model, which can then be used for performance prediction. The danger of this for newcomers is that they can easily be daunted by the apparent complexity of the circuit, and as a result lose sight of the key messages that ought to emerge. And although equivalent circuits can provide qualitative answers to some of the questions we will be addressing, there are other matters (such as the fact that the rotor frequency is different from the stator frequency) that are disguised in the circuit approach.

Experience has shown that to get the most benefit from an equivalent circuit, a good grasp of why the machine behaves as it does is an essential prerequisite. Armed with this knowledge the power of the equivalent circuit can be properly appreciated, so readers are urged to come to grips with the material in this chapter before exploring to Chapter 7, which can be regarded as an 'extra' for those seeking a different viewpoint.

The fundamental aspects we have explored so far (magnetic flux, MMF, reluctance, electromagnetic force, motional e.m.f.) will be needed again here, just as they were in the study of the d.c. machine. But despite their basic similarities, most reader will probably find that the induction motor is more difficult to understand than the d.c. machine. This is because we are now dealing with alternating rather than steady quantities (so, for example, inductive reactance becomes very significant), and also because (as mentioned earlier) a single winding acts simultaneously as the producer of the working flux and the supplier of the converted energy. Readers who are unfamiliar with routine a.c. circuit theory, including reactance, impedance, phasor diagrams (but not, at this stage, 'j' notation) and basic ideas about 3-phase systems will have to do some preparatory work before venturing further in this chapter.

THE ROTATING MAGNETIC FIELD

Before we look at how the rotating field is produced, we should be clear what it actually is. Because both the rotor and stator iron surfaces are smooth (apart from the regular slotting), and are separated by a small air gap, the flux produced by the stator windings crosses the air gap radially. The behaviour of the motor is dictated by this radial flux, so we will concentrate first on establishing a mental picture of what is meant by the 'flux wave' in an induction motor.

The pattern of flux in an ideal 4-pole induction motor supplied from a balanced 3-phase source is shown in Figure 5.1(a). The top sketch corresponds to time $t = 0$; the middle one shows the flux pattern one quarter of a cycle of the mains supply later (i.e. 5 ms if the frequency is 50 Hz) and the lower one corresponds to a further quarter cycle later. We note that the pattern of flux lines is repeated in each case, except that the middle and lower ones are rotated by 45° and 90°, respectively, with respect to the top sketch.

The term '4-pole' reflects the fact that the flux leaves the stator from two N poles, and returns at two S poles. Note, however, that there are no physical features of the stator iron mark it out as being 4-pole, rather than say 2-pole or 6-pole. As we will see, it is the layout and intercon-nection of the stator coils that sets the pole number.

If we plot the variation of the radial air-gap flux density with respect to distance round the stator, at each of the three instants of time, we get the patterns shown in Figure 5.1(b). The first feature to note is that the radial flux density varies sinusoidally in space. There are two N peaks and two S peaks, but the transition from N to S occurs in a smooth sinusoidal way, giving rise to the term 'flux wave'. The distance from the centre of one N pole to the centre of the adjacent S pole is called the pole-pitch, for obvious reasons.

Staying with Figure 5.1(b), we note that after one quarter of a cycle of the mains frequency, the flux wave retains its original shape, but has moved round the stator by half a pole-pitch, while after half a cycle it has moved round by a full pole-pitch. If we had plotted the patterns at intermediate times, we would have discovered that the wave maintained a constant shape, and progressed smoothly, advancing at a uniform rate of two pole-pitches per cycle of the mains. The term 'travelling flux wave' is thus an appropriate one to describe the air-gap field.

For the 4-pole wave here, one complete revolution takes two cycles of the supply, so the speed is 25 rev/s (1500 rev/min) with a 50 Hz supply, or 30 rev/s (1800 rev/min) at 60 Hz. The general expression for the speed of the field (which is known as the synchronous speed) N_s, in rev/min is

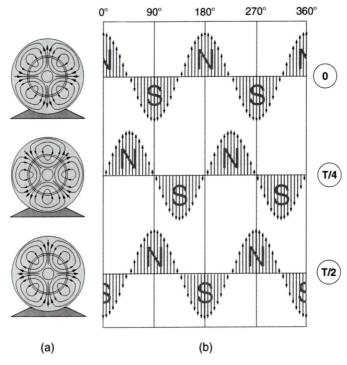

(a)   (b)

Figure 5.1 (a) *Flux pattern in a 4-pole induction motor at three successive instants of time, each one-quarter of a cycle apart;* (b) *radial flux density distribution in the air-gap at the three instants shown in Figure 5.1(a)*

$$N_s = \frac{120f}{p} \tag{5.1}$$

where p is the pole number. The pole number must be an even integer, since for every N pole there must be a S pole. Synchronous speeds for commonly used pole numbers are given in the table below.

	Synchronous speeds, in revs/min	
Pole number	50 Hz	60 Hz
2	3000	3600
4	1500	1800
6	1000	1200
8	750	900
10	600	720
12	500	600

We can see from the table that if we want the field to rotate at inter-mediate speeds, we will have to be able to vary the supply frequency, and this is what happens in inverter-fed motors, which are dealt in Chapter 8.

Production of rotating magnetic field

Now that we have a picture of the field, we turn to how it is produced. If we inspect the stator winding of an induction motor we find that it consists of a uniform array of identical coils, located in slots. The coils are in fact connected to form three identical groups or phase windings, distributed around the stator, and symmetrically displaced with respect to one another. The three-phase windings are connected either in star (wye) or delta (mesh), as shown in Figure 5.2.

The three-phase windings are connected directly to a three-phase a.c. supply, and so the currents (which produce the MMF that sets up the flux) are of equal amplitude but differ in time phase by one-third of a cycle (120°), forming a balanced three-phase set.

Field produced by each phase winding

The aim of the winding designer is to arrange the layout of the coils so that each phase winding, acting alone, produces an MMF wave (and hence an air-gap flux wave) of the desired pole number, and with a sinusoidal variation of amplitude with angle. Getting the desired pole number is not difficult: we simply have to choose the right number and pitch of coils, as shown by the diagrams of an elementary 4-pole winding in Figure 5.3.

In Figure 5.3(a) we see that by positioning two coils (each of which spans one pole-pitch) 180° apart we obtain the correct number of poles (i.e. 4). However, the air-gap field – shown by only two flux lines per pole for the sake of clarity – is uniform between each go and return coil side, not sinusoidal.

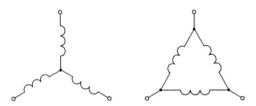

Figure 5.2   *Star (wye) and Delta connection of the three phase windings of a 3-phase induction motor*

Plate 5.1   *Stator of three-phase induction motor. The semi-closed slots of the stator core obscure the active sides of the stator coils, but the ends of the coils are just visible beneath the binding tape. (Photograph by courtesy of Brook Crompton)*

Plate 5.2   *Cage rotor for induction motor. The rotor conductor bars and end rings are cast in aluminium, and the blades attached to the end rings serve as a fan for circulating internal air. An external fan will be mounted on the non-drive end. (Photograph by courtesy of Brook Crompton)*

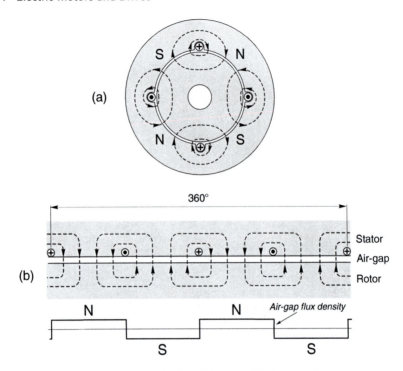

Figure 5.3   *Arrangement* (a) *and developed diagram* (b) *showing elementary 4-pole, single-layer stator winding consisting of 4 conductors spaced by 90°. The 'go' side of each coil (shown by the plus symbol) carries current into the paper at the instant shown, while the 'return' side (shown by the dot) carries current out of the paper*

   A clearer picture of the air-gap flux wave is presented in the developed view in Figure 5.3(b), where more equally spaced flux lines have been added to emphasise the uniformity of the flux density between the go and return sides of the coils. Finally, the plot of the air-gap flux density underlines the fact that this very basic arrangement of coils produces a rectangular flux density wave, whereas what we are seeking is a sinusoidal wave.

   We can improve matters by adding more coils in the adjacent slots, as shown in Figure 5.4. All the coils have the same number of turns, and carry the same current. The addition of the extra slightly displaced coils gives rise to the stepped waveform of MMF and air-gap flux density shown in Figure 5.4. It is still not sinusoidal, but is much better than the original rectangular shape.

   It turns out that if we were to insist on having a perfect sinusoidal flux density waveform, we would have to distribute the coils of one phase in a smoothly varying sinusoidal pattern over the whole periphery of the stator. This is not a practicable proposition, firstly because we would

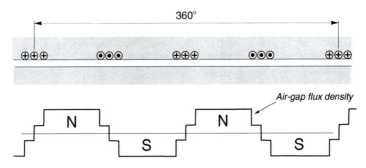

Figure 5.4   *Developed diagram showing flux density produced by one phase of a single-layer winding having three slots per pole per phase*

also have to vary the number of turns per coil from point to point, and secondly because we want the coils to be in slots, so it is impossible to avoid some measure of discretisation in the layout. For economy of manufacture we are also obliged to settle for all the coils being identical, and we must make sure that the three identical phase windings fit together in such a way that all the slots are fully utilised.

Despite these constraints we can get remarkably close to the ideal sinusoidal pattern, especially when we use a 'two-layer' winding. A typical arrangement of one phase is shown in Figure 5.5. The upper expanded sketch shows how each coil sits with its go side in the top of a slot while the return side occupies the bottom of a slot rather less than one pole-pitch away. Coils which span less than a full pole-pitch are known as short-pitch or short-chorded: in this particular case the coil pitch is six slots, the pole-pitch is nine slots, so the coils are short-pitched by three slots.

This type of winding is almost universal in all but small induction motors, the coils in each phase being grouped together to form 'phase-bands' or 'phase-belts'. Since we are concentrating on the field produced by only one of the phase windings (or 'phases'), only one-third of the coils in Figure 5.5 are shown carrying current. The remaining two-thirds of the coils form the other two phase windings, as discussed in Section 5.2.3.

Returning to the flux density plot in Figure 5.5 we see that the effect of short-pitching is to increase the number of steps in the waveform, and that as a result the field produced by one phase is a fair approximation to a sinusoid.

The current in each phase pulsates at the supply frequency, so the field produced by, say, phase A, pulsates in sympathy with the current in phase A, the axis of each 'pole' remaining fixed in space, but its polarity

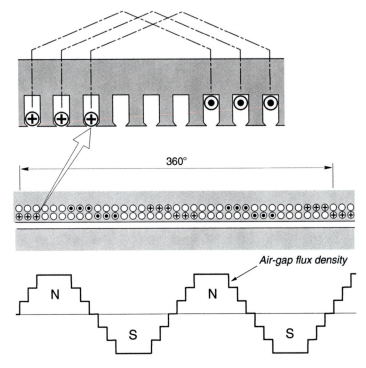

Figure 5.5 *Developed diagram showing layout of windings in a 3-phase, 4-pole, two-layer induction motor winding, together with the flux density wave produced by one phase acting alone. The upper detail shows how the coil sides form upper and lower layers in the slots*

changing from N to S and back once per cycle. There is no hint of any rotation in the field of one phase, but when the fields produced by each of the three-phases are combined, matters change dramatically.

Resultant field

The layout of coils for the complete 4-pole winding is shown in Figure 5.6(a). The go sides of each coil are represented by the capital letters (A, B, C) and the return sides are identified by bars over the letters (Ā, B̄, C̄). (For the sake of comparison, a 6-pole winding layout that uses the same stator slotting is shown in Figure 5.6(b): here the pole-pitch is six slots and the coils are short-pitched by one slot.)

Returning to the 4-pole winding, we can see that the windings of phases B and C are identical with that of phase A apart from the fact that they are displaced in space by plus and minus two-thirds of a pole-pitch respectively.

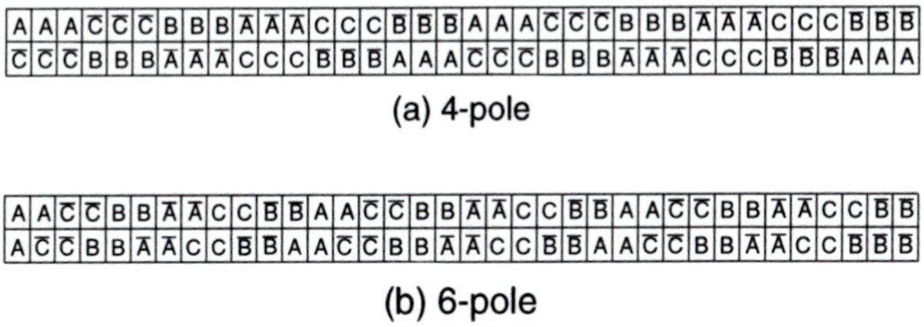

A	A	A	C̄	C̄	C̄	B	B	B	Ā	Ā	Ā	C	C	C	B̄	B̄	B̄	A	A	A	C̄	C̄	C̄	B	B	B	Ā	Ā	Ā	C	C	C	B̄	B̄	B̄
C̄	C̄	C̄	B	B	B	Ā	Ā	Ā	C	C	C	B̄	B̄	B̄	A	A	A	C̄	C̄	C̄	B	B	B	Ā	Ā	Ā	C	C	C	B̄	B̄	B̄	A	A	A

(a) 4-pole

A	A	C̄	C̄	B	B	Ā	Ā	C	C	B	B	A	A	C̄	C̄	B	B	Ā	Ā	C	C	B	B	A	A	C̄	C̄	B	B	Ā	Ā	C	C	B	B
A	C̄	C̄	B	B	Ā	Ā	C	C	B	B	A	A	C̄	C̄	B	B	Ā	Ā	C	C	B	B	A	A	C̄	C̄	B	B	Ā	Ā	C	C	B	B	B

(b) 6-pole

Figure 5.6 *Developed diagram showing arrangement of 3-phase, two-layer windings in a 36-slot stator. A 4-pole winding with three slots per pole per phase is shown in* (a), *and a 6-pole winding with two slots per pole per phase is shown in* (b)

Phases B and C therefore also produce pulsating fields, along their own fixed axes in space. But the currents in phases B and C also differ in time phase from the current in phase A, lagging by one-third and two-thirds of a cycle, respectively. To find the resultant field we must therefore superimpose the fields of the three phases, taking account not only of the spatial differences between windings, but also the time differences between the currents. This is a tedious process, so the intermediate steps have been omitted and instead we move straight to the plot of the resultant field for the complete 4-pole machine, for three discrete times during one complete cycle, as shown in Figure 5.7.

We see that the three pulsating fields combine beautifully and lead to a resultant 4-pole field, which rotates at a uniform rate, advancing by two pole-pitches for every cycle of the mains. The resultant field is not exactly sinusoidal in shape (though it is actually more sinusoidal than the field produced by the individual phase windings), and its shape varies a little from instant to instant; but these are minor worries. The resultant field is amazingly close to the ideal travelling wave and yet the winding layout is simple and easy to manufacture. This is an elegant engineering achievement, however one looks at it.

Direction of rotation

The direction of rotation depends on the order in which the currents reach their maxima, i.e. on the phase-sequence of the supply. Reversal of direction is therefore simply a matter of interchanging any two of the lines connecting the windings to the supply.

Main (air-gap) flux and leakage flux

Broadly speaking the motor designer shapes the stator and rotor teeth to encourage as much as possible of the flux produced by the stator to pass

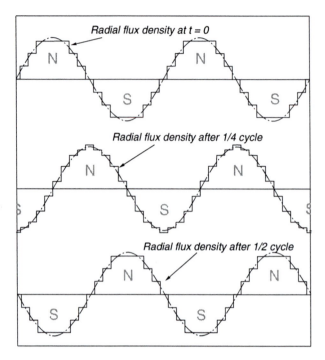

Figure 5.7   *Resultant air-gap flux density wave produced by a complete 3-phase, 4-pole winding at three successive instants in time*

right down the rotor teeth, so that before completing its path back to the stator it is fully linked with the rotor conductors (see later) which are located in the rotor slots. We will see later that this tight magnetic coupling between stator and rotor windings is necessary for good running performance, and the field which provides the coupling is of course the main or air-gap field, which we are in the midst of discussing.

In practice the vast majority of the flux produced by the stator is indeed main or 'mutual' flux. But there is some flux which bypasses the rotor conductors, linking only with the stator winding, and known as stator leakage flux. Similarly not all the flux produced by the rotor currents links the stator, but some (the rotor leakage flux) links only the rotor conductors.

The use of the perjorative-sounding term 'leakage' suggests that these leakage fluxes are unwelcome imperfections, which we should go out of our way to minimise. However, whilst the majority of aspects of performance are certainly enhanced if the leakage is as small as possible, others (notably the large and unwelcome current drawn from the mains when the motor is started from rest) are made much worse if the coupling is too good. So we have the somewhat paradoxical situation in which the designer finds it comparatively easy to layout the windings to produce a good main flux, but is then obliged to juggle the detailed design of the slots to obtain just the right amount of leakage flux to give acceptable all-round performance.

The weight which attaches to the matter of leakage flux is reflected in the prominent part played by the associated leakage reactance in equivalent circuit models of the induction motor, and is discussed in Chapter 7. However, such niceties are of limited importance to the user, so in this and the next chapters we will limit references to leakage reactance to well-defined contexts, and in general, where the term 'flux' is used, it will refer to the main air-gap field.

Magnitude of rotating flux wave

We have already seen that the speed of the flux wave is set by the pole number of the winding and the frequency of the supply. But what is it that determines the amplitude of the field?

To answer this question we can continue to neglect the fact that under normal conditions there will be induced currents in the rotor. We might even find it easier to imagine that the rotor conductors have been removed altogether: this may seem a drastic assumption, but will prove justified later. The stator windings are assumed to be connected to a balanced 3-phase a.c. supply so that a balanced set of currents flows in the windings. We denote the phase voltage by V, and the current in each phase by I_m, where the subscript m denotes 'magnetising' or flux-producing current.

From the discussion in Chapter 1, we know that the magnitude of the flux wave (B_m) is proportional to the winding MMF, and is thus proportional to I_m. But what we really want to know is how the flux density depends on the supply voltage and frequency, since these are the only two parameters over which we have control.

To guide us to the answer, we must first ask what effect the travelling flux wave will have on the stator winding. Every stator conductor will of course be cut by the rotating flux wave, and will therefore have an e.m.f. induced in it. Since the flux wave varies sinusoidally in space, and cuts each conductor at a constant velocity, a sinusoidal e.m.f. is induced in each conductor. The magnitude of the e.m.f. is proportional to the magnitude of the flux wave (B_m), and to the speed of the wave (i.e. to the supply frequency f). The frequency of the induced e.m.f. depends on the time taken for one N pole and one S pole to cut the conductor. We have already seen that the higher the pole number, the slower the field rotates, but we found that the field always advances by two pole-pitches for every cycle of the mains. The frequency of the e.m.f. induced in the stator conductors is therefore the same as the supply frequency, regardless of the pole number. (This conclusion is what we would have reached intuitively, since we would expect any linear system to react at the same frequency at which we excited it.)

The e.m.f. in each complete phase winding (E) is the sum of the e.m.f.'s in the phase coils, and thus will also be at supply frequency. (The alert reader will realise that whilst the e.m.f. in each coil has the same magnitude, it will differ in time phase, depending on the geometrical position of the coil. Most of the coils in each phase band are close together, however, so their e.m.f.'s – though slightly out of phase – will more or less add up directly.)

If we were to compare the e.m.f.'s in the three complete phase windings, we would find that they were of equal amplitude, but out of phase by one-third of a cycle (120°), thereby forming a balanced 3-phase set. This result could have been anticipated from the overall symmetry. This is very helpful, as it means that we need only consider one of the phases in the rest of the discussion.

So we find that when an alternating voltage V is applied, an alternating e.m.f., E, is induced. We can represent this state of affairs by the a.c. equivalent circuit for one phase shown in Figure 5.8.

The resistance shown in Figure 5.8 is the resistance of one complete phase winding. Note that the e.m.f. E is shown as opposing the applied voltage V. This must be so, otherwise we would have a runaway situation in which the voltage V produced the magnetising current I_m which in turn set up an e.m.f. E, which added to V, which further increased I_m and so on *ad infinitum*.

Applying Kirchoff's law to the a.c. circuit in Figure 5.8 yields

$$V = I_m R + E \qquad (5.2)$$

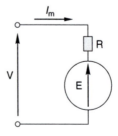

Figure 5.8   *Simple equivalent circuit for the induction motor under no-load conditions*

We find in practice that the term $I_m R$ (which represents the volt drop due to winding resistance) is usually very much less than the applied voltage V. In other words most of the applied voltage is accounted for by the opposing e.m.f., E. Hence, we can make the approximation

$$V \approx E \tag{5.3}$$

But we have already seen that the e.m.f. is proportional to B_m and to f, i.e.

$$E \propto B_m f \tag{5.4}$$

So by combining equations (5.3) and (5.4) we obtain

$$B_m = k \frac{V}{f} \tag{5.5}$$

where the constant k depends on the number of turns per coil, the number of coils per phase and the distribution of the coils.

Equation (5.5) is of fundamental importance in induction motor operation. It shows that if the supply frequency is constant, the flux in the air-gap is directly proportional to the applied voltage, or in other words the voltage sets the flux. We can also see that if we raise or lower the frequency (to increase or reduce the speed of rotation of the field), we will have to raise or lower the voltage in proportion if, as is usually the case, we want the magnitude of the flux to remain constant.

It may seem a paradox that having originally homed-in on the magnetising current I_m as being the source of the MMF which in turn produces the flux, we find that the actual value of the flux is governed only by the applied voltage and frequency, and I_m does not appear at all in equation (5.5). We can see why this is by looking again at Figure 5.8 and asking what would happen if, for some reason, the e.m.f. E were to reduce. We would find that I_m would increase, which in turn would lead to a higher MMF, more flux, and hence to an increase in E. There is clearly a negative feedback effect taking place, which continually tries to keep E equal to V. It is rather like the d.c. motor (see Chapter 3) where the speed of the unloaded motor always adjusted itself so that the back e.m.f. equalled the applied voltage. Here, the magnetising current always adjusts itself so that the induced e.m.f. is almost equal to the applied voltage.

Needless to say this does not mean that the magnetising current is arbitrary, but to calculate it we would have to know the number of turns in the winding, the length of the air-gap (from which we could calculate the gap reluctance) and the reluctance of the iron paths. From a user point of view there is no need to delve further in this direction. We should however recognise that the reluctance will be dominated by the air-gap, and that the magnitude of the magnetising current will therefore depend mainly on the size of the gap. The larger the gap, the bigger the magnetising current. Since the magnetising current contributes to stator copper loss, but not to useful output power, we would like it to be as small as possible, so we find that induction motors usually have the smallest air-gap, which is consistent with providing the necessary mechanical clearances. Despite the small air-gap the magnetising current can be appreciable: in a 4-pole motor, it may be typically 50% of the full-load current, and even higher in 6-pole and 8-pole designs.

Excitation power and VA

The setting up of the travelling wave by the magnetising current amounts to the provision of 'excitation' for the motor. Some energy is stored in the magnetic field, but since the amplitude remains constant once the field has been established, no power input is needed to sustain the field. We therefore find that under the conditions discussed so far, i.e. in the absence of any rotor currents, the power input to the motor is very small. (We should perhaps note that the rotor currents in a real motor are very small when it is running light, so the hypothetical situation we are looking at is not so far removed from reality as we may have supposed.)

Ideally the only source of power losses would be the copper losses in the stator windings, but to this must be added the 'iron losses' which arise from eddy currents and hysteresis in the laminated steel cores of rotor and stator. However, we have seen that the magnetising current can be quite large, its value being largely determined by the air-gap, so we can expect an unloaded induction motor to draw appreciable current from the supply, but very little real power. The VA will therefore be substantial, but the power factor will be very low, the magnetising current lagging the supply voltage by almost 90°, as shown in the phasor diagram (see Figure 5.9).

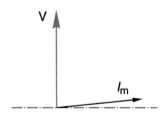

Figure 5.9   *Phasor diagram for the induction motor under no-load conditions, showing magnetising current I_m*

Viewed from the supply the stator looks more or less like a pure inductance, a fact which we would expect intuitively given that – having ignored the rotor circuit – we are left with only an arrangement of flux-producing coils surrounded by a good magnetic circuit. This matter is explored further in Chapter 7.

Summary

When the stator is connected to a 3-phase supply, a sinusoidally distributed, radially directed rotating magnetic flux density wave is set up in the air-gap. The speed of rotation of the field is directly proportional to the frequency of the supply, and inversely proportional to the pole number of the winding. The magnitude of the flux wave is proportional to the applied voltage, and inversely proportional to the frequency.

When the rotor circuits are ignored (i.e. under no-load conditions), the real power drawn from the mains is small, but the magnetising current itself can be quite large, giving rise to a significant reactive power demand from the mains.

TORQUE PRODUCTION

In this section we begin with a brief description of rotor types, and introduce the notion of 'slip', before moving onto explore how the torque is produced, and investigate the variation of torque with speed. We will find that the behaviour of the rotor varies widely according to the slip, and we therefore look separately at low and high values of slip. Throughout this section we will assume that the rotating magnetic field is unaffected by anything which happens on the rotor side of the air-gap. Later, we will see that this assumption is pretty well justified.

Rotor construction

Two types of rotor are used in induction motors. In both the rotor 'iron' consists of a stack of steel laminations with evenly spaced slots punched around the circumference. As with the stator laminations, the surface is coated with an oxide layer, which acts as an insulator, preventing unwanted axial eddy currents from flowing in the iron.

The cage rotor is by far the most common: each rotor slot contains a solid conductor bar and all the conductors are physically and electrically joined together at each end of the rotor by conducting 'end-rings' (see Figure 5.10). The conductors may be of copper, in which case the

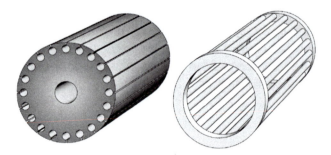

Figure 5.10   *Cage rotor construction. The stack of pre-punched laminations is shown on the left, with the copper or aluminium rotor bars and end-rings on the right*

end-rings are brazed-on. Or, in small and medium sizes, the rotor conductors and end-rings can be die cast in aluminium.

The term squirrel cage was widely used at one time and the origin should be clear from Figure 5.10. The rotor bars and end-rings are reminiscent of the rotating cages used in bygone days to exercise small rodents (or rather to amuse their human captors).

The absence of any means for making direct electrical connection to the rotor underlines the fact that in the induction motor the rotor currents are induced by the air-gap field. It is equally clear that because the rotor cage comprises permanently short-circuited conductor bars, no external control can be exercised over the resistance of the rotor circuit once the rotor has been made. This is a significant drawback that can be avoided in the second type of rotor, which is known as the 'wound-rotor' or 'slipring' type.

In the wound rotor, the slots accommodate a set of three phase-windings very much like those on the stator. The windings are connected in star, with the three ends brought out to three sliprings (see Figure 5.11). The rotor circuit is thus open, and connection can be made via brushes bearing on the sliprings. In particular, the resistance of each phase of the rotor circuit can be increased by adding external resistances as indicated in Figure 5.11. Adding resistance in appropriate circumstances can be beneficial, as we will see.

Cage rotors are usually cheaper to manufacture, and are very robust and reliable. Until the advent of variable-frequency inverter supplies, however, the superior control which was possible from the slipring type meant that the extra expense of the wound rotor and its associated control gear were frequently justified, especially for high-power machines. Nowadays comparatively few are made, and then only in large sizes. But many old motors remain in service, so they are included in Chapter 6.

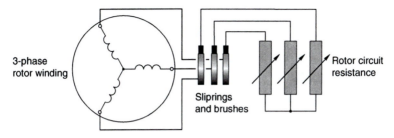

Figure 5.11   *Schematic diagram of wound rotor for induction motor, showing sliprings and brushes to provide connection to the external (stationary) 3-phase resistance*

Slip

A little thought will show that the behaviour of the rotor depends very much on its relative velocity with respect to the rotating field. If the rotor is stationary, for example, the rotating field will cut the rotor conductors at synchronous speed, thereby inducing a high e.m.f. in them. On the other hand, if the rotor was running at the synchronous speed, its relative velocity with respect to the field would be zero, and no e.m.f.'s would be induced in the rotor conductors.

The relative velocity between the rotor and the field is known as the slip. If the speed of the rotor is N, the slip speed is $N_s - N$, where N_s is the synchronous speed of the field, usually expressed in rev/min. The slip (as distinct from slip speed) is the normalised quantity defined by

$$s = \frac{N_s - N}{N_s} \tag{5.6}$$

and is usually expressed either as a ratio as in equation (5.6), or as a percentage. A slip of 0 therefore indicates that the rotor speed is equal to the synchronous speed, while a slip of 1 corresponds to zero speed. (When tests are performed on induction motors with their rotor deliberately held stationary so that the slip is 1, the test is said to be under 'locked-rotor' conditions. The same expression is often used loosely to mean zero speed, even when the rotor is free to move, e.g. when it is started from rest.)

Rotor induced e.m.f., current and torque

The rate at which the rotor conductors are cut by the flux – and hence their induced e.m.f. – is directly proportional to the slip, with no induced e.m.f. at synchronous speed ($s = 0$) and maximum induced e.m.f. when the rotor is stationary ($s = 1$).

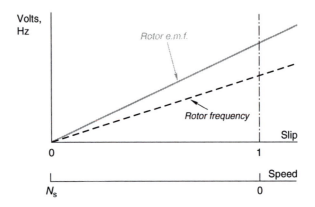

Figure 5.12   *Variation of rotor induced e.m.f and frequency with speed and slip*

The frequency of rotor e.m.f. is also directly proportional to slip, since the rotor effectively slides with respect to the flux wave, and the higher the relative speed, the more times in a second each rotor conductor is cut by a N and a S pole. At synchronous speed (slip = 0) the frequency is zero, while at standstill (slip = 1), the rotor frequency is equal to the supply frequency. These relationships are shown in Figure 5.12.

Although the e.m.f. induced in every rotor bar will have the same magnitude and frequency, they will not be in phase. At any particular instant, bars under the peak of the N poles of the field will have maximum positive voltage in them, those under the peak of the S poles will have maximum negative voltage (i.e. 180° phase shift), and those in between will have varying degrees of phase shift. The pattern of instantaneous voltages in the rotor is thus a replica of the flux density wave, and the rotor induced 'voltage wave' therefore moves relative to the rotor at slip speed, as shown in Figure 5.13.

Since all the rotor bars are short-circuited by the end-rings, the induced voltages will drive currents along the rotor bars, the currents

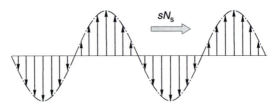

Figure 5.13   *Pattern of induced e.m.f.'s in rotor conductors. The rotor 'voltage wave' moves at a speed of sN_s with respect to the rotor surface*

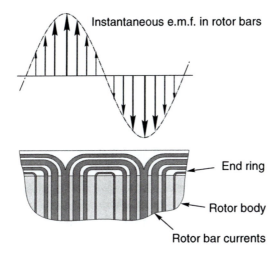

Instantaneous e.m.f. in rotor bars

End ring

Rotor body

Rotor bar currents

Figure 5.14   *Instantaneous sinusoidal pattern of rotor currents in rotor bars and end-rings. Only one pole-pitch is shown, but the pattern is repeated*

forming closed paths through the end-rings, as shown in the developed diagram (see Figure 5.14).

In Figure 5.14 the variation of instantaneous e.m.f. in the rotor bars is shown in the upper sketch, while the corresponding instantaneous currents flowing in the rotor bars and end-rings are shown in the lower sketch. The lines representing the currents in the rotor bars have been drawn so that their width is proportional to the instantaneous currents in the bars.

The axial currents in the rotor bars will interact with the radial flux wave to produce the driving torque of the motor, which will act in the same direction as the rotating field, the rotor being dragged along by the field. We note that slip is essential to this mechanism, so that it is never possible for the rotor to catch up with the field, as there would then be no rotor e.m.f., no current and no torque. Finally, we can see that the cage rotor will automatically adapt to whatever pole number is impressed by the stator winding, so that the same rotor can be used for a range of different stator pole numbers.

Rotor currents and torque – small slip

When the slip is small (say between 0 and 10%), the frequency of induced e.m.f. is also very low (between 0 and 5 Hz if the supply frequency is 50 Hz). At these low frequencies the impedance of the rotor circuits is predominantly resistive, the inductive reactance being small because the rotor frequency is low.

The current in each rotor conductor is therefore in time phase with the e.m.f. in that conductor, and the rotor current wave is therefore in space phase with the rotor e.m.f. wave, which in turn is in space phase with the flux wave. This situation was assumed in the previous discussion, and is represented by the space waveforms shown in Figure 5.15.

To calculate the torque we first need to evaluate the 'BIl_r' product (see equation (1.2)) to obtain the tangential force on each rotor conductor. The torque is then given by the total force multiplied by the rotor radius. We can see from Figure 5.15 that where the flux density has a positive peak, so does the rotor current, so that particular bar will contribute a high tangential force to the total torque. Similarly, where the flux has its maximum negative peak, the induced current is maximum and negative, so the tangential force is again positive. We don't need to work out the torque in detail, but it should be clear that the resultant will be given by an equation of the form

$$T = kBI_r \qquad (5.7)$$

where B and I_r denote the amplitudes of the flux density wave and the rotor current wave, respectively. Provided that there are a large number of rotor bars (which is a safe bet in practice), the waves shown in Figure 5.15 will remain the same at all instants of time, so the torque remains constant as the rotor rotates.

If the supply voltage and frequency are constant, the flux will be constant (see equation (5.5)). The rotor e.m.f. (and hence I_r) is then proportional to slip, so we can see from equation (5.7) that the torque is directly proportional to slip. We must remember that this discussion relates to low values of slip only, but since this is the normal running condition, it is extremely important.

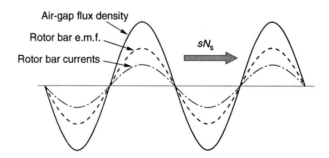

Figure 5.15   *Pattern of air-gap flux density, induced e.m.f. and current in cage rotor bars at low values of slip*

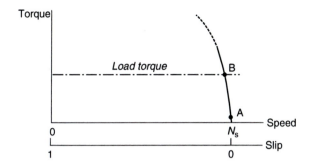

Figure 5.16 *Torque–speed relationship for low values of slip*

The torque–speed (and torque/slip) relationship for small slips is thus approximately a straight-line, as shown by the section of line AB in Figure 5.16.

If the motor is unloaded, it will need very little torque to keep running – only enough to overcome friction in fact – so an unloaded motor will run with a very small slip at just below the synchronous speed, as shown at A in Figure 5.16.

When the load is increased, the rotor slows down, and the slip increases, thereby inducing more rotor e.m.f. and current, and thus more torque. The speed will settle when the slip has increased to the point where the developed torque equals the load torque – e.g. point B in Figure 5.16.

Induction motors are usually designed so that their full-load torque is developed for small values of slip. Small ones typically have a full-load slip of 8%, large ones around 1%. At the full-load slip, the rotor conductors will be carrying their safe maximum continuous current, and if the slip is any higher, the rotor will begin to overheat. This overload region is shown by the dotted line in Figure 5.16.

The torque–slip (or torque–speed) characteristic shown in Figure 5.16 is a good one for most applications, because the speed only falls a little when the load is raised from zero to its full value. We note that, in this normal operating region, the torque–speed curve is very similar to that of a d.c. motor (see Figure 3.9), which explains why both d.c. and induction motors are often in contention for constant-speed applications.

Rotor currents and torque – large slip

As the slip increases, the rotor e.m.f. and rotor frequency both increase in direct proportion to the slip. At the same time the rotor inductive reactance, which was negligible at low slip (low rotor frequency) begins to be appreciable in comparison with the rotor resistance. Hence, although the

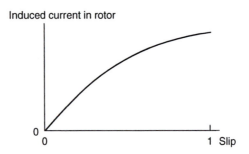

Figure 5.17   *Magnitude of current induced in rotor over the full range of slip*

induced current continues to increase with slip, it does so more slowly than at low values of slip, as shown in Figure 5.17.

At high values of slip, the rotor current also lags behind the rotor e.m.f. because of the inductive reactance. The alternating current in each bar reaches its peak well after the induced voltage, and this in turn means that the rotor current wave has a space-lag with respect to the rotor e.m.f. wave (which is in space phase with the flux wave). This space-lag is shown by the angle ϕ_r in Figure 5.18.

The space-lag means that the peak radial flux density and peak rotor currents no longer coincide, which is bad news from the point of view of torque production, because although we have high values of both flux density and current, they do not occur simultaneously at any point around the periphery. What is worse is that at some points we even have flux density and currents of opposite sign, so over those regions of the rotor surface the torque contributed will actually be negative. The overall torque will still be positive, but is much less than it would be if

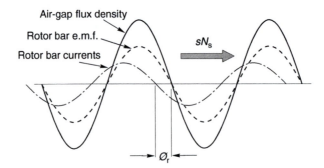

Figure 5.18   *Pattern of air-gap flux density, induced e.m.f. and current in cage rotor bars at high values of slip. (These waveforms should be compared with the corresponding ones when the slip is small, see Figure 5.15.)*

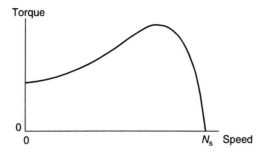

Figure 5.19   *Typical complete torque–speed characteristic for cage induction motor*

the flux and current waves were in phase. We can allow for the unwelcome space-lag by modifying equation (5.7), to obtain a more general expression for torque as

$$T = kBI_r \, \cos \phi_r \qquad (5.8)$$

Equation (5.7) is merely a special case of equation (5.8), which only applies under low-slip conditions where $\cos \phi_r \approx 1$.

For most cage rotors, it turns out that as the slip increases the term $\cos \phi_r$ reduces more quickly than the current (I_r) increases, so that at some slip between 0 and 1 the developed torque reaches a maximum value. This is illustrated in the typical torque–speed characteristic shown in Figure 5.19. The peak torque actually occurs at a slip at which the rotor inductive reactance is equal to the rotor resistance, so the motor designer can position the peak torque at any slip by varying the reactance to resistance ratio.

We will return to the torque–speed curve after we have checked that when we allow for the interaction of the rotor with the stator, our interim conclusions regarding torque production remain valid.

INFLUENCE OF ROTOR CURRENT ON FLUX

Up to now all our discussion has been based on the assumption that the rotating magnetic field remains constant, regardless of what happens on the rotor. We have seen how torque is developed, and that mechanical output power is produced. We have focused attention on the rotor, but the output power must be provided from the stator winding, so we must turn attention to the behaviour of the whole motor, rather than just the rotor. Several questions spring to mind.

Firstly, what happens to the rotating magnetic field when the motor is working? Won't the MMF of the rotor currents cause it to change?

Secondly, how does the stator know when to start supplying real power across the air-gap to allow the rotor to do useful mechanical work? And finally, how will the currents drawn by the stator vary as the slip is changed?

These are demanding questions, for which full treatment is beyond our scope. But we can deal with the essence of the matter without too much difficulty. Further illumination can be obtained from study of the equivalent circuit, and this is dealt with in Chapter 7.

Reduction of flux by rotor current

We should begin by recalling that we have already noted that when the rotor currents are negligible ($s = 0$), the e.m.f. that the rotating field induces in the stator winding is very nearly equal to the applied voltage. Under these conditions a reactive current (which we termed the magnetising current) flows into the windings, to set up the rotating flux. Any slight tendency for the flux to fall is immediately detected by a corresponding slight reduction in e.m.f., which is reflected in a disproportionately large increase in magnetising current, which thus opposes the tendency for the flux to fall.

Exactly the same feedback mechanism comes into play when the slip increases from zero, and rotor currents are induced. The rotor currents are at slip frequency, and they give rise to a rotor MMF wave, which therefore rotates at slip speed (sN_s) relative to the rotor. But the rotor is rotating at a speed of $(1 - s)N_s$, so that when viewed from the stator, the rotor MMF wave always rotates at synchronous speed, regardless of the speed of the rotor.

The rotor MMF wave would, if unchecked, cause its own 'rotor flux wave', rotating at synchronous speed in the air-gap, in much the same way that the stator magnetising current originally set up the flux wave. The rotor flux wave would oppose the original flux wave, causing the resultant flux wave to reduce.

However, as soon as the resultant flux begins to fall, the stator e.m.f. reduces, thereby admitting more current to the stator winding, and increasing its MMF. A very small drop in the e.m.f. induced in the stator is sufficient to cause a large increase in the current drawn from the mains because the e.m.f. E (see Figure 5.8) and the supply voltage V are both very large in comparison with the stator resistance volt drop, IR. The 'extra' stator MMF produced by the large increase in stator current effectively 'cancels' the MMF produced by the rotor currents, leaving the resultant MMF (and hence the rotating flux wave) virtually unchanged.

There *must* be a *small* drop in the resultant MMF (and flux) of course, to alert the stator to the presence of rotor currents. But because of the delicate balance between the applied voltage and the induced e.m.f. in the stator the change in flux with load is very small, at least over the normal operating speed range, where the slip is small. In large motors, the drop in flux over the normal operating region is typically less than 1%, rising to perhaps 10% in a small motor.

The discussion above should have answered the question as to how the stator knows when to supply mechanical power across the air-gap. When a mechanical load is applied to the shaft, the rotor slows down, the slip increases, rotor currents are induced and their MMF results in a modest (but vitally important) reduction in the air-gap flux wave. This in turn causes a reduction in the e.m.f. induced in the stator windings and therefore an increase in the stator current drawn form the supply. We can anticipate that this is a stable process (at least over the normal operating range) and that the speed will settle when the slip has increased sufficiently that the motor torque equals the load torque.

As far as our conclusions regarding torque are concerned, we see that our original assumption that the flux was constant is near enough correct when the slip is small. We will find it helpful and convenient to continue to treat the flux as constant (for given stator voltage and frequency) when we turn later to methods of controlling the normal running speed.

It has to be admitted, however, that at high values of slip (i.e. low rotor speeds), we cannot expect the main flux to remain constant, and in fact we would find in practice that when the motor was first switched-on, with the rotor stationary, the main flux might typically be only half what it was when the motor was at full speed. This is because at high slips, the leakage fluxes assume a much greater importance than under normal low-slip conditions. The simple arguments we have advanced to predict torque would therefore need to be modified to take account of the reduction of main flux if we wanted to use them quantitatively at high slips. There is no need for us to do this explicitly, but it will be reflected in any subsequent curves portraying typical torque–speed curves for real motors. Such curves are of course used when selecting a motor, since they provide the easiest means of checking whether the starting and run-up torque is adequate for the job in hand.

STATOR CURRENT-SPEED CHARACTERISTICS

In the previous section, we argued that as the slip increased, and the rotor did more mechanical work, the stator current increased. Since the extra current is associated with the supply of real (i.e. mechanical

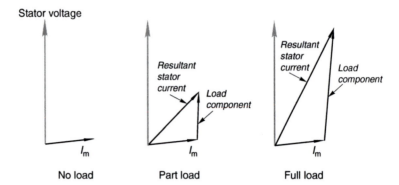

Figure 5.20   *Phasor diagrams showing stator current at no-load, part-load and full-load. The resultant current in each case is the sum of the no-load (magnetising) current and the load component*

output) power (as distinct from the original magnetising current which was seen to be reactive), this additional 'work' component of current is more or less in phase with the supply voltage, as shown in the phasor diagrams (Figure 5.20).

The resultant stator current is the sum of the magnetising current, which is present all the time, and the load component, which increases with the slip. We can see that as the load increases, the resultant stator current also increases, and moves more nearly into phase with the voltage. But because the magnetising current is appreciable, the difference in magnitude between no-load and full-load currents may not be all that great. (This is in sharp contrast to the d.c. motor, where the no-load current in the armature is very small in comparison with the full-load current. Note, however, that in the d.c. motor, the excitation (flux) is provided by a separate field circuit, whereas in the induction motor the stator winding furnishes both the excitation and the work currents. If we consider the behaviour of the work components of current only, both types of machine look very similar.)

The simple ideas behind Figure 5.20 are based on an approximation, so we cannot push them too far: they are fairly close to the truth for the normal operating region, but breakdown at higher slips, where the rotor and stator leakage reactances become significant. A typical current locus over the whole range of slips for a cage motor is shown in Figure 5.21. We note that the power factor becomes worse again at high slips, and also that the current at standstill (i.e. the 'starting' current) is perhaps five times the full-load value.

Very high starting currents are one of the worst features of the cage induction motor. They not only cause unwelcome volt drops in the

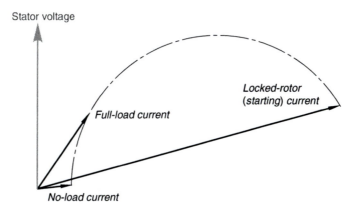

Figure 5.21   *Phasor diagram showing the locus of stator current over the full range of speeds from no-load (full speed) down to the locked-rotor (starting) condition*

supply system, but also call for heavier switchgear than would be needed to cope with full-load conditions. Unfortunately, for reasons discussed earlier, the high starting currents are not accompanied by high starting torques, as we can see from Figure 5.22, which shows current and torque as functions of slip for a general-purpose cage motor.

We note that the torque per ampere of current drawn from the mains is typically very low at start up, and only reaches a respectable value in the normal operating region, i.e. when the slip is small. This matter is explored further in Chapter 6, and also by means of the equivalent circuit in Chapter 7.

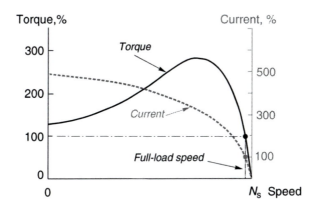

Figure 5.22   *Typical torque–speed and current–speed curves for a cage induction motor. The torque and current axes are scaled so that 100% represents the continuously rated (full-load) value*

REVIEW QUESTIONS

The first ten questions are designed to reinforce understanding by means of straightforward calculations, while the remainder are more demanding and involve some extension of the basic ideas.

1) At what frequency must a 4-pole motor be supplied so that its synchronous speed is 1200 rev/min?

2) The nameplate of a standard 50 Hz induction motor quotes full-load speed as 2950 rev/min. Find the pole number and the rated slip.

3) A 4-pole, 60 Hz induction motor runs with a slip of 4%. Find: (a) the speed; (b) the rotor frequency; (c) the speed of rotation of the rotor current wave relative to the rotor surface; (d) the speed of rotation of the rotor current wave relative to the stator surface.

4) Choose a suitable pole number for an induction motor to cover the speed range from 400 rev/min to 800 rev/min when supplied from a 30–75 Hz variable-frequency source.

5) The r.m.s. current in the rotor bars of an induction motor running with a slip of 1% is 25 A, and the torque produced is 20 N m. Estimate the rotor current and torque when the load is increased so that the motor slip is 3%.

6) An induction motor designed for operation from 440 V is supplied at 380 V instead. What effect will the reduced voltage have on the following: (a) the synchronous speed; (b) the magnitude of the air-gap flux; (c) the induced current in the rotor when running at rated speed; (d) the torque produced at rated speed.

7) The rotor from a 6-pole induction motor is to be used with a 4-pole stator having the same bore and length as the 6-pole stator. What modifications would be required to the rotor if it was (a) a squirrel-cage type, and (b) a wound-rotor type?

8) A 440 V, 60 Hz induction motor is to be used on a 50 Hz supply. What voltage should be used?

9) The stator of a 220 V induction motor is to be rewound for operation from a 440 V supply. The original coils each had 15 turns of 1 mm diameter wire. Estimate the number of turns and diameter of wire for the new stator coils.

10) The slip of an induction motor driving a constant torque load is 2.0%. If the voltage is reduced by 5%, estimate: (a) the new steady-

state rotor current, expressed in terms of its original value; (b) the new steady-state slip.

11) Why can an induction motor never run at its synchronous speed?

12) Why is it important to maintain the ratio of voltage:frequency at the correct value for an induction motor? What would be the consequences of making the ratio (a) higher, and (b) lower, than the specified value?

13) An induction motor with a 2 mm air-gap has a no-load (magnetising) current of 5 A. Assuming that the air-gap represents the only significant reluctance in the path of the main flux, explain why, if the rotor was reground so that the air-gap became 3 mm, the magnetising current would be expected to increase to 7.5 A. What effect would the increase in air-gap be expected to have on the magnitude of the air-gap flux wave?

14) As the slip of an induction motor increases, the current in the rotor increases, but beyond a certain slip the torque begins to fall. Why is this?

15) For a given rotor diameter, the stator diameter of a 2-pole motor is much greater than the stator diameter of, say, an 8-pole motor. By sketching and comparing the magnetic flux patterns for machines with low and high pole numbers, explain why more stator iron is required as the pole number reduces.

16) The layout of coils for 4-pole and 6-pole windings in a 36-slot stator are shown in Figure 5.6. All the coils have the same number of turns of the same wire, the only real difference being that the 6-pole coils are of shorter pitch.

Sketch the MMF wave produced by one phase, for each pole number, assuming that the same current flows in every coil. Hence show that to achieve the same amplitude of flux wave, the current in the 6-pole winding would have to be about 50% larger than in the 4-pole.

How does this exercise relate to the claim that the power factor of a low-speed induction motor is usually lower than a high-speed one?

6

OPERATING CHARACTERISTICS OF INDUCTION MOTORS

This chapter is concerned with how the induction motor behaves when connected to a constant frequency supply. This is by far the most widely used and important mode of operation, the motor running directly connected to a constant voltage mains supply. 3-phase motors are the most important, so they are dealt with first.

METHODS OF STARTING CAGE MOTORS

Direct Starting – Problems

Our everyday domestic experience is likely to lead us to believe that there is nothing more to starting a motor than closing a switch, and indeed for most low-power machines (say up to a few kW) – of whatever type – that is indeed the case. By simply connecting the motor to the supply we set in train a sequence of events which sees the motor draw power from the supply while it accelerates to its target speed. When it has absorbed and converted sufficient energy from electrical to kinetic form, the speed stabilises and the power drawn falls to a low level until the motor is required to do useful mechanical work. In these low-power applications acceleration to full speed may take less than a second, and we are seldom aware of the fact that the current drawn during the acceleration phase is often higher than the continuous rated current.

For motors over a few kW, however, it is necessary to assess the effect on the supply system before deciding whether or not the motor can be started simply by switching directly onto the supply. If supply systems were ideal (i.e. the supply voltage remained unaffected regardless of how much current was drawn) there would be no problem starting any induction motor, no matter how large. The problem is that the heavy current drawn while the motor is running up to speed may cause a large drop in the supply system voltage, annoying other customers on the same supply and perhaps taking it outside statutory limits.

It is worthwhile reminding ourselves about the influence of supply impedance at this point, as this is at the root of the matter, so we begin by noting that any supply system, no matter how complicated, can be modelled by means of the delightfully simple Thevenin equivalent circuit shown in Figure 6.1. (We here assume a balanced 3-phase operation, so a 1-phase equivalent circuit will suffice.)

The supply is represented by an ideal voltage source (V_s) in series with the supply impedance Z_s. When no load is connected to the supply, and the current is zero, the terminal voltage is V_s; but as soon as a load is connected the load current (I) flowing through the source impedance results in a volt drop, and the output voltage falls from V_s to V, where

$$V = V_s - IZ_s \tag{6.1}$$

For most industrial supplies the source impedance is predominantly inductive, so that Z_s is simply an inductive reactance, X_s. Typical phasor diagrams relating to a supply with a purely inductive reactance are shown in Figure 6.2: in (a) the load is also taken to be purely reactive, while the load current in (b) has the same magnitude as in (a) but the load is resistive. The output (terminal) voltage in each case is represented by the phasor labelled V.

For the inductive load (a) the current lags the terminal voltage by 90° while for the resistive load (b) the current is in phase with the terminal voltage. In both cases the volt drop across the supply reactance (IX_s) leads the current by 90°.

The first point to note is that, for a given magnitude of load current, the volt drop is in phase with V_s when the load is inductive, whereas with

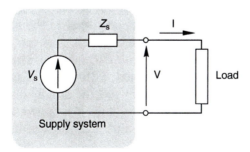

Figure 6.1   *Equivalent circuit of supply system*

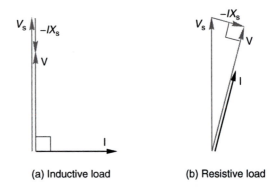

(a) Inductive load   (b) Resistive load

Figure 6.2   *Phasor diagrams showing the effect of supply-system impedance on the output voltage with* (a) *inductive load and* (b) *resistive load*

a resistive load the volt drop is almost at $90°$ to V_s. This results in a much greater fall in the magnitude of the output voltage when the load is inductive than when it is resistive. The second, obvious, point is that the larger the current, the more the drop in voltage.

Unfortunately, when we try to start a large cage induction motor we face a double-whammy because not only is the starting current typically five or six times rated current, but it is also at a low-power factor, i.e. the motor looks predominantly inductive when the slip is high. (In contrast, when the machine is up to speed and fully loaded, its current is perhaps only one fifth of its starting current and it presents a predominantly resistive appearance as seen by the supply. Under these conditions the supply voltage is hardly any different from at no-load.)

Since the drop in voltage is attributable to the supply impedance, if we want to be able to draw a large starting current without upsetting other consumers it would be clearly best for the supply impedance to be as low as possible, and preferably zero. But from the supply authority view-point a very low supply impedance brings the problem of how to scope in the event of an accidental short-circuit across the terminals. The short-circuit current is inversely proportional to the supply impedance, and tends to infinity as Z_s approaches zero. The cost of providing the switch-gear to clear such a large fault current would be prohibitive, so a compromise always has to be reached, with values of supply impedances being set by the supply authority to suit the anticipated demands.

Systems with a low internal impedance are known as 'stiff' supplies, because the voltage is almost constant regardless of the current drawn.

(An alternative way of specifying the nature of the supply is to consider the fault current that would flow if the terminals were short-circuited: a system with a low impedance would have a high fault current or 'fault level'.) Starting on a stiff supply requires no special arrangements and the three motor leads are simply switched directly onto the mains. This is known as 'direct-on-line' (DOL) or 'direct-to-line' (DTL) starting. The switching will usually be done by means of a relay or contactor, incorporating fuses and other overload protection devices, and operated manually by local or remote pushbuttons, or interfaced to permit operation from a programmable controller or computer.

In contrast, if the supply impedance is high (i.e. a low-fault level) an appreciable volt drop will occur every time the motor is started, causing lights to dim and interfering with other apparatus on the same supply. With this 'weak' supply, some form of starter is called for to limit the current at starting and during the run-up phase, thereby reducing the magnitude of the volt drop imposed on the supply system. As the motor picks up speed, the current falls, so the starter is removed as the motor approaches full speed. Naturally enough the price to be paid for the reduction in current is a lower starting torque, and a longer run-up time.

Whether or not a starter is required depends on the size of the motor in relation to the capacity or fault level of the supply, the prevailing regulations imposed by the supply authority, and the nature of the load.

The references above to 'low' and 'high' supply impedances must therefore be interpreted in relation to the impedance of the motor when it is stationary. A large (and therefore low impedance) motor could well be started quite happily DOL in a major industrial plant, where the supply is 'stiff', i.e. the supply impedance is very much less than the motor impedance. But the same motor would need a starter when used in a rural setting remote from the main power system, and fed by a relatively high impedance or 'weak' supply. Needless to say, the stricter the rules governing permissible volt drop, the more likely it is that a starter will be needed.

Motors which start without significant load torque or inertia can accelerate very quickly, so the high starting current is only drawn for a short period. A 10 kW motor would be up to speed in a second or so, and the volt drop may therefore be judged as acceptable. Clutches are sometimes fitted to permit 'off-load' starting, the load being applied after the motor has reached full speed. Conversely, if the load torque and/or inertia are high, the run-up may take many seconds, in which case a starter may prove essential. No strict rules can be laid down, but obviously the bigger the motor, the more likely it is to require a starter.

Star/delta (wye/mesh) starter

This is the simplest and most widely used method of starting. It provides for the windings of the motor to be connected in star (wye) to begin with, thereby reducing the voltage applied to each phase to 58% $(1/\sqrt{3})$ of its DOL value. Then, when the motor speed approaches its running value, the windings are switched to delta (mesh) connection. The main advantage of the method is its simplicity, while its main drawbacks are that the starting torque is reduced (see below), and the sudden transition from star to delta gives rise to a second shock – albeit of lesser severity – to the supply system and to the load. For star/delta switching to be possible both ends of each phase of the motor windings must be brought out to the terminal box. This requirement is met in the majority of motors, except small ones which are usually permanently connected in delta.

With a star/delta starter the current drawn from the supply is approximately one third of that drawn in a DOL start, which is very welcome, but at the same time the starting torque is also reduced to one third of its DOL value. Naturally we need to ensure that the reduced torque will be sufficient to accelerate the load, and bring it up to a speed at which it can be switched to delta without an excessive jump in the current.

Various methods are used to detect when to switch from star to delta. In manual starters, the changeover is determined by the operator watching the ammeter until the current has dropped to a low level, or listening to the sound of the motor until the speed becomes steady. Automatic versions are similar in that they detect either falling current or speed rising to a threshold level, or else they operate after a preset time.

Autotransformer starter

A 3-phase autotransformer is usually used where star/delta starting provides insufficient starting torque. Each phase of an autotransformer consists of a single winding on a laminated core. The mains supply is connected across the ends of the coils, and one or more tapping points (or a sliding contact) provide a reduced voltage output, as shown in Figure 6.3.

The motor is first connected to the reduced voltage output, and when the current has fallen to the running value, the motor leads are switched over to the full voltage.

If the reduced voltage is chosen so that a fraction α of the line voltage is used to start the motor, the starting torque is reduced to approximately α^2 times its DOL value, and the current drawn from the mains is

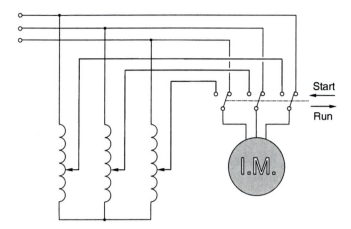

Figure 6.3   *Autotransformer starter for cage induction motor*

also reduced to α^2 times its direct value. As with the star/delta starter, the torque per ampere of supply current is the same as for a direct start.

The switchover from the starting tap to the full voltage inevitably results in mechanical and electrical shocks to the motor. In large motors the transient overvoltages caused by switching can be enough to damage the insulation, and where this is likely to pose a problem a modified procedure known as the Korndorfer method is used. A smoother changeover is achieved by leaving part of the winding of the autotransformer in series with the motor winding all the time.

Resistance or reactance starter

By inserting three resistors or inductors of appropriate value in series with the motor, the starting current can be reduced by any desired extent, but only at the expense of a disproportionate reduction in starting torque.

For example, if the current is reduced to half its DOL value, the motor voltage will be halved, so the torque (which is proportional to the square of the voltage – see later) will be reduced to only 25% of its DOL value. This approach is thus less attractive in terms of torque per ampere of supply current than the star/delta method. One attractive feature, however, is that as the motor speed increases and its effective impedance rises, the volt drop across the extra impedance reduces, so the motor voltage rises progressively with the speed, thereby giving more torque. When the motor is up to speed, the added impedance is shorted-out by means of a contactor. Variable-resistance starters (manually or motor operated) are sometimes used with small motors where a smooth jerk-free start is required, for example in film or textile lines.

Solid-state soft starting

This method is now the most widely used. It provides a smooth build-up of current and torque, the maximum current and acceleration time are easily adjusted, and it is particularly valuable where the load must not be subjected to sudden jerks. The only real drawback over conventional starters is that the mains currents during run-up are not sinusoidal, which can lead to interference with other equipment on the same supply.

The most widely used arrangement comprises three pairs of back-to-back thyristors connected in series with three supply lines, as shown in Figure 6.4(a).

Each thyristor is fired once per half-cycle, the firing being synchronised with the mains and the firing angle being variable so that each pair conducts for a varying proportion of a cycle. Typical current waveforms are shown in Figure 6.4(b): they are clearly not sinusoidal but the motor will tolerate them quite happily.

A wide variety of control philosophies can be found, with the degree of complexity and sophistication being reflected in the price. The cheapest open-loop systems simply alter the firing angle linearly with time, so that the voltage applied to the motor increases as it accelerates. The 'ramp-time' can be set by trial and error to give an acceptable start, i.e. one in which the maximum allowable current from the supply is not exceeded at any stage. This approach is reasonably satisfactory when the load remains the same, but requires resetting each time the load changes. Loads with high static friction are a problem because nothing happens for the first part of the ramp, during which time the motor torque is insufficient to move the load. When the load finally moves, its acceleration is often too rapid. The more advanced open-loop versions allow the level of current at the start of the ramp to be chosen, and this is helpful with 'sticky' loads.

More sophisticated systems – usually with on-board digital controllers – provide for tighter control over the acceleration profile by incorporating closed-loop current feedback. After an initial ramping up to the start level (over the first few cycles), the current is held constant at the desired level throughout the accelerating period, the firing angle of the thyristors being continually adjusted to compensate for the changing effective impedance of the motor. By keeping the current at the maximum value, which the supply can tolerate the run-up time, is minimised.

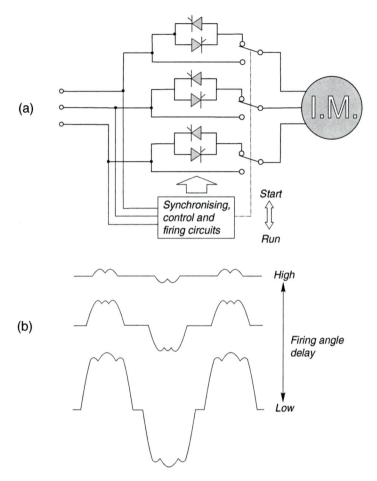

Figure 6.4   (a) *Thyristor soft-starter,* (b) *typical motor current waveforms*

Alternatively, if a slow run-up is desirable, a lower accelerating current can be selected.

As with the open-loop systems the velocity–time profile is not necessarily ideal, since with constant current the motor torque exhibits a very sharp rise as the pullout slip is reached, resulting in a sudden surge in speed.

Prospective users need to be wary of some of the promotional literature, where naturally enough the virtues are highlighted while the shortcomings are played down. Claims are sometimes made that massive reductions in starting current can be achieved without corresponding reductions in starting torque. This is nonsense: the current can certainly be limited, but as far as torque per line amp is concerned soft-start systems are no better than series reactor systems, and not as good as the autotransformer and star/delta methods.

Starting using a variable-frequency inverter

Operation of induction motors from variable-frequency inverters is discussed in Chapter 8, but it is appropriate to mention here that one of the advantages of inverter-fed operation is that starting is not a problem because it is usually possible to obtain at least rated torque at zero speed without drawing an excessive current from the mains supply. None of the other starting methods we have looked at have this ability, so in some applications it may be that the comparatively high cost of the inverter is justified solely on the grounds of its starting and run-up potential.

RUN-UP AND STABLE OPERATING REGIONS

In addition to having sufficient torque to start the load it is obviously necessary for the motor to bring the load up to full speed. To predict how the speed will rise after switching on we need the torque–speed curves of the motor and the load, and the total inertia.

By way of example, we can look at the case of a motor with two different loads (see Figure 6.5) . The solid line is the torque–speed curve of the motor, while the dotted lines represent two different load charac-teristics. Load (A) is typical of a simple hoist, which applies constant torque to the motor at all speeds, while load (B) might represent a fan. For the sake of simplicity, we will assume that the load inertias (as seen at the motor shaft) are the same.

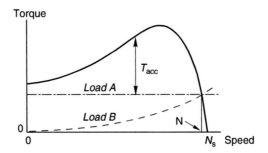

Figure 6.5   *Typical torque–speed curve showing two different loads which have the same steady running speed* (N)

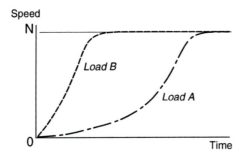

Figure 6.6   *Speed–time curves during run-up, for motor and loads shown in Figure 6.5*

The speed–time curves for run-up are shown in Figure 6.6. Note that the gradient of the speed–time curve (i.e. the acceleration) is obtained by dividing the accelerating torque T_{acc} (which is the difference between the torque developed by the motor and the torque required to run the load at that speed) by the total inertia.

In this example, both loads ultimately reach the same steady speed, N (i.e. the speed at which motor torque equals load torque), but B reaches full speed much more quickly because the accelerating torque is higher during most of the run-up. Load A picks up speed slowly at first, but then accelerates hard (often with a characteristic 'whoosh' produced by the ventilating fan) as it passes through the peak torque–speed and approaches equilibrium conditions.

It should be clear that the higher the total inertia, the slower the acceleration and vice versa. The total inertia means the inertia as seen at the motor shaft, so if gearboxes or belts are employed the inertia must be 'referred' as discussed in Chapter 11.

An important qualification ought to be mentioned in the context of the motor torque–speed curves shown by the solid line in Figure 6.5. This is that curves like this represent the torque developed by the motor when it has settled down at the speed in question, i.e. they are the true steady-state curves. In reality, a motor will generally only be in a steady-state condition when it settles at its normal running speed, so for most of the speed range the motor will be accelerating. In particular, when the motor is first switched on, there will be a transient period of a few cycles as the three currents gradually move towards a balanced 3-phase pattern. During this period the torque can fluctuate wildly and the motor can pick up significant speed, so the actual torque may be very different from that shown by the steady-state curve, and as a result the instantaneous speeds can fluctuate about the mean value. Fortunately, the average torque during run-up can be fairly reliably obtained from the steady-state curves, particularly if the inertia is high and the motor takes many cycles to reach full speed, in which case we would consider the torque–speed curve as being 'quasi-steady state'.

Harmonic effects – skewing

A further cautionary note in connection with the torque–speed curves shown in this and most other books relate to the effects of harmonic air-gap fields. In Chapter 5, it was explained that despite the limitations imposed by slotting, the stator winding magnetic flux (MMF) is remarkably close to the ideal of a pure sinusoid. Unfortunately, because it is not a perfect sinusoid, Fourier analysis reveals that in addition to the predominant fundamental component, there are always additional unwanted 'space harmonic' fields. These harmonic fields have synchronous speeds that are inversely proportional to their order. For example a 4-pole, 50 Hz motor will have a main field rotating at 1500 rev/min, but in addition there may be a fifth harmonic (20-pole) field rotating in the reverse direction at 300 rev/min, a seventh harmonic (28-pole) field rotating forwards at 214 rev/min, etc. These space harmonics are minimised by stator winding design, but can seldom be eliminated.

If the rotor has a very large number of bars it will react to the harmonic field in much the same way as to the fundamental, producing additional induction motor torques centred on the synchronous speed of the harmonic, and leading to unwanted dips in the torque speed, typically as shown in Figure 6.7.

Users should not be too alarmed as in most cases the motor will ride through the harmonic during acceleration, but in extreme cases a motor might, for example, stabilise on the seventh harmonic, and 'crawl' at about 214 rev/min, rather than running up to 4-pole speed (1500 rev/min at 50 Hz), as shown by the dot in Figure 6.7.

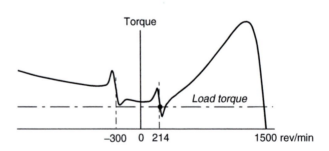

Figure 6.7   *Torque–speed curve showing the effect of space harmonics, and illustrating the possibility of a motor 'crawling' on the seventh harmonic*

To minimise the undesirable effects of space harmonics the rotor bars in the majority of induction motors are not parallel to the axis of rotation, but instead they are skewed (typically by around one or two slot pitches) along the rotor length. This has very little effect as far as the fundamental field is concerned, but can greatly reduce the response of the rotor to harmonic fields.

Because the overall influence of the harmonics on the steady-state curve is barely noticeable, and their presence might worry users, they are rarely shown, the accepted custom being that 'the' torque–speed curve represents the behaviour due to the fundamental component only.

High inertia loads – overheating

Apart from accelerating slowly, high inertia loads pose a particular problem of rotor heating, which can easily be overlooked by the unwary user. Every time an induction motor is started from rest and brought up to speed, the total energy dissipated as heat in the motor windings is equal to the stored kinetic energy of the motor plus load. (This matter is explored further via the equivalent circuit in Chapter 7.) Hence with high inertia loads, very large amounts of energy are released as heat in the windings during run-up, even if the load torque is negligible when the motor is up to speed. With totally enclosed motors the heat ultimately has to find its way to the finned outer casing of the motor, which is cooled by air from the shaft-mounted external fan. Cooling of the rotor is therefore usually much worse than the stator, and the rotor is thus most likely to overheat during high inertia run-ups.

No hard and fast rules can be laid down, but manufacturers usually work to standards which specify how many starts per hour can be tolerated. Actually, this information is useless unless coupled with reference to the total inertia, since doubling the inertia makes the problem twice as bad. However, it is usually assumed that the total inertia is not likely to be more than twice the motor inertia, and this is certainly the case for most loads. If in doubt, the user should consult the manufacturer who may recommend a larger motor than might seem necessary simply to supply the full-load power requirements.

Steady-state rotor losses and efficiency

The discussion above is a special case, which highlights one of the less attractive features of induction machines. This is that it is never possible for all the power crossing the air-gap from the stator to be converted to mechanical output, because some is always lost as heat in the rotor circuit resistance. In fact, it turns out that at slip s the total power (P_r) crossing the air-gap always divides so that a fraction sP_r is lost as heat, while the remainder $(1 - s)P_r$ is converted to useful mechanical output (see also Chapter 7.).

Hence, when the motor is operating in the steady state the energy conversion efficiency of the rotor is given by

$$\eta_r = \frac{\text{Mechanical output power}}{\text{Rated power input to rotor}} = (1 - s) \qquad (6.2)$$

This result is very important, and shows us immediately why operating at small values of slip is desirable. With a slip of 5% (or 0.05), for example, 95% of the air-gap power is put to good use. But if the motor was run at half the synchronous speed ($s = 0.5$), 50% of the air-gap power would be wasted as heat in the rotor.

We can also see that the overall efficiency of the motor must always be significantly less than $(1-s)$, because in addition to the rotor copper losses there are stator copper losses, iron losses and windage and friction losses. This fact is sometimes forgotten, leading to conflicting claims such as 'full-load slip = 5%, overall efficiency = 96%', which is clearly impossible.

Steady-state stability – pullout torque and stalling

We can check stability by asking what happens if the load torque suddenly changes for some reason. The load torque shown by the dotted line in Figure 6.8 is stable at speed X, for example: if the load torque increased from T_a to T_b, the load torque would be greater than the motor torque, so the motor torque would decelerate. As the speed dropped, the motor torque would rise, until a new equilibrium was reached, at the slightly

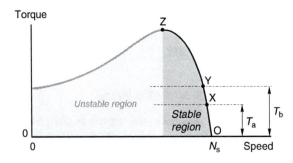

Figure 6.8   *Torque–speed curve illustrating stable operating region (0XYZ)*

lower speed (Y). The converse would happen if the load torque were reduced, leading to a higher stable running speed.

But what happens if the load torque is increased more and more? We can see that as the load torque increases, beginning at point X, we eventually reach point Z, at which the motor develops its maximum torque. Quite apart from the fact that the motor is now well into its overload region, and will be in danger of overheating, it has also reached the limit of stable operation. If the load torque is further increased, the speed falls (because the load torque is more than the motor torque), and as it does so the shortfall between motor torque and load torque becomes greater and greater. The speed therefore falls faster and faster, and the motor is said to be 'stalling'. With loads such as machine tools (a drilling machine, for example), as soon as the maximum or 'pullout' torque is exceeded, the motor rapidly comes to a halt, making an angry humming sound. With a hoist, however, the excess load would cause the rotor to be accelerated in the reverse direction, unless it was prevented from doing so by a mechanical brake.

TORQUE–SPEED CURVES – INFLUENCE OF ROTOR PARAMETERS

We saw earlier that the rotor resistance and reactance influenced the shape of the torque–speed curve. The designer can vary both of these parameters, and we will explore the pros and cons of the various alternatives. To limit the mathematics the discussion will be mainly qualitative, but it is worth mentioning that the whole matter can be dealt rigorously using the equivalent circuit approach, as discussed in Chapter 7.

We will deal with the cage rotor first because it is the most important, but the wound rotor allows a wider variation of resistance to be obtained, so it is discussed later.

Cage rotor

For small values of slip, i.e. in the normal running region, the lower we make the rotor resistance the steeper the slope of the torque–speed curve becomes, as shown in Figure 6.9. We can see that at the rated torque (shown by the horizontal dotted line in Figure 6.9) the full-load slip of the low-resistance cage is much lower than that of the high-resistance cage. But we saw earlier that the rotor efficiency is equal to $(1 - s)$, where s is the slip. So, we conclude that the low-resistance rotor not only gives better speed holding, but is also much more efficient. There is of course a limit to how low we can make the resistance: copper allows us

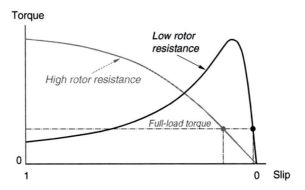

Figure 6.9   *Influence of rotor resistance on torque–speed curve of cage motor. The full-load running speeds are indicated by the vertical dotted lines*

to achieve a lower resistance than aluminium, but we can't do anything better than fill the slots with solid copper bars.

As we might expect there are drawbacks with a low-resistance rotor. The starting torque is reduced (see Figure 6.9), and worse still the starting current is increased. The lower starting torque may prove insufficient to accelerate the load, while increased starting current may lead to unacceptable volt drops in the supply.

Altering the rotor resistance has little or no effect on the value of the peak (pullout) torque, but the slip at which the peak torque occurs is directly proportional to the rotor resistance. By opting for a high enough resistance (by making the cage from bronze, brass or other relatively high resistivity material) we can if we wish to arrange for the peak torque to occur at or close to starting, as shown in Figure 6.9. The snag in doing this is that the full-load efficiency is inevitably low because the full-load slip will be high (see Figure 6.9).

There are some applications for which high-resistance motors are well suited, an example being for metal punching presses, where the motor accelerates a flywheel, which is used to store energy. To release a significant amount of energy, the flywheel slows down appreciably during impact, and the motor then has to accelerate it back up to full speed. The motor needs a high torque over a comparatively wide speed range, and does most of its work during acceleration. Once up to speed the motor is effectively running light, so its low efficiency is of little consequence. High-resistance motors are also used for speed control of fan-type loads, and this is taken up again in Section 6.6, where speed control is explored.

To sum up, a high-rotor resistance is desirable when starting and at low speeds, while a low resistance is preferred under normal running conditions. To get the best of both worlds, we need to be able to alter the resistance from a high value at starting to a lower value at full speed. Obviously we cannot change the actual resistance of the cage once it has been manufactured, but it is possible to achieve the desired effect with either a 'double cage' or a 'deep bar' rotor. Manufacturers normally offer a range of designs, which reflect these trade-offs, and the user then selects the one which best meets his particular requirements.

Double cage rotors

Double cage rotors have an outer cage made up of relatively high resistivity material such as bronze, and an inner cage of low resistivity, usually copper, as shown in Figure 6.10.

The inner cage is sunk deep into the rotor, so that it is almost completely surrounded by iron. This causes the inner bars to have a much higher leakage inductance than if they were near the rotor surface, so that under starting conditions (when the induced rotor frequency is high) their inductive reactance is very high and little current flows in them. In contrast, the bars of the outer cage are placed so that their leakage fluxes face a much higher reluctance path, leading to a low-leakage inductance. Hence, under starting conditions, rotor current is concentrated in the outer cage, which, because of its high resistance, produces a high starting torque.

At the normal running speed the roles are reversed. The rotor frequency is low, so both cages have low reactance and most of the current therefore flows in the low-resistance inner cage. The torque–speed curve is therefore steep, and the efficiency is high.

Considerable variation in detailed design is possible to shape the torque–speed curve to particular requirements. In comparison with a single-cage rotor, the double cage gives much higher starting torque, substantially less starting current, and marginally worse running performance.

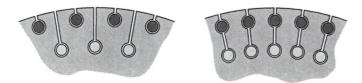

Figure 6.10   *Alternative arrangements of double cage rotors. The outer cage has a high resistance (e.g. bronze) while the inner cage has a low resistance (e.g. copper)*

Deep bar rotors

The deep bar rotor has a single cage, usually of copper, formed in slots which are deeper and narrower than in a conventional single-cage design. Construction is simpler and therefore cheaper than in a double cage rotor, as shown in Figure 6.11.

The deep bar approach ingeniously exploits the fact that the effective resistance of a conductor is higher under a.c. conditions than under d.c. conditions. With a typical copper bar of the size used in an induction motor rotor, the difference in effective resistance between d.c. and say 50 or 60 Hz (the so-called 'skin-effect') would be negligible if the conductor was entirely surrounded by air. But when it is almost completely surrounded by iron, as in the rotor slots, its effective resistance at mains frequency may be two or three times its d.c. value.

Under starting conditions, when the rotor frequency is equal to the supply frequency, the skin effect is very pronounced, and the rotor current is concentrated towards the top of the slots. The effective resistance is therefore increased, resulting in a high-starting torque from a low-starting current. When the speed rises and the rotor frequency falls, the effective resistance reduces towards its d.c. value, and the current distributes itself more uniformly across the cross section of the bars. The normal running performance thus approaches that of a low-resistance single-cage rotor, giving a high efficiency and stiff torque–speed curve. The pullout torque is, however, somewhat lower than for an equivalent single-cage motor because of the rather higher leakage reactance.

Most small and medium motors are designed to exploit the deep bar effect to some extent, reflecting the view that for most applications the slightly inferior running performance is more than outweighed by the much better starting behaviour. A typical torque–speed curve for a general-purpose medium-size (55 kW) motor is shown in Figure 6.12. Such motors are unlikely to be described by the maker specifically as 'deep-bar' but they nevertheless incorporate a measure of the skin effect

Figure 6.11   *Typical deep-bar rotor construction*

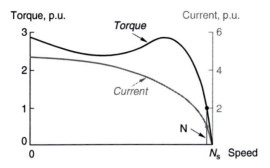

Figure 6.12   *Typical torque–speed and current–speed curves for a general-purpose industrial cage motor*

and consequently achieve the 'good' torque–speed characteristic shown by the solid line in Figure 6.12.

The current–speed relationship is shown by the dotted line in Figure 6.12, both torque and current scales being expressed in per unit (p.u.). This notation is widely used as a shorthand, with 1 p.u. (or 100%) representing rated value. For example, a torque of 1.5 p.u. simply means one and a half times rated value, while a current of 400% means a current of four times rated value.

Starting and run-up of slipring motors

By adding external resistance in series with the rotor windings the starting current can be kept low but at the same time the starting torque is high. This is the major advantage of the wound-rotor or slipring motor, and makes it well suited for loads with heavy starting duties such as stone-crushers, cranes and conveyor drives.

The influence of rotor resistance is shown by the set of torque–speed curves in Figure 6.13. The curve on the right corresponds to no added rotor resistance, with the other six curves showing the influence of progressively increasing the external resistance.

A high-rotor resistance is used when the motor is first switched on, and depending on the value chosen any torque up to the pullout value (perhaps twice full load) can be obtained. Typically, the resistance will be selected to give full-load torque at starting, together with rated current from the mains. The starting torque is then as indicated by point A in Figure 6.13.

As the speed rises, the torque would fall more or less linearly if the resistance remained constant, so to keep close to full-load torque the resistance is gradually reduced, either in steps, in which case the

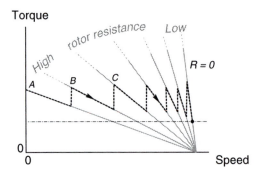

Figure 6.13 *Torque–speed curves for a wound-rotor (slipring) motor showing how the external rotor-circuit resistance* (R) *can be varied in steps to provide an approximately constant torque during acceleration*

trajectory ABC etc. is followed (see Figure 6.13), or continuously so that maximum torque is obtained throughout. Ultimately the external resistance is made zero by shorting-out the sliprings, and thereafter the motor behaves like a low-resistance cage motor, with a high running efficiency.

As mentioned earlier, the total energy dissipated in the rotor circuit during run-up is equal to the final stored kinetic energy of the motor and load. In a cage motor this energy ends up in the rotor, and can cause overheating. In the slipring motor, however, most of the energy goes into the external resistance. This is a good thing from the motor point of view, but means that the external resistance has to absorb the thermal energy without overheating.

Fan-cooled grid resistors are often used, with tappings at various resistance values. These are progressively shorted-out during run-up, either by a manual or motor-driven drum-type controller, or with a series of timed contactors. Alternatively, where stepless variation of resistance is required, a liquid resistance controller is often employed. It consists of a tank of electrolyte (typically caustic soda) into which three electrodes can be raised or lowered. The resistance between the electrodes depends on how far they are immersed in the liquid. The electrolyte acts as an excellent short-term reservoir for the heat released, and by arranging for convection to take place via a cooling radiator, the equipment can also be used continuously for speed control (see later).

Attempts have been made to vary the effective rotor circuit resistance by means of a fixed external resistance and a set of series connected thyristors, but this approach has not gained wide acceptance.

INFLUENCE OF SUPPLY VOLTAGE ON TORQUE–SPEED CURVE

We established earlier that at any given slip, the air-gap flux density is proportional to the applied voltage, and the induced current in the rotor is proportional to the flux density. The torque, which depends on the product of the flux and the rotor current, therefore depends on the square of the applied voltage. This means that a comparatively modest fall in the voltage will result in a much larger reduction in torque capability, with adverse effects which may not be apparent to the unwary until too late.

To illustrate the problem, consider the torque–speed curves for a cage motor shown in Figure 6.14. The curves (which have been expanded to focus attention on the low-slip region) are drawn for full voltage (100%), and for a modestly reduced voltage of 90%. With full voltage and full-load torque the motor will run at point X, with a slip of say 5%. Since this is the normal full-load condition, the rotor and stator currents will be at their rated values.

Now suppose that the voltage falls to 90%. The load torque is assumed to be constant so the new operating point will be at Y. Since the air-gap flux density is now only 0.9 of its rated value, the rotor current will have to be about 1.1 times rated value to develop the same torque, so the rotor e.m.f. is required to increase by 10%. But the flux density has fallen by 10%, so an increase in slip of 20% is called for. The new slip is therefore 6%.

The drop in speed from 95% of synchronous to 94% may well not be noticed, and the motor will apparently continue to operate quite happily. But the rotor current is now 10% above its rated value, so the rotor heating will be 21% more than is allowable for continuous running.

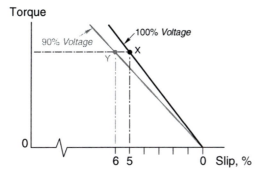

Figure 6.14 *Influence of stator supply voltage on torque–speed curves*

The stator current will also be above rated value, so if the motor is allowed to run continuously, it will overheat. This is one reason why all large motors are fitted with protection, which is triggered by over-temperature. Many small and medium motors do not have such protection, so it is important to guard against the possibility of undervoltage operation.

GENERATING AND BRAKING

Having explored the torque–speed curve for the normal motoring region, where the speed lies between zero and just below synchronous, we must ask what happens if the speed is above the synchronous speed, or is negative.

A typical torque–speed curve for a cage motor covering the full range of speeds, which are likely to be encountered in practice, is shown in Figure 6.15.

We can see from Figure 6.15 that the decisive factor as far as the direction of the torque is concerned is the slip, rather than the speed. When the slip is positive the torque is positive, and vice versa. The torque therefore always acts so as to urge the rotor to run at zero slip, i.e. at the synchronous speed. If the rotor is tempted to run faster than the field it will be slowed down, whilst if it is running below synchronous speed it will be urged to accelerate forwards. In particular, we note that for slips greater than 1, i.e. when the rotor is running backwards (i.e. in

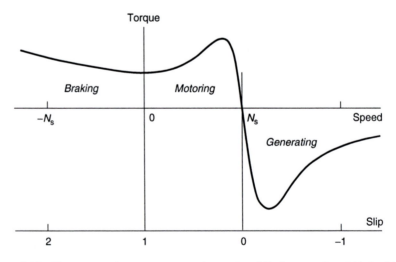

Figure 6.15   *Torque–speed curve over motoring region (slip between 0 and 1), braking region (slip greater than 1) and generating region (negative slip)*

the opposite direction to the field), the torque will remain positive, so that if the rotor is unrestrained it will first slow down and then change direction and accelerate in the direction of the field.

Generating region – overhauling loads

For negative slips, i.e. when the rotor is turning in the same direction, but at a higher speed than the travelling field, the 'motor' torque is in fact negative. In other words the machine develops a torque that which opposes the rotation, which can therefore only be maintained by applying a driving torque to the shaft. In this region the machine acts as an induction generator, converting mechanical power from the shaft into electrical power into the supply system. Many cage induction machines are used in this way in wind power generation schemes because their inherently robust construction leads to low maintenance.

It is worth stressing that, just as with the d.c. machine, we do not have to make any changes to an induction motor to turn it into an induction generator. All that is needed is a source of mechanical power to turn the rotor faster than the synchronous speed. On the other hand, we should be clear that the machine can only generate when it is connected to the supply. If we disconnect an induction motor from the mains and try to make it generate simply by turning the rotor we will not get any output because there is nothing to set up the working flux: the flux (excitation) is not present until the motor is supplied with magnetising current from the supply.

There are comparatively few applications in which mains-fed motors find themselves in the generating region, though as we will see later it is quite common in inverter-fed drives. We will, however, look at one example of a mains-fed motor in the so-called 'regenerative' mode to underline the value of the motor's inherent ability to switch from motoring to generating automatically, without the need for any external intervention.

Consider a cage motor driving a simple hoist through a reduction gearbox, and suppose that the hook (unloaded) is to be lowered. Because of the static friction in the system, the hook will not descend on its own, even after the brake is lifted, so on pressing the 'down' button the brake is lifted and power is applied to the motor so that it rotates in the lowering direction. The motor quickly reaches full speed and the hook descends. As more and more rope winds off the drum, a point is reached where the lowering torque exerted by the hook and rope is greater than the running friction, and a restraining torque is then needed to prevent a runaway. The necessary stabilising torque is automatically provided by the motor acting as a generator as soon as the synchronous speed is exceeded, as shown in Figure 6.15. The speed will therefore be held at just above the synchronous speed, provided of course that the peak generating torque (see Figure 6.15) is not exceeded.

Plug reversal and plug braking

Because the rotor always tries to catch up with the rotating field, it can be reversed rapidly simply by interchanging any two of the supply leads. The changeover is usually obtained by having two separate 3-pole contactors, one for forward and one for reverse. This procedure is known as plug reversal or plugging, and is illustrated in Figure 6.16.

The motor is initially assumed to be running light (and therefore with a very small positive slip) as indicated by point A on the dotted torque–speed curve in Figure 6.16(a). Two of the supply leads are then reversed, thereby reversing the direction of the field, and bringing the mirror-image torque–speed curve shown by the solid line into play. The slip of the motor immediately after reversal is approximately 2, as shown by point B on the solid curve. The torque is thus negative, and the motor decelerates, the speed passing through zero at point C and then rising in the reverse direction before settling at point D, just below the synchronous speed.

The speed–time curve is shown in Figure 6.16(b). We can see that the deceleration (i.e. the gradient of the speed–time graph) reaches a maximum as the motor passes through the peak torque (pullout) point, but thereafter the final speed is approached gradually, as the torque tapers down to point D.

Very rapid reversal is possible using plugging; for example a 1 kW motor will typically reverse from full speed in under 1 s. But large cage

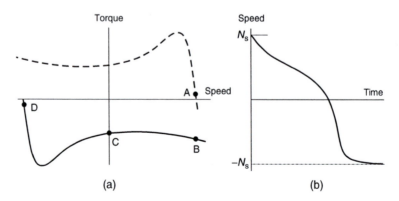

Figure 6.16   *Torque–speed and speed–time curves for plug reversal of cage motor*

motors can only be plugged if the supply can withstand the very high currents involved, which are even larger than when starting from rest. Frequent plugging will also cause serious overheating, because each reversal involves the 'dumping' of four times the stored kinetic energy as heat in the windings.

Plugging can also be used to stop the rotor quickly, but obviously it is then necessary to disconnect the supply when the rotor comes to rest, otherwise it will run-up to speed in reverse. A shaft-mounted reverse-rotation detector is therefore used to trip out the reverse contactor when the speed reaches zero.

We should note that, whereas, in the regenerative mode (discussed in the previous section) the slip was negative, allowing mechanical energy from the load to be converted to electrical energy and fed back to the mains, plugging is a wholly dissipative process in which all the kinetic energy ends up as heat in the motor.

Injection braking

This is the most widely used method of electrical braking. When the 'stop' button is pressed, the 3-phase supply is interrupted, and a d.c. current is fed into the stator via two of its terminals. The d.c. supply is usually obtained from a rectifier fed via a low-voltage high-current transformer.

We saw earlier that the speed of rotation of the air-gap field is directly proportional to the supply frequency, so it should be clear that since d.c. is effectively zero frequency, the air-gap field will be stationary. We also saw that the rotor always tries to run at the same speed as the field. So, if the field is stationary, and the rotor is not, a braking torque will be exerted. A typical torque–speed curve for braking a cage motor is shown in Figure 6.17, from which we see that the braking (negative) torque falls to zero as the rotor comes to rest.

This is in line with what we would expect, since there will be induced currents in the rotor (and hence torque) only when the rotor is 'cutting' the flux. As with plugging, injection (or dynamic) braking is a dissipative process, all the kinetic energy being turned into heat inside the motor.

SPEED CONTROL

We have seen that to operate efficiently an induction motor must run with a small slip. It follows that any efficient method of speed control must be based on varying the synchronous speed of the field, rather than

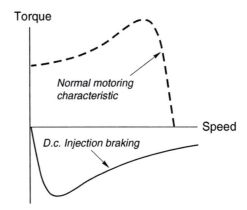

Figure 6.17   *Torque–speed curve for d.c. injection braking of cage motor*

the slip. The two factors that determine the speed of the field, are the supply frequency and the pole number (see equation (5.1)).

The pole number has to be an even integer, so where continuously adjustable speed control over a wide range is called for, the best approach is to provide a variable-frequency supply. This method is very important, and is discussed in Chapter 8. In this chapter we are concerned with constant frequency mains operation, so we have a choice between pole-changing, which can provide discrete speeds only, or slip control which can provide continuous speed control, but is inherently inefficient.

Pole-changing motors

For some applications continuous speed control may be an unnecessary luxury, and it may be sufficient to be able to run at two discrete speeds. Among many instances where this can be acceptable and economic are pumps, lifts and hoists, fans and some machine tool drives.

We established in Chapter 5 that the pole number of the field was determined by the layout and interconnection of the stator coils, and that once the winding has been designed, and the frequency specified, the synchronous speed of the field is fixed. If we wanted to make a motor, that could run at either of two different speeds, we could construct it with two separate stator windings (say 4-pole and 6-pole), and energise the appropriate one. There is no need to change the cage rotor since the pattern of induced currents can readily adapt to suit the stator pole number. Early 2-speed motors did have 2 distinct stator windings, but were bulky and inefficient.

It was soon realised that if half of the phase belts within each phase-winding could be reversed in polarity, the effective pole number could be halved. For example, a 4-pole MMF pattern (N-S-N-S) would become (N-N-S-S), i.e. effectively a 2-pole pattern with one large N pole and one large S pole. By bringing out six leads instead of three, and providing switching contactors to effect the reversal, two discrete speeds in the ratio 2:1 are therefore possible from a single winding. The performance at the high (e.g. 2-pole) speed is relatively poor, which is not surprising in view of the fact that the winding was originally optimised for 4-pole operation.

It was not until the advent of the more sophisticated pole amplitude modulation (PAM) method in the 1960s that 2-speed single-winding high-performance motors with more or less any ratio of speeds became available from manufacturers. This subtle technique allows close ratios such as 4/6, 6/8, 8/10 or wide ratios such as 2/24 to be achieved. Close ratios are used in pumps and fans, while wide ratios are used for example in washing machines where a fast spin is called for.

The beauty of the PAM method is that it is not expensive. The stator winding has more leads brought out, and the coils are connected to form non-uniform phase belts, but otherwise construction is the same as for a single-speed motor. Typically six leads will be needed, three of which are supplied for one speed, and three for the other, the switching being done by contactors. The method of connection (star or delta) and the number of parallel paths within the winding are arranged so that the air-gap flux at each speed matches the load requirement. For example, if constant torque is needed at both speeds, the flux needs to be made the same, whereas if reduced torque is acceptable at the higher speed the flux can obviously be lower.

Voltage control of high-resistance cage motors

Where efficiency is not of paramount importance, the torque (and hence the running speed) of a cage motor can be controlled simply by altering the supply voltage. The torque at any slip is approximately proportional to the square of the voltage, so we can reduce the speed of the load by reducing the voltage. The method is not suitable for standard low-resistance cage motors, because their stable operating speed range is very restricted, as shown in Figure 6.18(a) . But if special high-rotor resistance motors are used, the slope of the torque–speed curve in the stable region is much less, and a rather wider range of steady-state operating speeds is available, as shown in Figure 6.18(b).

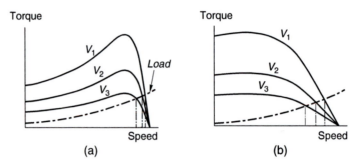

Figure 6.18   *Speed control of cage motor by stator voltage variation; (a) low-resistance rotor, (b) high-resistance rotor*

The most unattractive feature of this method is the low efficiency, which is inherent in any form of slip control. We recall that the rotor efficiency at slip s is $(1 - s)$, so if we run at say 70% of synchronous speed (i.e. $s = 0.3$), 30% of the power crossing the air-gap is wasted as heat in the rotor conductors. The approach is therefore only practicable where the load torque is low at low speeds, so that at high slips the heat in the rotor is tolerable. A fan-type characteristic is suitable, as shown in Figure 6.18(b), and many ventilating systems therefore use voltage control.

Voltage control became feasible only when relatively cheap thyristor a.c. voltage regulators arrived on the scene during the 1970s. Previously the cost of autotransformers or induction regulators to obtain the variable voltage supply was simply too high. The thyristor hardware required is essentially the same as discussed earlier for soft starting, and a single piece of kit can therefore serve for both starting and speed control. Where accurate speed control is needed, a tachogenerator must be fitted to the motor to provide a speed feedback signal, and this naturally increases the cost significantly.

Applications are numerous, mainly in the range 0.5–10 kW, with most motor manufacturers offering high-resistance motors specifically for use with thyristor regulators.

Speed control of wound-rotor motors

The fact that the rotor resistance can be varied easily allows us to control the slip from the rotor side, with the stator supply voltage and frequency constant. Although the method is inherently inefficient it is still used in many medium and large drives such as hoists, conveyors and crushers because of its simplicity and comparatively low cost.

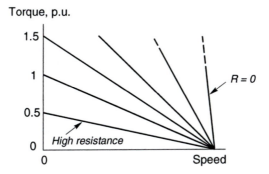

Figure 6.19   *Influence of external rotor resistance (R) on torque–speed curve of wound-rotor motor*

A set of torque–speed characteristics is shown in Figure 6.19, from which it should be clear that by appropriate selection of the rotor circuit resistance, any torque up to typically 1.5 times full-load torque can be achieved at any speed.

POWER FACTOR CONTROL AND ENERGY OPTIMISATION

Voltage control

In addition to their use for soft-start and speed control, thyristor voltage regulators are sometimes marketed as power factor controllers and/or energy optimisers for cage motors. Much of the hype surrounding their introduction has evaporated, but users should remain sceptical of some of the more extravagant claims, which can still be found.

The fact is that there are comparatively few situations where considerations of power factor and/or energy economy alone are sufficient to justify the expense of a voltage controller. Only when the motor operates for very long periods running light or at low load can sufficient savings be made to cover the outlay. There is certainly no point in providing energy economy when the motor spends most of its time working at or near full load.

Both power factor control and energy optimisation rely on the fact that the air-gap flux is proportional to the supply voltage, so that by varying the voltage, the flux can be set at the best level to cope with the prevailing load. We can see straightaway that nothing can be achieved at full load, since the motor needs full flux (and hence full voltage) to operate as intended. Some modest savings in losses can be achieved at reduced load, as we will see.

If we imagine the motor to be running with a low-load torque and full voltage, the flux will be at its full value, and the magnetising component of the stator current will be larger than the work component, so the input power factor ($\cos\phi_a$) will be very low, as shown in Figure 6.20(a).

Now suppose that the voltage is reduced to say half (by phasing back the thyristors), thereby halving the air-gap flux and reducing the magnetising current by at least a factor of two. With only half the flux, the rotor current must double to produce the same torque, so the work current reflected in the stator will also double. The input power factor ($\cos\phi_b$) will therefore improve considerably (see Figure 6.20(b)). Of course the slip with 'half-flux' operation will be higher (by a factor of four), but with a low-resistance cage it will still be small, and the drop in speed will therefore be slight.

The success (or otherwise) of the energy economy obtained depends on the balance between the iron losses and the copper losses in the motor. Reducing the voltage reduces the flux, and hence reduces the eddy current and hysteresis losses in the iron core. But as we have seen above, the rotor current has to increase to produce the same torque, so the rotor copper loss rises. The stator copper loss will reduce if (as in Figure 6.20) the magnitude of the stator current falls. In practice, with average general-purpose motors, a net saving in losses only occurs for

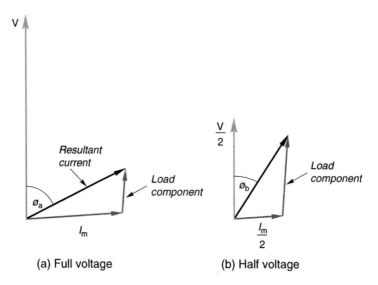

(a) Full voltage   (b) Half voltage

Figure 6.20   *Phasor diagram showing improvement of power factor by reduction of stator voltage*

light loads, say at or below 25% of full load, though the power factor will always increase.

Slip energy recovery (wound rotor motors)

Instead of wasting rotor circuit power in an external resistance, it can be converted and returned to the mains supply. Frequency conversion is necessary because the rotor circuit operates at slip frequency, so it cannot be connected directly to the mains.

In a slip energy recovery system, the slip frequency a.c. from the rotor is first rectified in a 3-phase diode bridge and smoothed before being returned to the mains supply via a 3-phase thyristor bridge converter operating in the inverting mode (see Chapter 4). A transformer is usually required to match the output from the controlled bridge to the mains voltage.

Since the cost of both converters depends on the slip power they have to handle, this system (which is known as the static Kramer drive) is most often used where only a modest range of speeds (say from 80% of synchronous and above) is required, such as in large pump and compressor drives. Speed control is obtained by varying the firing angle of the controlled converter, the torque–speed curves for each firing angle being fairly steep (i.e. approximating to constant speed), thereby making closed-loop speed control relatively simple.

SINGLE-PHASE INDUCTION MOTORS

Single-phase induction motors are simple, robust and reliable, and are used in enormous numbers especially in domestic and commercial applications where 3-phase supplies are not available. Although outputs of up to a few kW are possible, the majority are below 0.5 kW, and are used in such applications such as refrigeration compressors, washing machines and dryers, pumps and fans, small machine tools, tape decks, printing machines, etc.

Principle of operation

If one of the leads of a 3-phase motor is disconnected while it is running light, it will continue to run with a barely perceptible drop in speed, and a somewhat louder hum. With only two leads remaining there can only be one current, so the motor must be operating as a single-phase machine. If load is applied the slip increases more quickly than under

Plate 6.1 *Single-phase capacitor-run induction motor. Output power range is typically from about 70 W to 2.2 kW, with pole-numbers from 2 to 8. (Photo courtesy of ABB)*

3-phase operation, and the stall torque is much less, perhaps one third. When the motor stalls and comes to rest it will not restart if the load is removed, but remains at rest drawing a heavy current and emitting an angry hum. It will burn out if not disconnected rapidly.

It is not surprising that a truly single-phase cage induction motor will not start from rest, because as we saw in Chapter 5 the single winding, fed with a.c., simply produces a pulsating flux in the air-gap, without any suggestion of rotation. It is, however, surprising to find that if the motor is given a push in either direction it will pick up speed, slowly at first but then with more vigour, until it settles with a small slip, ready to take-up load. Once turning, a rotating field is evidently brought into play to continue propelling the rotor.

We can understand how this comes about by first picturing the pulsating MMF set up by the current in the stator winding as being the resultant of two identical travelling waves of MMF, one in the forward direction and the other in reverse. (This equivalence is not self-evident but is easily proved; the phenomenon is often discussed in physics textbooks under the heading of standing waves.) When the rotor is stationary, it reacts equally to both travelling waves, and no torque is developed. When the rotor is turning, however, the induced rotor currents are such that their MMF opposes the reverse stator MMF to a greater extent than they oppose the forward stator MMF. The result is that the forward flux wave (which is what develops the forward torque) is bigger than the reverse flux wave (which exerts a drag). The difference widens as the speed increases, the forward flux wave becoming progressively bigger as the speed rises while the reverse flux wave simultaneously reduces. This 'positive feedback' effect explains why the speed builds slowly at first, but later zooms up to just below synchronous speed. At the normal running speed (i.e. small slip), the forward flux is many times larger than the backward flux, and the drag torque is only a small percentage of the forward torque.

As far as normal running is concerned, a single winding is therefore sufficient. But all motors must be able to self-start, so some mechanism has to be provided to produce a rotating field even when the rotor is at rest. Several methods are employed, all of them using an additional winding.

The second winding usually has less copper than the main winding, and is located in the slots which are not occupied by the main winding, so that its MMF is displaced in space relative to that of the main winding. The current in the second winding is supplied from the same single-phase source as the main winding current, but is caused to have a phase-lag, by various means which are discussed later. The combination of a space displacement between the two windings together with a time displacement between the currents produces a 2-phase machine. If the two windings were identical, displaced by 90°, and fed with currents with 90° phase-shift, an ideal rotating field would be produced. In practice we can never achieve a 90° phase-shift between the currents, and it turns out to be more economic not to make the windings identical. Nevertheless, a decent rotating field is set up, and entirely satisfactory starting torque can be obtained. Reversal is simply a matter of reversing the polarity of one of the windings, and performance is identical in both directions.

The most widely used methods are described below. At one time it was common practice for the second or auxiliary winding to be energised only during start and run-up, and for it to be disconnected by means of a centrifugal switch mounted on the rotor, or sometimes by a time switch. This practice gave rise to the term 'starting winding'. Nowadays it is more common to find both windings in use all the time.

Capacitor-run motors

A capacitor is used in series with the auxiliary winding (see Figure 6.21) to provide a phase-shift between the main and auxiliary winding currents. The capacitor (usually of a few μF, and with a voltage rating

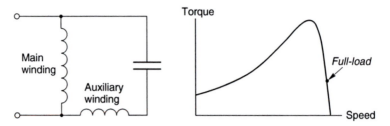

Figure 6.21   *Single-phase capacitor-run induction motor*

which may well be higher than the mains voltage) may be mounted
piggyback fashion on the motor, or located elsewhere. Its value represents a compromise between the conflicting requirements of high-
starting torque and good running performance.

A typical torque–speed curve is also shown in Figure 6.21; the modest
starting torque indicates that the capacitor-run motors are generally best
suited to fan-type loads. Where higher starting torque is needed, two
capacitors can be used, one being switched out when the motor is up to
speed.

As mentioned above, the practice of switching out the starting winding altogether is no longer favoured for new machines, but many old
ones remain, and where a capacitor is used they are known as 'capacitor
start' motors.

Split-phase motors

The main winding is of thick wire, with a low resistance and high
reactance, while the auxiliary winding is made of fewer turns of thinner
wire with a higher resistance and lower reactance (see Figure 6.22). The
inherent difference in impedance is sufficient to give the required phase-

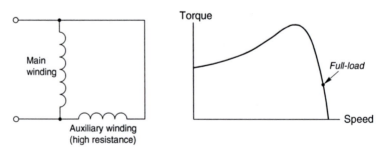

Figure 6.22   *Single-phase split-phase induction motor*

shift between the two currents without needing any external elements in series. Starting torque is good at typically 1.5 times full-load torque, as also shown in Figure 6.22. As with the capacitor type, reversal is accomplished by changing the connections to one of the windings.

Shaded-pole motors

There are several variants of this extremely simple, robust and reliable cage motor, which predominate for low-power applications such as hairdryers, oven fans, tape decks, office equipment, display drives, etc. A 2-pole version from the cheap end of the market is shown in Figure 6.23.

The rotor, typically between 1 and 4 cm diameter, has a die-cast aluminium cage, while the stator winding is a simple concentrated coil wound round the laminated core. The stator pole is slotted to receive the 'shading ring', which is a single short-circuited turn of thick copper or aluminium.

Most of the pulsating flux produced by the stator winding bypasses the shading ring and crosses the air-gap to the rotor. But some of the flux passes through the shading ring, and because it is alternating it induces an e.m.f. and current in the ring. The opposing MMF of the ring current diminishes and retards the phase of the flux through the ring, so that the flux through the ring reaches a peak after the main flux, thereby giving what amounts to a rotation of the flux across the face of the pole. This far from perfect travelling wave of flux produces the motor torque by interaction with the rotor cage. Efficiencies are low because of the rather poor magnetic circuit and the losses caused by the induced

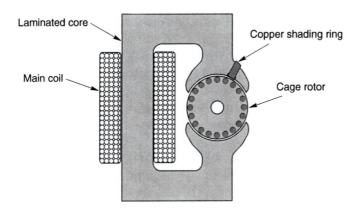

Figure 6.23   *Shaded-pole induction motor*

currents in the shading ring, but this is generally acceptable when the aim is to minimise first cost. Series resistance can be used to obtain a crude speed control, but this is only suitable for fan-type loads. The direction of rotation depends on whether the shading ring is located on the right or left side of the pole, so shaded pole motors are only suitable for unidirectional loads.

SIZE RANGE

Having praised the simplicity and elegance of the induction motor, and noting that they can be provided up to multi-MW powers, we might wonder why there are no very small ones. Industrial (3-phase) induction motors are rarely found below about 200 W, and even single-phase versions, which might be expected to dominate the domestic scene, rarely extend below about 50 W.

We will see that when we scale down a successful design, the excitation or flux-producing function of the windings becomes more and more demanding until eventually the heat produced in the windings by the excitation current causes the permissible temperature to be reached. There is then no spare capacity for the vital function of supplying the mechanical output power, so the machine is of no use.

Scaling down – the excitation problem

We can get to the essence of the matter by imagining that we take a successful design and scale all the linear dimensions by half. We know that to fully utilise the iron of the magnetic circuit we would want the air-gap flux density to be the same as in the original design, so because the air-gap length has been halved the stator MMF needs to be half of what it was. The number of coils and the turns in each coil remains as before, so if the original magnetising current was I_m, the magnetising current of the half-scale motor will be $I_m/2$.

Turning now to what happens to the resistance of the winding, we will assume that the resistance of the original winding was R. In the half-scale motor, the total length of wire is half of what it was, but the cross-sectional area of the wire is only a quarter of the original. As a result, the new resistance is twice as great, i.e. $2R$.

The power dissipated in providing the air-gap flux in the original motor is given by $I_m^2 R$, while the corresponding excitation power in the half-scale motor is given by $(I_m/2)^2 \times 2R = 1/2 I_m^2 R$.

When we consider what determines the steady temperature rise of a body in which heat is dissipated, we find that the equilibrium condition is reached when the rate of loss of heat to the surroundings is equal to the rate of production of heat inside the body. And, not surprisingly, the rate of loss of heat to the surroundings depends on the temperature difference between the body and its surroundings, and on the surface area through which the heat escapes. In the case of copper windings in a motor, the permissible temperature rise depends on the quality of insulation, so we will make a reasonable assumption that the same insulation is used for the scaled motor as for the original.

We have worked out that the power dissipation in the new motor is half of that in the original. However, the surface area of the new winding is only one quarter, so clearly the temperature rise will be higher, and if all other things were equal, it will double. We might aim to ease matters by providing bigger slots so that the current density in the copper could be reduced, but as explained in Chapter 1 this means that there is less iron in the teeth to carry the working flux. A further problem arises because it is simply not practicable to go on making the air-gap smaller because the need to maintain clearances between the moving parts would require unacceptably tight manufacturing tolerances.

Obviously, there are other factors that need to be considered, not least that a motor is designed to reach its working temperature when the full current (not just the magnetising current) is flowing. But the fact is that the magnetisation problem we have highlighted is the main obstacle in small sizes, not only in induction motors but also in any motor that derives its excitation from the stator windings. Permanent magnets therefore become attractive for small motors, because they provide the working flux without producing unwelcome heat. Several examples of small permanent magnet d.c. motors were included in Chapter 3, notably the toy motor discussed in Section 3.8.

REVIEW QUESTIONS

1) Why do large induction motors sometimes cause a dip in the supply system voltage when they are switched direct-on-line?

2) Why, for a given induction motor, might it be possible to employ direct-on-line starting in one application, but be necessary to employ a starter in another application?

3) What is meant by the term 'stiff' in relation to an industrial electrical supply?

4) Why would a given motor take longer to run-up to speed when started from a weak supply, as compared with the time it would take to run-up to speed on a stiff supply?

5) When a particular induction motor is started at its full-rated voltage, its starting torque is 20% greater than the load torque. By how much could the supply voltage be reduced before the motor would not start?

6) The voltage applied to each phase of a motor when it is star-connected is $1/\sqrt{3}$ times the voltage applied when in delta connection. Using this information, explain briefly why the line current and starting torque in star are both 1/3 of their values in delta.

7) A thyristor soft starter claims to achieve 50% starting torque with only 25% current in the supply lines. Explain why this claim does not make sense.

8) Explain briefly why induction motors are often described as 'constant speed' machines.

9) Explain briefly why, in many cage rotors, the conductor bars are not insulated from the laminated core.

10) How could the pole number of an induction motor be determined by inspection of the stator windings?

11) What determines the direction of rotation of an induction motor? How is the direction reversed?

12) Choose suitable pole numbers of cage induction motors for the following applications: (a) a grindstone to run at about 3500 rev/min when the supply is 60 Hz; (b) a pump drive to run at approximately 700 rev/min from a 50 Hz supply; (c) a turbocompressor to run at 8000 rev/min from 60 Hz; (d) a direct-drive turntable for theatre stage set to rotate at approximately 20 rev/min.

13) The full-load speed of a 4-pole, 60 Hz induction motor is 1700 rev/min. Why is it unlikely that the full-load efficiency could be as high as 94%?

14) Sketch a typical cage motor torque–speed curve and indicate: (a) the synchronous speed; (b) the starting torque; (c) the stable operating region; (d) the stall speed.

15) Discuss the pros and cons of low-resistance and high-resistance rotors in induction motors.

16) Increasing the external rotor resistance of a slipring motor reduces the rotor current at standstill, but can increase the torque. How can this apparent paradox be explained?

17) The full-load speed of a 2-pole, 60 Hz, low-resistance cage induction motor is 1740 rev/min. Estimate the speed under the following conditions: (a) Half-rated torque, full voltage and (b) Full torque, 85% voltage. Why would prolonged operation in condition (b) be unwise?

18) What voltage would you recommend to allow a 25 kW, 550 V, 60 Hz, 3-phase, 4-pole, 1750 rev/min induction motor to be used without modification on a 50 Hz supply? What would be the new rated power and speed?

19) Why might the rotor of an induction motor become very hot if it was switched on and off repeatedly, even though it was not connected to any mechanical load?

20) Sketch a typical torque–speed curve for an induction machine, covering the range of slips from 2 to −1. Identify the motoring, generating and braking regions. Which quadrants of the torque–speed plane are accessible when an induction machine is operating from a constant voltage and constant frequency supply?

21) The torque–speed curve of a particular 10-pole induction motor is approximately linear for low values of slip either side of its synchronous speed. When used as a motor on a 50 Hz supply, its efficiency is 90% and it produces a mechanical output power of 25 kW at a speed of 550 rev/min. Estimate the power produced when the machine is connected to the same 50 Hz mains and driven by a wind turbine at 650 rev/min.

22) The book explains that the speed of rotation of the space harmonics of the air-gap field in an induction motor rotate at a speed that is inversely proportional to their order. For example, the fifth harmonic rotates forward at one fifth of synchronous speed, while the seventh rotates backward at one seventh of the synchronous speed. Calculate the frequencies of the e.m.f.'s induced by these two harmonic fields in the stator winding.

7

INDUCTION MOTOR EQUIVALENT CIRCUIT

INTRODUCTION

It is important to begin by stressing that although readers who can absorb the material in this chapter will undoubtedly be better versed in induction motor matters than those who decide to skip it, it should be seen as a bonus in terms of the added understanding it can provide, rather than an essential.

Nothing on the equivalent circuit was included in the first two editions of the book, because it was feared that readers might find it too daunting. But on reflection there are three reasons why it is logical to include such a potentially illuminating topic. Firstly, the parameters that appear in the equivalent circuit (e.g. leakage reactance, magnetising reactance) are common currency in any serious discussion of induction motors, and understanding what they mean is necessary if we wish to engage effectively with those who use the language. Secondly, a knowledge of the structure and behaviour of the circuit brings a new perspective to support the 'physical' basis followed in the rest of the book, and in this respect it can be an excellent aide-memoire when attempting to recall aspects of motor performance. And finally, it provides the only simple method for quantitative performance predictions and thereby permits us to justify some of the qualitative statements made in Chapters 5 and 6.

As elsewhere in the book a knowledge of elementary circuit theory is required, together with an understanding of the material on magnetic circuits introduced in Chapter 1. In addition, readers whose basic knowledge of steady-state a.c. circuits (reactance, impedance, power-factor and phasors) and introductory-level calculus have become rusty will find it helpful to brush-up on these topics before reading on.

Outline of approach

So far in this book we have not referred to the parallels between the induction motor and the transformer, not least because the former is designed to convert energy from electrical to mechanical form, while the latter converts electrical power from one voltage to another. Physically, however, the construction of the wound-rotor induction motor has striking similarities with that of the 3-phase transformer, with the stator and rotor windings corresponding to the primary and secondary windings of a transformer. In the light of these similarities, which are discussed in Section 7.2, it is not surprising that the induction motor equivalent circuit is derived from that of the transformer.

The behaviour of the transformer is covered in Section 7.3, beginning with the 'ideal' transformer, for which the governing equations are delightfully simple. We then extend the equivalent circuit so that it includes the modest imperfections of the real transformer. In the course of this discussion we will establish the meaning of the terms magnetising reactance and leakage reactance, which also feature in the induction motor equivalent circuit. And perhaps even more importantly for what comes later, we will be in a position to appreciate the benefit of being able to assess what effect a load connected to the secondary winding has on the primary winding by making use of the concept of a 'referred' secondary load impedance. This concept will be central when we reach the induction motor.

The emphasis throughout will be on how good a transformer is, and how for most purposes a very simple equivalent circuit is more than adequate. The approach taken differs from that taken in many textbooks, which begin with the all-singing, all-dancing circuit which can not only look frighteningly complicated to a newcomer, but also tends to give the erroneous impression that the transformer is riddled with serious imperfections.

We will reach the induction motor equivalent circuit in Section 7.4. This is where a really clever leap of the imagination is revealed. We have already seen that the magnitude and frequency of the rotor currents depend on the slip, and that they interact with the air-gap flux to produce torque. So how are we to represent what is going on in the rotor in a single equivalent circuit that must necessarily also include the stator variables, and in which all the voltages and currents are at mains frequency? We will discover that, despite the apparent complexity, all of the electromechanical interactions can be represented by means of a transformer equivalent circuit, with a hypothetical slip-dependent 'electromechanical resistance' connected where the secondary 'load' would normally be. The equivalent circuit must have created quite a stir when it was first developed, but is now taken for granted. This is a pity because it represents a major intellectual achievement.

SIMILARITY BETWEEN INDUCTION MOTOR AND TRANSFORMER

The development of a wound-rotor induction motor from a single-phase transformer is depicted in Figure 7.1.

In Figure 7.1(a), we see a section through one of several possible arrangements of a single-phase iron-cored transformer, with primary and secondary windings wound concentrically on the centre limb. In most transformers there will be many turns on both windings, but for the sake of simplicity only four coils are shown for each winding.

Operation of the transformer is explored in Section 7.3, but here we should recall that the purpose of a transformer is to take in electrical power at one voltage and supply it at a different voltage. When an a.c. supply is connected to the primary winding, a pulsating magnetic flux (shown by the dotted lines in Figure 7.1) is set up. The pulsating flux links the secondary winding, inducing a voltage in each turn, so by choosing the number of turns in series the desired output voltage is obtained. Because no mechanical energy conversion is involved there is no need for an air-gap in the magnetic circuit, which therefore has an extremely low reluctance.

In Figure 7.1(b), we see a hypothetical set-up in which two small air-gaps have been introduced into the magnetic circuit to allow for the motion that is essential in the induction motor. Needless to say we would not do this deliberately in a transformer as it would cause an unnecessary increase in the reluctance of the flux path (though the effect on performance would be much less than we might fear, as discussed in Section 7.3). The central core and winding space have also been enlarged somewhat (anticipating the need for two more phases), without materially altering the functioning of the transformer.

Finally, in Figure 7.1(c) we see the two sets of coils arranged as they would be in a 2-pole wound-rotor induction motor with full-pitched coils.

The most important points to note are:

- Flux produced by the stator (primary) winding links the rotor (secondary) winding in much the same way as it did in Figure 7.1(a), i.e. the two windings remain tightly coupled by the magnetic field.

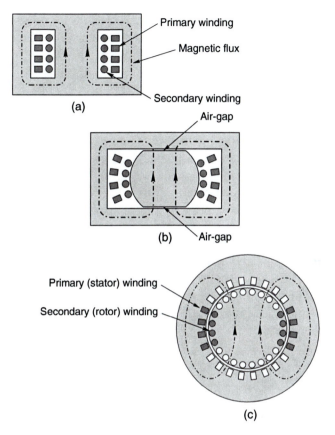

Figure 7.1   *Sketch showing similarity between transformer and induction motor*

- Only one-third of the slots are taken up because the remaining two-thirds will be occupied by the windings of the other two phases: these have been omitted for the sake of clarity.
- The magnetic circuit has two air-gaps, and because of the slots that accommodate the coils, the flux threads its way down the teeth, so that it not only fully links the aligned winding, but also partially links the other two phase-windings.
- When the rotor turns, the rotor winding also turns and it therefore links less of the flux produced by the stator. If the rotor in Figure 7.1(c) turns through 90°, there will be no mutual flux linkage, i.e. the degree to which the two windings are magnetically coupled depends on the rotor position.

If the two windings highlighted in Figure 7.1(c) were used as primary and secondary, this set-up would work perfectly well as a single-phase transformer.

We now turn to the theory of the transformer, to develop its equivalent circuit and in so doing lay the foundations for the induction motor equivalent circuit that is our ultimate objective.

THE IDEAL TRANSFORMER

Because we are dealing with balanced 3-phase motors we can achieve considerable simplification by developing single-phase models, it being understood that any calculations using the equivalent circuit (e.g. torque or power) will yield 'per phase' values which will be multiplied by three to give the total torque or power.

A quasi-circuit model of an iron-cored transformer is shown diagrammatically in Figure 7.2. This represents the most common application of the transformer, with the primary drawing power from an a.c. constant-voltage source (V_1) and supplying it to a load impedance (Z_2) at a different voltage (V_2). (In the real transformer there would not be a big hole in the middle (as in Figure 7.2) because the primary and secondary windings would fill the space, each winding having the same total volume of copper.)

The primary winding has N_1 turns with total resistance R_1, and the secondary winding has N_2 turns with total resistance R_2, and they share a common 'iron' magnetic circuit with no air-gap and therefore very low reluctance.

Ideal transformer – no-load condition, flux and magnetising current

We will begin by asking how the ideal transformer behaves when its primary winding is connected to the voltage source as shown in Figure 7.2, but the secondary is open circuited. This is known as the no-load

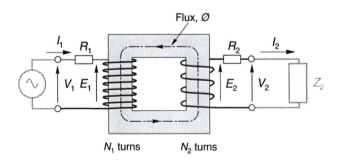

Figure 7.2   *Single-phase transformer supplying secondary load Z_2*

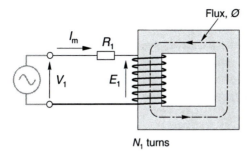

Figure 7.3   *No-load condition (i.e. secondary open-circuited), with secondary winding omitted for the sake of clarity*

condition. Since the secondary has no current it cannot influence matters, so we can temporarily remove it to yield the simpler model shown in Figure 7.3. What we want to know is what determines the magnitude of the flux in the core, and how much current (known for obvious reasons as the magnetising current, I_m) will be drawn from the voltage source to set up the flux.

We apply Kirchoff's voltage law to relate the applied voltage (V_1) to the voltage induced in the primary winding by the pulsating flux (E_1) and to the volt-drop across the primary resistance, yielding

$$V_1 = E_1 + I_m R_1 \qquad (7.1)$$

We also apply Faraday's law to obtain the induced e.m.f. in terms of the flux, i.e.

$$E_1 = N_1 \frac{d\phi}{dt} \qquad (7.2)$$

We have of course seen similar equations several times before.

We can simplify equation (7.1) further in the case of the 'ideal' transformer, which not unexpectedly is assumed to have windings made of wire with zero resistance, i.e. $R_1 = 0$. So we deduce that for an ideal transformer, the induced e.m.f. is equal to the applied voltage, i.e. $E_1 = V_1$. This is very important and echoes the discussion in Chapter 5: but what we really want is to find out what the flux is doing, so we must now use the result that $E_1 = V_1$ to recast equation (7.2) in the form

$$\frac{d\phi}{dt} = \frac{E_1}{N_1} = \frac{V_1}{N_1} \qquad (7.3)$$

Equation (7.3) shows that the rate of change of flux at any instant is determined by the applied voltage, so if we want to know how the flux behaves in time we must specify the nature of the applied voltage. We will look at two cases: the first voltage waveform is good for illustrative purposes because it is easy to derive the flux waveform, while the second represents the commonplace situation, i.e. use on an a.c. supply.

Firstly, we suppose that the applied voltage is a square wave. The rate of change of flux is positive and constant while V_1 is positive, so the flux increases linearly with time. Conversely, when the applied voltage is negative the flux ramps down, so the overall voltage and flux waves are as shown in Figure 7.4(a).

It should be evident from Figure 7.4 that the maximum flux in the core (φ_m) is determined not only by the magnitude of the applied voltage (which determines the slope of the flux/time plot), but also by the frequency (which determines for how long the positive slope continues). Normally we want to utilise the full capacity of the magnetic circuit, so we must adjust the voltage and frequency together to keep φ_m at its rated value. For example, in Figure 7.4(b) the voltage has been doubled, but so has the frequency, to keep φ_m the same as in Figure 7.4(a): if we had left the frequency the same, the flux would have tried to reach twice its rated value, and in the process the core would have become saturated, as explained in Chapter 1.

Turning now to the everyday situation in which the transformer is connected to a sinusoidal voltage given by $V_1 = \hat{V} \sin \omega t$, we can

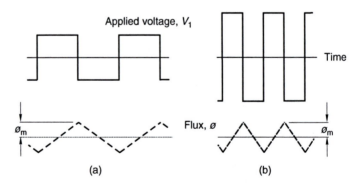

(a)   (b)

Figure 7.4   *Flux and voltage waveforms for ideal transformer operating with square-wave primary voltage. (The voltage and frequency in (b) are doubled compared with (a), but the peak flux remains the same.)*

integrate equation (7.3) to obtain an expression for the flux. The flux is then given by

$$\phi = -\frac{\hat{V}}{\omega N_1} \cos \omega t = \phi_{\mathrm{m}} \sin\left(\omega t - \frac{\pi}{2}\right) \tag{7.4}$$

where

$$\phi_{\mathrm{m}} = \frac{\hat{V}}{\omega N_1} = \frac{\hat{V}}{2\pi f N_1} \tag{7.5}$$

Typical primary voltage and flux waves are shown in Figure 7.5. A special feature of a sine function is that its differential (gradient) is basically the same shape as the function itself, i.e. differentiating a sine yields a cosine, which is simply a sine shifted by 90°; and similarly, as here, differentiating a (−)cosine wave of flux yields a sinewave of voltage.

Equation (7.5) shows that the amplitude of the flux wave is proportional to the applied voltage, and inversely proportional to the frequency. As mentioned above, we normally aim to keep the peak flux constant in order to fully utilise the magnetic circuit, and this means that changes to voltage or frequency must be done so that the ratio of voltage to frequency is maintained. This is shown in Figure 7.5 where, to keep the peak flux (φ_{m}) in Figure 7.5(a) the same when the frequency is doubled to that in

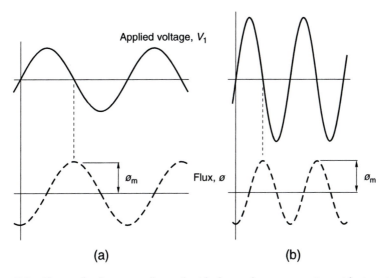

(a)　　　　　　　　　　　　　　(b)

Figure 7.5   *Flux and voltage waveforms for ideal transformer operating with sinusoidal primary voltage. (The voltage and frequency in (b) are doubled compared with (a), but the peak flux remains the same.)*

Figure 7.5(b), the voltage must also be doubled. We drew the same conclusion in relation to the induction motor in Chapters 5 and 6.

Although equation (7.5) was developed for an ideal transformer, it is also applicable with very little error to the real transformer, and is in fact a basic design equation. For example, suppose we have a transformer core with a cross-sectional area 5 cm × 5 cm, and we decide we want to use it as a 240 V, 50 Hz mains transformer. How many turns will be required on the primary winding?

We can assume that, as discussed in Chapter 1, the flux density in the core will have to be limited to say 1.4 T to avoid saturation. Hence the peak flux in the core is given by

$$\phi_m = B_m \times A = 1.4 \times 0.05 \times 0.05 = 3.5 \times 10^{-3} \text{ Wb} = 3.5 \text{ m Wb}$$

The peak voltage ($\hat{V}$) is the r.m.s (240) multiplied by $\sqrt{2}$; the frequency (f) is 50, so we can substitute these together with ϕ_m in equation (7.5) to obtain the number of turns of the primary winding as

$$N_1 = \frac{240\sqrt{2}}{2\pi \times 50 \times 3.5 \times 10^{-3}} = 308.7 \text{ turns}$$

We cannot have a fraction of a turn, so we choose 309 turns for the primary winding. If we used fewer turns the flux would be too high, and if we used more, the core would be under-utilised.

The important message to take from this analysis is that under sinusoidal conditions at a fixed frequency, the flux in a given transformer is determined by the applied voltage. Interestingly, however, the only assumption we have to make to arrive at this result is that the resistance of the windings is zero: the argument so far is independent of the magnetic circuit, so we must now see how the reluctance of the transformer core makes its presence felt.

We know that although the amplitude of the flux waveform is determined by the applied voltage, frequency and turns, there will need to be an MMF (i.e. a current in the primary winding) to drive the flux around the magnetic circuit: the magnetic Ohm's law tells us that the MMF required to drive flux φ around a magnetic circuit that has reluctance R is given by MMF $= R\varphi$.

But in this section we are studying an ideal transformer, so we can assume that the magnetic circuit is made of infinitely permeable material, and therefore has zero reluctance. This means that no MMF is required, so the current drawn from the supply (the 'magnetising current', I_m) in the ideal transformer is zero.

To sum up, the flux in the ideal transformer is determined by the applied voltage, and the no-load current is zero. This hypothetical situation is never achieved in practice, but real transformers (especially large ones) come close to it.

Viewed from the supply, the ideal transformer at no-load looks like an open circuit, as it draws no current. We will see later that a real transformer at no-load draws a small current, lagging the applied voltage by almost 90°, and that from the supply viewpoint it therefore has a high inductive reactance, known for obvious reasons as the 'magnetising reactance'. An ideal transformer is thus seen to have an infinite magnetising reactance.

Ideal transformer – no-load condition, voltage ratio

We now consider the secondary winding to be restored, but leave it disconnected from the load so that its current is zero, in which case it can clearly have no influence on the flux. Because the magnetic circuit is perfect, none of the flux set up by the primary winding leaks out, and all of it therefore links the secondary winding. We can therefore apply Faraday's law and make use of equation (7.3) to obtain the secondary induced e.m.f. as

$$E_2 = N_2 \frac{d\phi}{dt} = N_2 \frac{V_1}{N_1} = \frac{N_2}{N_1} V_1 \qquad (7.6)$$

There is no secondary current, so there is no volt-drop across R_2 and therefore the secondary terminal voltage V_2 is equal to the induced e.m.f. E_2. Hence the voltage ratio is given by

$$\frac{V_1}{V_2} = \frac{N_1}{N_2} \qquad (7.7)$$

This equation shows that any desired secondary voltage can be obtained simply by choosing the number of turns on the secondary winding. For example, if we wish to obtain a secondary voltage of 28 V in the mains transformer discussed in the previous section, the number of turns of the secondary winding is given by

$$N_2 = \frac{V_2}{V_1} N_1 = \frac{28}{240} \times 309 = 36 \text{ turns.}$$

It is worth mentioning that equation (7.7) applies regardless of the nature of the waveform, so if we apply a square wave voltage to the primary, the secondary voltage would also be square wave with amplitude scaled according to equation (7.7).

We will see later that when the transformer supplies a load the primary and secondary currents are inversely proportional to their respective voltages: so if the secondary voltage is lower than the primary there will be fewer secondary turns but the current will be higher and therefore the cross-sectional area of the wire used will be greater. The net result is that the total volumes of copper in primary and secondary are virtually the same, as is to be expected since they both handle the same power.

Ideal transformer on load

We now consider what happens when we connect the secondary winding to a load impedance Z_2. We have already seen that the flux is determined solely by the applied primary voltage, so when current flows to the load it can have no effect on the flux, and hence because the secondary winding resistance is zero, the secondary voltage remains as it was at no-load, given by equation (7.7). (If the voltage were to change when we connected the load we could be forgiven for beginning to doubt the validity of the description 'ideal' for such a device!)

The current drawn by the load will be given by $I_2 = V_2/Z_2$, and the secondary winding will therefore produce an MMF of $N_2 I_2$ acting around the magnetic circuit. If this MMF went unchecked it would tend to reduce the flux in the core, but, as we have seen, the flux is determined by the applied voltage, and cannot change. We have also seen that because the core is made of ideal magnetic material it has no reluctance, and therefore the resultant MMF (due to both the primary winding and the secondary winding) is zero. These two conditions are met by the primary winding drawing a current such that the sum of the primary MMF and the secondary MMF is zero, i.e.

$$N_1 I_1 + N_2 I_2 = 0, \quad \text{or} \quad \frac{I_1}{I_2} = -\frac{N_2}{N_1} \tag{7.8}$$

In other words, as soon as the secondary current begins to flow, a primary current automatically springs up to neutralise the demagnetising effect of the secondary MMF.

The minus sign in equation (7.8) serves to remind us that primary and secondary MMFs act in opposition. It has no real meaning until we define what we mean by the positive direction of winding the turns around the core, so because we are not concerned with transformer manufacture we can afford to ignore it from now on.

The current ratio in equation (7.8) is seen to be the inverse of the voltage ratio in equation (7.7). We could have obtained the current ratio by a different approach if we had argued from a power basis, by saying that in an ideal transformer, the instantaneous input power and the instantaneous output power must be equal. This would lead us to conclude that $V_1 I_1 = V_2 I_2$, and hence from equation (7.7) that

$$\frac{V_1}{V_2} = \frac{N_1}{N_2} = \frac{I_2}{I_1} \qquad (7.9)$$

To conclude our look at the ideal transformer, we should ask what the primary winding of an ideal transformer 'looks like' when the secondary is connected to a load impedance Z_2. As far as the primary supply is concerned, the apparent impedance looking into the primary of the transformer is simply the ratio V_1/I_1, which can be expressed in secondary terms as

$$\frac{V_1}{I_1} = \frac{(N_1/N_2)V_2}{(N_2/N_1)I_2} = \left(\frac{N_1}{N_2}\right)^2 \frac{V_2}{I_2} = \left(\frac{N_1}{N_2}\right)^2 Z_2 = Z_2' \qquad (7.10)$$

So, when we connect an impedance Z_2 to the secondary, it appears from the primary side as if we have connected an impedance Z_2' across the primary terminals. This equivalence is summed-up diagrammatically in Figure 7.6.

The ideal transformer effectively 'scales' the voltages by the turns ratio and the currents by the inverse turns ratio, and from the point of view of the input terminals, the ideal transformer and its secondary load (inside the shaded area in Figure 7.6(a)) is indistinguishable from the circuit in Figure 7.6(b), in which the impedance Z_2' (known as the 'referred' impedance) is connected across the supply. We should note that when we use referred impedances, the equivalent circuit of the ideal transformer simply reduces to a link between primary and referred secondary circuits: this point has been stressed by showing the primary and secondary terminals in Figure 7.6, even though there is nothing between them. We will make use of the idea of referring impedance from secondary to primary when we model the imperfections of the real transformer, and then we will find that there are circuit elements between the input (V_1) and output (V_2') terminals (Figure 7.10).

Finally, we should note that although both windings of an ideal transformer have infinite inductance, there is not even a vestige of

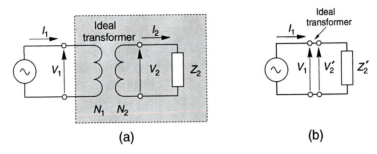

Figure 7.6   *Equivalent circuits of ideal transformer*

inductance in the equivalent circuit. This remarkable result is due to the perfect magnetic coupling between the windings. As we will see shortly, the real transformer can come very close to the ideal, but for reasons that will also become apparent, ultimate perfection is not usually what we seek.

THE REAL TRANSFORMER

We turn now to the real transformer, with the aim of developing its equivalent circuit. Real transformers behave much like ideal ones (except in very small sizes), and the approach is therefore to extend the model of the real transformer to allow for the imperfections of the real one.

For the sake of completeness we will establish the so-called 'exact' equivalent circuit first on a step-by-step basis. As its name implies the exact circuit can be used to predict all aspects of transformer performance, but we will find that it looks rather fearsome, and is not well adapted to offering simple insights into transformer behaviour. Fortunately, given that we are not seeking great accuracy, we can retreat from the complexity of the exact circuit and settle instead for the much less daunting 'approximate' circuit which is more than adequate for yielding answers to the questions we need to pursue, not only for the transformer itself but also when we model the induction motor.

Real transformer – no-load condition, flux and magnetising current

In modelling the real transformer at no-load we take account of the finite resistances of the primary and secondary windings; the finite reluctance of the magnetic circuit; and the losses due to the pulsating flux in the iron core.

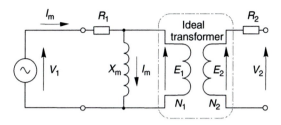

Figure 7.7 *Equivalent circuit of real transformer under no-load conditions, allowing for presence of magnetising current and winding resistances*

The winding resistances are included by adding resistances R_1 and R_2, respectively in series with the primary and secondary of the ideal transformer, as shown in Figure 7.7.

The transformer in Figure 7.7 is shown without a secondary load i.e. it is under no-load conditions. As previously explained the current drawn from the supply in this condition is known as the magnetising current, and is therefore denoted by I_m.

Because the ideal transformer (within the dotted line in Figure 7.7) is on open-circuit, the secondary current is zero and so the primary current must also be zero. So to allow for the fact that the real transformer has a no-load magnetising current we add an inductive branch (known as the 'magnetising reactance') in parallel with the primary of the ideal transformer, as shown in Figure 7.7. In circuit theory terms, the magnetising reactance is simply the reactance due to the self-inductance of the primary of the transformer. Its value reflects the magnetising current, as discussed later.

We now need to recall that when we apply Kirchoff's law to the primary winding, we obtain equation (7.1), which includes the term $I_m R_1$, the volt-drop across the primary resistance due to the magnetising current. It turns out that when a transformer is operated at its rated voltage, the term $I_m R_1$ is very much less than V_1, so we can make the approximation $V_1 \approx E_1$, i.e. in a real transformer the induced e.m.f. is equal to the applied voltage, just as in the ideal transformer. This is a welcome news as it allows us to use the simple design equation linking flux, voltage and turns (developed in Section 7.3.1 for the ideal transformer) for the case of the real transformer. More importantly, we can continue to say that the flux will be determined by the applied voltage, and it will not depend on the reluctance of the magnetic circuit.

However, the magnetic circuit of the real transformer clearly has some reluctance, so it is to be expected that it will require a current to provide the MMF needed to set up the flux in the core. If the reluctance is $\mathcal{R}$ and the peak flux is ϕ_m, the magnetic Ohm's law gives

$$\text{MMF} = N_1 I_m = \mathcal{R}\,\phi_m, \quad \therefore I_m = \frac{\mathcal{R}\,\phi_m}{N_1} \qquad (7.11)$$

The magnetising reactance X_m is given by

$$X_m = \frac{V_1}{I_m} \qquad (7.12)$$

In most transformers the reluctance of the core is small, and as a result the magnetising current (I_m) is much smaller than the normal full-load (rated) primary current. The magnetising reactance is therefore high, and we will see that it can be neglected for many purposes.

However, it was pointed out in Section 7.2 that the induction motor (our ultimate goal!) resembles a transformer with two air-gaps in its magnetic circuit. If we were to simulate the motor magnetic circuit by making a couple of saw-cuts across our transformer core, the reluctance of the magnetic circuit would clearly be increased because, relative to iron, air is a very poor magnetic medium. Indeed, unless the saw-blade was exceptionally thin we would probably find that the reluctance of the two air-gaps greatly exceeded the reluctance of the iron core.

We might have thought that making saw-cuts and thereby substantially increasing the reluctance would reduce the flux in the core, but we need to recall that provided the term $I_m R_1$ is very much less than V_1, the flux is determined by the applied voltage, and the magnetising current adjusts to suit, as given by equation (7.11). So when we make the saw-cuts and greatly increase the reluctance, there is a compensating large increase in the magnetising current, but the flux remains virtually the same.

We can see that there must be a *small* reduction in the flux when we increase the reluctance by rearranging equation (7.1) in the form:

$$I_m = \frac{V_1 - E_1}{R_1}$$

From which we can see that the only way that I_m can increase to compensate for an increase in the reluctance is for E_1 to reduce. However, since E_1 is very nearly equal to V_1, a very slight reduction in E_1 (and hence a very slight reduction in flux) is sufficient to yield the required large increase in the magnetising current.

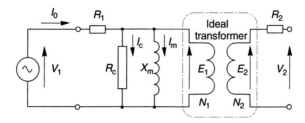

Figure 7.8   *Development of no-load equivalent circuit to include presence of losses in iron core*

The final refinement of the no-load model takes account of the fact that the pulsating flux induces eddy current and hysteresis losses in the core. The core is laminated to reduce the eddy current losses, but the total loss is often significant in relation to overall efficiency and cooling, and we need to allow for it in our model. The iron losses depend on the square of the peak flux density, and they are therefore proportional to the square of the induced e.m.f. This means that we can model the losses simply by including a resistance (R_c – where the suffix denotes 'core loss') in parallel with the magnetising reactance, as shown in Figure 7.8.

The no-load current (I_0) consists of the reactive magnetising component (I_m) lagging the applied voltage by 90°, and the core-loss (power) component (I_c) that is in phase with the applied voltage. The magnetising component is usually much greater than the core-loss component, so the real transformer looks predominantly reactive under open-circuit (no-load) conditions.

Real transformer – leakage reactance

In the ideal transformer it was assumed that all the flux produced by the primary winding linked the secondary, but in practice some of the primary flux will exist outside the core (see Figure 1.7) and will not link with the secondary. This leakage flux, which is proportional to the primary current, will induce a voltage in the primary winding whenever the primary current changes, and it can therefore be represented by a 'primary leakage inductance' (l_1) in series with the primary winding of the ideal transformer. We will normally be using the equivalent circuit under steady-state operating conditions at a given frequency (f), in which case we are more likely to refer to the 'primary leakage reactance' ($X_1 = \omega l_1 = 2\pi f l_1$) as in Figure 7.9.

The induced e.m.f. in the secondary winding arising from the secondary leakage flux (and thus proportional to the secondary current) is modelled

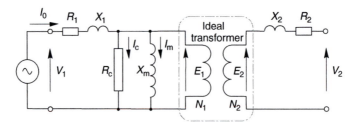

Figure 7.9   *Development of no-load equivalent circuit to include leakage reactances*

in a similar way, and this is represented by the 'secondary leakage react-ance' (X_2) shown in Figure 7.9.

It is important to point out that under no-load conditions when the primary current is small the volt-drop across X_1 will be much less than V_1, so all our previous arguments relating to the relationship between flux, turns and frequency can be used in the design of the real transformer.

Because the leakage reactances are in series, their presence is felt when the currents are high, i.e. under normal full-load conditions, when the volt-drops across X_1 and X_2 may not be negligible, and especially under extreme conditions (e.g. when the secondary is short-circuited) when they provide a vitally important current-limiting function.

Looking at Figure 7.9 the conscientious reader could be forgiven for beginning to feel a bit downhearted to discover how complex matters appear to be getting, so it will come as good news to learn that we have reached the turning point. The circuit in Figure 7.9 represents all aspects of transformer behaviour if necessary, but it is more elaborate than we need, so from now on things become simpler, particularly when we take the usual step and refer all secondary parameters to the primary.

Real transformer on load – exact equivalent circuit

The equivalent circuit showing the transformer supplying a secondary load impedance Z_2 is shown in Figure 7.10(a). This diagram has been annotated to show how the ideal transformer at the centre imposes the relationships between primary and secondary currents. Provided that we know the values of the transformer parameters we can use this circuit to calculate all the voltages, currents and powers when either the primary or secondary voltage are specified.

However, we seldom use the circuit in this form, as it is usually much more convenient to work with the referred impedances. We saw in Section 7.3.3 that, from the primary viewpoint, we could model the

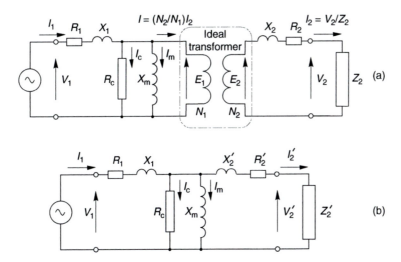

Figure 7.10 'Exact' equivalent circuit of real transformer supplying load impedance Z_2. (The circuit in (a) includes the actual secondary parameters, while in (b) the parameters have been 'referred' to the primary side.)

combination of ideal transformer and its load by 'referring' the second-ary impedance to the primary. We saw that an impedance Z_2 connected to the secondary could be modelled by a referred impedance Z_2' con-nected at the primary, where $Z_2' = (N_1/N_2)^2 Z_2$.

We can use the same approach to refer not only the secondary load impedance Z_2 but also the secondary resistance and leakage reactance of the transformer itself, as shown in Figure 7.10(b). As before, the referred or effective load impedance, secondary resistance and leakage reactance as seen at the primary are denoted by Z_2', R_2' and X_2', respectively. When circuit calculations have been done using this circuit the actual second-ary voltage and current are obtained from their referred counterparts using the equations

$$V_2 = V_2' \times \frac{N_2}{N_1} \quad \text{and} \quad I_2 = I_2' \times \frac{N_1}{N_2} \tag{7.13}$$

At this point we should recall the corresponding circuit of the ideal transformer, as shown in Figure 7.6(b). There we saw that the referred secondary voltage V_2' was equal to the primary voltage V_1. For the real transformer, however, (Figure 7.10(b)) we see that the imperfections of the transformer are reflected in circuit elements lying between the input voltage V_1 and the secondary voltage V_2', i.e. between the input and output terminals.

In the ideal transformer the series elements (resistances and leakage reactances) are zero, while the parallel elements (magnetising reactance and core-loss resistance) are infinite. The principal effect of the non-zero series elements of the real transformer is that because of the volt-drops across them, the secondary voltage V_2' will be less than it would be if the transformer were ideal. And the principal effect of the parallel elements is that the real transformer draws a (magnetising) current and consumes (a little) power even when the secondary is open-circuited.

Although the circuit of Figure 7.10(b) represents a welcome simplification compared with that in Figure 7.10(a), calculations are still cumbersome, because for a given primary voltage, every volt-drop and current changes whenever the load on the secondary alters. Fortunately, for most practical purposes a further gain in terms of simplicity (at the expense of only a little loss of accuracy) is obtained by simplifying the 'exact' circuits in Figure 7.10 to obtain the so-called 'approximate' equivalent circuit.

Real transformer – approximate equivalent circuit

The justification for the approximate equivalent circuit (see Figure 7.11(a)) rests on the fact that for all transformers except very small ones (i.e. for all transformers that would cause serious harm if they fell on ones foot) the series elements in Figure 7.10(b) are of low impedance and the parallel elements are of high impedance.

Actually, to talk of low or high impedances without qualification is nonsense. What the rather loose language in the paragraph above really means is that under normal conditions, the volt-drop across R_1 and X_1 will be a small fraction of the applied voltage V_1. Hence the voltage across the magnetising branch (X_m in parallel with R_c) is almost equal to V_1: so by moving the magnetising branch to the left-hand side, the magnetising current and the core-loss current will be almost unchanged.

Moving the magnetising branch brings massive simplification in terms of circuit calculations. Firstly because the current and power in the magnetising branch are independent of the load current, and secondly because primary and referred secondary impedances now carry the same current (I_2') so it is easy to calculate V_2' from V_1 or vice-versa.

Further simplification results if we ignore the resistance (R_c) that models iron losses in the magnetic circuit, and we combine the primary and secondary resistances and leakage reactances to yield total resistance (R_T) and total leakage reactance (X_T) as shown in Figure 7.11(b).

The justification for ignoring the iron-loss resistance is that the eddy current and hysteresis losses in the iron are only of interest when we are

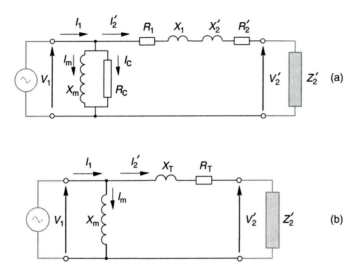

Figure 7.11   *'Approximate' equivalent circuit of real transformer*

calculating efficiency, which is not our concern here, and in most transformers the associated current (I_c) is small, so ignoring it has little impact on the total input current.

Given the difference in the numbers of turns and wire thickness between primary and secondary windings it is not surprising that the actual primary and secondary resistances are sometimes very different. But when referred to the primary, the secondary resistance is normally of similar value to the primary resistance, so both primary and secondary contribute equally to the total effective resistance. This is in line with commonsense: since both windings contain the same quantities of copper and handle the same power, both would be expected to have similar influence on the overall performance.

When leakage reactances are quoted separately for primary and secondary they too may have very different ohmic values, but again when the secondary is referred its value will normally be the same as that of the primary. (In fact, the validity of regarding leakage reactance as something that can be uniquely apportioned between primary and secondary is questionable, as they cannot be measured separately, and for most purposes all that we need to know is the total leakage reactance X_T.)

The equivalent circuit in Figure 7.11(b) will serve us well when we extend the treatment to the induction motor. It is delightfully simple, having only three parameters, i.e. the magnetising reactance X_m, the total leakage reactance X_T and the total effective resistance R_T.

Measurement of parameters

Two simple tests are used to measure the transformer parameters – the open-circuit or no-load test and the short-circuit test. We will see later that very similar tests are used to measure induction motor parameters.

In the open-circuit test, the secondary is left disconnected and with rated voltage (V_1) applied to the primary winding, the input current (I_0) and power (W_0) are measured.

If we are seeking values for the three-element model in Figure 7.11(b), we would expect no power at no load, because the only circuit is that via the magnetising reactance X_m. So if, as is usually the case, the no-load input power W_0 is very much less than $V_1 I_0$ (i.e. the transformer looks predominantly reactive) we assume that the input current I_0 is entirely magnetising current and deduce the value of the magnetising reactance from

$$X_m = \frac{V_1}{I_0} \tag{7.14}$$

(If we want to use the more accurate model in Figure 7.11(a), we note that the input power will be solely that due to the iron loss in the resistor R_c. Hence $W_0 = V_1^2/R_c$, i.e. $R_c = V_1^2/W_0$.)

In the short-circuit test the secondary terminals are shorted together and a low voltage is applied to the primary – just sufficient to cause rated current in both windings. The voltage (V_{sc}), current (I_{sc}) and input power (W_{sc}) are measured.

In this test, X_m is in parallel with the impedance Z_T (i.e. the series combination of X_T and R_T), so the input current divides between the two paths. However, the reactance X_m is invariably very much larger than the impedance Z_T, so we usually neglect the small current through X_m and assume that the current I_{sc} flows only through Z_T.

The resistance can then be derived from the input power using the relationship:

$$W_{sc} = I_{sc}^2 \, R_T, \therefore R_T = \frac{W_{sc}}{I_{sc}^2} \tag{7.15}$$

And the leakage reactance can be derived from

$$Z_T = \frac{V_{sc}}{I_{sc}} = \sqrt{R_T^2 + X_T^2} \tag{7.16}$$

To split the resistances, the primary and/or secondary winding resistances are measured under d.c. conditions.

A great advantage of the open-circuit and short-circuit tests is that neither require more than a few per cent of the rated power, so both can be performed from a modestly rated supply.

Significance of equivalent circuit parameters

If the study of transformers was our aim, we would now turn to some numerical examples to show how the equivalent circuit was used to predict performance: for example, we might want to know how much the secondary voltage dropped when we applied the load, or what the efficiency was at various loads. But our aim is to develop an equivalent circuit to illuminate induction motor behaviour, not to become experts at transformer analysis, so we will avoid quantitative study at this point. On the other hand, as we will see, the similarities with the induction motor circuit mean that there are several qualitative observations that are worth making at this point.

Firstly, if it were possible to choose whatever values we wished for the parameters, we would probably look back at the equivalent circuit of the ideal transformer, and try to emulate it by making the parallel elements infinite, and the series elements zero. And if we only had to think about normal operation (from no-load up to rated load) this would be a sensible aim, as there would then be no magnetising current, no iron losses, no copper losses and no drop in secondary voltage when the load is applied.

Most transformers do come close to this ideal when viewed from an external perspective. At full load the input power is only slightly greater than the output power (because the losses in the resistive elements are a small fraction of the rated power); and the secondary voltage falls only slightly from no-load to full-load conditions (because the volt-drop across Z_T at full-load is only a small fraction of the input voltage).

But in practice we have to cope with the abnormal as well as the normal, and in particular we need to be aware of the risk of a short-circuit at the secondary terminals. If the transformer primary is supplied from a constant-voltage source, the only thing that limits the secondary current if the secondary is inadvertently short-circuited is the series impedance Z_T. But we have already seen that the volt-drop across Z_T at rated current is only a small fraction of the rated voltage, which means that if rated voltage is applied across Z_T, the current will be many times a rated value. So the lower the impedance of the transformer, the higher the short-circuit current.

From the transformer viewpoint, a current of many times rated value will soon cause the windings to melt, to say nothing of the damaging inter-turn electromagnetic forces produced. And from the supply viewpoint the fault current must be cleared quickly to avoid damaging the supply cables, so a circuit breaker will be required with the ability to interrupt the short-circuit current. The higher the fault current, the higher the cost of the protective equipment, so the supply authority will normally specify the maximum fault current that can be tolerated.

As a transformer designer we then face a dilemma: if we do what seems obvious and aim to reduce the series impedance to improve the steady-state performance, we find that the short-circuit current increases and there is the problem of exceeding the permissible levels dictated by the supply authority. Clearly the so-called ideal transformer is not what we want after all, and we are obliged to settle for a compromise. We deliberately engineer sufficient impedance (principally via the control of leakage reactance) to ensure that under abnormal (short-circuit) conditions the system does not self-destruct!

In Chapters 5 and 6, we mentioned something very similar in relation to the induction motor: we saw that when the induction motor is started direct-on-line it draws a heavy current, and limiting the current requires compromises in the motor design. We will see shortly that this behaviour is exactly like that of a short-circuited transformer.

DEVELOPMENT OF THE INDUCTION MOTOR EQUIVALENT CIRCUIT

Stationary conditions

In Section 7.2 it was shown that, on a per-phase basis, the stationary induction motor is very much like a transformer, so to model the induction motor at rest we can use any of the transformer equivalent circuits we have looked at so far.

We can represent the motor at rest (the so-called 'locked rotor' condition) simply by setting $Z_2' = 0$ in Figure 7.11. (For a wound-rotor motor we would have to increase R_T to account for any external rotor-circuit resistance.) Given the applied voltage we can calculate the current and power that will be drawn from the supply, and if we know the effective turns ratio we can also calculate the rotor current and the power being supplied to the rotor.

But although our induction motor resembles a transformer, its purpose in life is very different because it is designed to convert electrical power to mechanical power, which of necessity involves movement. Our locked-rotor calculations will therefore be of limited use unless we can calculate the starting torque developed. Far more importantly, we need to be able to represent the electromechanical processes that take place over the whole speed range, so that we can predict the input current, power and developed torque at any speed. The remarkably simple and effective way that this can be achieved is discussed next.

Modelling the electromechanical energy conversion process

In Chapters 5 and 6 we saw that the behaviour of the motor was determined primarily by the slip. In particular we saw that if the motor was unloaded, it would settle at almost the synchronous speed (i.e. with a very small slip), with very little induced rotor current, at very low frequency. As the load torque was increased the rotor slowed relative to the travelling flux; the magnitude and frequency of the induced rotor currents increased; the rotor thereby produced more torque; and the stator current and power drawn from the supply automatically increased to furnish the mechanical output power.

A very important observation in relation to what we are now seeking to do is that we recognised earlier that although the rotor currents were at slip frequency, their effect (i.e. their MMF) was always reflected back at the stator windings at the supply frequency. This suggests that it must be possible to represent what takes place at slip-frequency on the rotor by referring the action to the primary (fixed-frequency) side, using a transformer-type model; and it turns out that we can indeed model the entire energy-conversion process in a very simple way. All that is required is to replace the referred rotor resistance (R_2') with a fictitious slip-dependent resistance (R_2'/s) in the short-circuited secondary of our transformer equivalent circuit.

Hence if we build from the exact transformer circuit in Figure 7.10(b), we obtain the induction motor equivalent circuit shown in Figure 7.12.

At any given slip, the power delivered to this 'load' resistance represents the power crossing the air-gap from rotor to stator. We can see straightaway that because the fictitious load resistance is inversely proportional to slip, it reduces as the slip increases, thereby causing the power across the air-gap to increase and resulting in more current and power being drawn in from the supply. This behaviour is of course in line with what we already know about the induction motor.

We will see how to use the equivalent circuit to predict and illuminate motor behaviour in the next section, but first there are two points worth making.

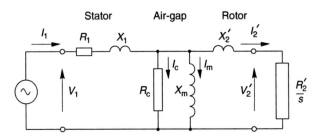

Figure 7.12 *'Exact' per-phase equivalent circuit of induction motor. The secondary (rotor) parameters have been referred to the primary (stator) side*

Firstly, given the complexity of the spatial and temporal interactions in the induction motor it is extraordinary that everything can be properly represented by such a simple equivalent circuit, and it has always seemed a pity to the author that something so elegant receives little by way of commendation in the majority of textbooks.

Secondly, the following brief discussion is offered for the benefit of readers who are seeking at least some justification for introducing the fictitious resistance R'_2/s, though it has to be admitted that full treatment is beyond our scope. Pragmatists who are content to accept that the method works can jump to the next section.

The key to developing the representation lies in ensuring that the magnitude and phase of the referred rotor current (at supply frequency) in the transformer model is in agreement with the actual current (at slip frequency) in the rotor. We argued in Chapter 5 that the induced e.m.f. in the rotor at slip s would be sE_2 at frequency sf, where E_2 is the e.m.f. induced under locked rotor ($s = 1$) conditions, when the rotor frequency is the same as the supply frequency, i.e. f. This e.m.f. acts on the series combination of the rotor resistance R_2 and the rotor leakage reactance, which at frequency sf is given by sX_2, where X_2 is the rotor leakage reactance at supply frequency. Hence the magnitude of the rotor current is given by

$$I_2 = \frac{sE_2}{\sqrt{R_2^2 + s^2 X_2^2}} \tag{7.17}$$

In the supply-frequency equivalent circuit, e.g. Figure 7.9, the secondary e.m.f. is E_2, rather than sE_2, so to obtain the same current in this model as given by equation (7.17), we require the slip-frequency rotor resistance and reactance to be divided by s, in which case the secondary current would be correctly given by

$$I_2 = \frac{E_2}{\sqrt{(R_2/s)^2 + X^2}} = \frac{sE_2}{\sqrt{R_2^2 + s^2 X_2^2}} \qquad (7.18)$$

PROPERTIES OF INDUCTION MOTORS

We have started with the exact circuit in Figure 7.12 because the air-gap in the induction motor causes its magnetising reactance to be lower than a transformer of similar rating, while its leakage reactance will be higher. We therefore have to be a bit more cautious before we make major simplifications, though we will find later that for many purposes the approximate circuit (with the magnetising branch on the left) is actually adequate. In this section, we concentrate on what can be learned from a study of the rotor section, which is the same in exact and approximate circuits, so our conclusions from this section are completely general.

Of the power that is fed across the air-gap into the rotor, some is lost as heat in the rotor resistance, and the remainder (hopefully the majority!) is converted to useful mechanical output power. To represent this in the referred circuit we split the fictitious resistance R_2'/s into two parts, R_2' and $R_2'((1-s)/s)$, as shown in Figure 7.13.

The rotor copper loss is represented by the power in R_2', and the useful mechanical output power is represented by the power in the 'electro-mechanical' resistance $R_2'((1-s)/s)$, shown shaded in Figure 7.13. To further emphasise the intended function of the motor – the production of mechanical power – the electromechanical element is shown in Figure 7.13 as the secondary 'load' would be in a transformer. For the motor to be a good electromechanical energy converter, most of the power entering the circuit on the left must appear in the electromechanical load resistance. This is equivalent to saying that for good performance, the output voltage V_2' must be as near as possible to the input voltage, V_1. And if we ignore the current in the centre magnetising branch, the 'good' condition simply requires that the load resistance is large compared with the other series elements. This desirable condition is met under normal running condition, when the slip is small and hence $R_2'((1-s)/s)$ is large.

Our next step is to establish some important general formulae, and draw some broad conclusions.

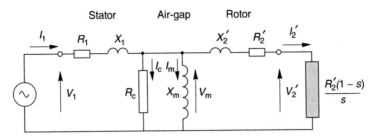

Figure 7.13 *Exact equivalent circuit with effective rotor resistance (R'_2/s) split into R'_2 and $R'_2((1-s)/s)$. The power dissipated in R'_2 represents the rotor copper loss per-phase, while the power in the shaded resistance $R'_2((1-s)/s)$ corresponds to the mechanical output power per-phase, when the slip is s*

Power balance

The power balance for the rotor can be derived as follows:

Power into rotor (P_2) = Power lost as heat in rotor conductors + Mechanical output power.

Hence using the notation in Figure 7.13 we obtain

$$P_2 = (I'_2)^2 R'_2 + (I'_2)^2 R'_2 \left(\frac{1-s}{s}\right) = (I'_2)^2 \frac{R'_2}{s} \qquad (7.19)$$

We can rearrange equation (7.19) to express the power loss and the mechanical output power in terms of the air-gap power, P_2 to yield

Power lost in rotor heating $= sP_2$

Mechanical output power $= (1-s)P_2$ $\qquad (7.20)$

Rotor efficiency $(P_{mech}/P_2) = (1-s) \times 100\%$

These relationships were mentioned in Chapters 5 and 6, and they are of fundamental importance and universal applicability. They show that of the power delivered across the air-gap fraction s is inevitably lost as heat, leaving the fraction $(1-s)$ as useful mechanical output. Hence an induction motor can only operate efficiently at low values of slip.

Torque

We can also obtain the relationship between the power entering the rotor and the torque developed. We know that mechanical power is torque times speed, and that when the slip is s the speed is $(1-s)\,\omega_s$, where ω_s is the synchronous speed. Hence from the power equations above we obtain

$$\text{Torque} = \frac{\text{Mechanical power}}{\text{Speed}} = \frac{(1-s)P_2}{(1-s)\omega_s} = \frac{P_2}{\omega_s} \qquad (7.21)$$

Again this is of fundamental importance, showing that the torque developed is proportional to the power entering the rotor.

All of the relationships derived in this section are universally applicable and do not involve any approximations. Further useful deductions can be made when we simplify the equivalent circuit, but first we will look at an example of performance prediction based on the exact circuit.

PERFORMANCE PREDICTION – EXAMPLE

The per-phase equivalent circuit parameters (referred to the stator) of a 4-pole, 60 Hz, 440 V three-phase delta-connected induction motor are as follows:

Stator resistance, $R_1 = 0.2\ \Omega$
Stator and rotor leakage reactances, $X_1 = X_2' = j1.0\ \Omega$
Rotor resistance, $R_2' = 0.3\ \Omega$
Magnetising reactance, $X_m = j40\ \Omega$
Iron-loss resistance, $R_c = 250\ \Omega$.

The mechanical frictional losses at normal speed amount to 2.5 kW.

We will calculate the input line current, the output power and the efficiency at the full-load speed of 1728 rev/min.

The per-phase equivalent circuit, with all values in Ohms, is shown in Figure 7.14. The synchronous speed is 1800 rev/min and the slip speed is thus $1800 - 1728 = 72$ rev/min, giving a slip of $72/1800 = 0.04$. The electromechanical resistance of 7.2 Ω is calculated from the expression $R_2'((1-s)/s)$, with $R_2' = 0.3\ \Omega$. Given that the power in this resistor

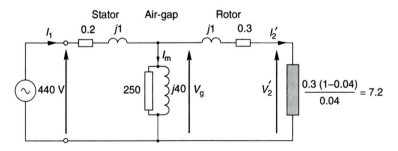

Figure 7.14 *Exact equivalent circuit for induction motor in Section 7.7, under full-load conditions (i.e. slip s = 0.04). All of the impedance values are in Ohms*

represents the mechanical output power, we will see that, as expected, it dominates our calculations.

Although the various calculations in this example are straightforward, they are not for the faint-hearted as they inevitably involve frequent manipulation of complex numbers, so most of the routine calculations will be omitted and we will concentrate on the results and their interpretation.

Line current

To find the line current we must find the effective impedance looking in from the supply, so we first find the impedances of the rotor branch and the magnetising (air-gap) branches. Expressed in 'real and imaginary' and also 'modulus and argument' form these are:

Rotor branch impedance, $Z_r = 7.5 + j1 = 7.57 \angle 7.6°$ Ω
Magnetising branch impedance, $Z_m = 6.25 + j39.0 = 39.5 \angle 80.9°$ Ω

The parallel combination of these two branches has an effective impedance of $6.74 + j2.12$ or $7.07 \angle 17.5°$. At roughly 7 Ω, and predominantly resistive, this is in line with what we would expect: the magnetising branch impedance is so much higher than the rotor branch that the latter remains dominant. We must now add the stator impedance of $0.2 + j1$ to yield the total motor impedance per phase as $6.94 + j3.12$, or $7.61 \angle 24.2°$ Ω.

The motor is mesh connected so the phase voltage is equal to the line voltage, i.e. 440 V. We will use the phase voltage as the reference for angles, so the phase current is given by

$$I_1 = \frac{440 \angle 0°}{7.61 \angle 24.2°} = 57.82 \angle -24.2° \text{ A.}$$

The line current is $\sqrt{3}$ times the phase current, i.e. 100 A.

Output power

The output power is the power converted to mechanical form, i.e. the power in the fictitious load resistance of 7.2 Ω, so we must first find the current through it.

The input current I_1 divides between the rotor and magnetising branches according to the expressions

$$I_2' = \frac{Z_m}{Z_m + Z_r} I_1 \quad \text{and} \quad I_m = \frac{Z_r}{Z_m + Z_r} I_1$$

Inserting values from above yields:

$$I_2' = 53.99\angle - 14.38°\,\text{A}$$
$$I_m = 10.35\angle - 87.7°\ \text{A}$$

Again we see that these currents line up with our expectations. Most of the input current flows in the rotor branch, which is predominantly resistive and the phase-lag is small; conversely relatively little current flows in the magnetising branch and its large phase-lag is to be expected from its predominantly inductive nature.

The power in the load resistor is given by

$P_{mech} = (I_2')^2 \times 7.2 = (53.99)^2 \times 7.2 = 20,987\,\text{W/phase}$.   There   are three phases so the total mechanical power developed is 62.96 kW.

The generated torque follows by dividing the power by the speed in rad/s, yielding torque as 348 Nm, or 116 Nm/phase.

We are told that the windage and friction power is 2.5 kW, so the useful output power is $62.96 - 2.5 = 60.46\,\text{kW}$.

Efficiency

We can proceed in two alternative ways, so we will do both to act as a check. Firstly, we will find the total loss per phase by summing the powers in the winding resistances and the iron-loss resistor in Figure 7.14.

The losses in the two winding resistances are easy to find because we already know the currents through both of them: the losses are $(I_1)^2 R_1 = (57.82)^2 \times 0.2 = 668$ W/phase, and $(I_2')^2 R_2' = (53.99)^2 \times 0.3 = 874.5$ W/phase for the stator and rotor, respectively.

The iron loss is a bit more tricky, as first we need to find the voltage (V_m) across the magnetising branch (sometimes called the gap voltage, V_g, as in Figure 7.14). We can get this from the product of the rotor current and the rotor impedance, which yields $V_g = 53.99\angle - 14.38° \times 7.57\angle 7.6° = 408.5\angle - 6.8$ V.

The power in the 250 Ω iron-loss resistor is now given by $(408.5)^2 / 250 = 667$ W/phase. The total loss (including the windage and friction) is then given by

$$\text{Total loss} = 3(668 + 874.5 + 667) + 2500 = 9.13\,\text{kW}.$$

The efficiency is given by

$$\text{Efficiency} = \frac{P_{mech}}{P_{in}} = \frac{P_{mech}}{P_{mech} + \text{losses}} = \frac{60.46}{60.46 + 9.13} = \frac{60.46}{69.59}$$
$$= 0.869 \text{ or } 86.9\%.$$

Alternatively we could have found the input power directly from the known input voltage and current, using $P_{in} = 3(V_1 I_1 \cos\phi_1) = 3 \times 440 \times 57.82 \times \cos 24.3° = 69.59\,\text{kW}$, which agrees with the result above.

Phasor diagram

It is instructive to finish by looking at the phasor diagrams showing the principal voltages and currents under full-load conditions, as shown in Figure 7.15, which is drawn to scale.

From the voltage phasor diagram, we note that in this motor the volt-drop due to the stator leakage reactance and resistance is significant, in that the input voltage of 440 V is reduced to 408.5 V, i.e. a reduction of just over 7%.

The voltage across the magnetising branch determines the magnetising current and hence the air-gap flux density, so in this motor the air-gap flux density will fall by about 7% between no-load and full-load. (At no-load the slip is very small and the impedance of the rotor branch is high, so that there is negligible rotor current and very little volt-drop due

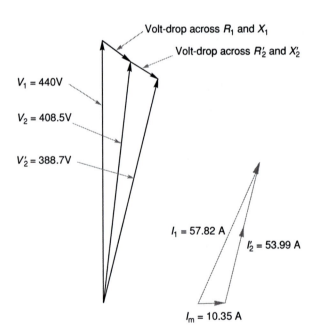

Figure 7.15 *Phasor diagrams, drawn to scale, for induction motor in Section 7.7 under full-load conditions*

to the stator impedance, leaving almost the full 440 V across the magnetising branch.)

A similar volt-drop occurs across the rotor impedance, leaving just under 390 V applied at the effective load. Note, however, that because the stator and rotor impedances are predominantly reactive, the appreciable volt-drops are not responsible for corresponding power losses.

These significant volt-drops reflect the fact that this motor is suitable for direct-on-line starting from a comparatively weak supply, where it is important to limit the starting current. The designer has therefore deliberately made the leakage reactances higher than they would otherwise be in order to limit the current when the motor is switched on.

To estimate the starting current we ignore the magnetising branch and put the slip $s = 1$, in which case the rotor branch impedance becomes $0.3 + j1$, so that the total impedance is $0.5 + j2$ or $2.06 \, \Omega$. When the full voltage is applied the phase current will be $440/2.06 = 214$ A. The full-load current is 57.8 A, so the starting current is 3.7 times the full-load current. This relatively modest ratio is required where the supply system is weak: a stiff supply might be happy with a ratio of 5 or 6.

We have already seen that the torque is directly proportional to the power into the rotor, so it should be clear that when we limit the starting current to ease the burden on the supply, the starting torque is unavoidably reduced. This is discussed further in the following section.

The current phasor diagram underlines the fact that when the motor is at full-load, the magnetising current is only a small fraction of the total. The total current lags the supply voltage by 24.3°, so the full-load power-factor is cos 24.3° $= 0.91$, i.e. very satisfactory.

APPROXIMATE EQUIVALENT CIRCUITS

This section is devoted to what can be learned from the equivalent circuit in simplified form, beginning with the circuit shown in Figure 7.16, in which the magnetising branch has been moved to the left-hand side. This makes calculations very much easier because the current and power in the magnetising branch are independent of the load branch. The approximation involved in doing this are greater than in the case of a transformer because for a motor the ratio of magnetising reactance to leakage reactance is lower, but algebraic analysis is much simpler and the results can be illuminating.

A cursory examination of electrical machines textbooks reveals a wide variety in the approaches taken to squeeze value from the study of the approximate equivalent circuit, but in the author's view there are often so many formulae that the reader becomes overwhelmed. So here we will

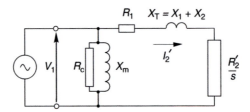

Figure 7.16   *Approximate equivalent circuit for induction motor*

focus on two simple messages. Firstly, we will develop an expression that neatly encapsulates the fundamental behaviour of the induction motor, and illustrates the trade-offs involved in design; and secondly we will examine how the relative values of rotor resistance and reactance influence the shape of the torque–speed curve.

Starting and full-load relationships

Straightforward circuit analysis of the circuit in Figure 7.16, together with equation (7.20) for the torque yields the following expressions for the load component of current (I_2') and for the torque per phase:

$$ I_2' = \frac{V_1}{\sqrt{\left(R_1 + \dfrac{R_2'}{s}\right)^2 + X_T^2}} \quad \text{and} \quad T = \frac{1}{\omega_s}\left\{ \frac{V_1^2}{\left(R_1 + \dfrac{R_2'}{s}\right)^2 + X_T^2} \right\}\frac{R_2'}{s} \tag{7.22} $$

The second expression (with the square of voltage in the numerator) reminds us of the sensitivity of torque to voltage variation, in that a 5% reduction in voltage gives a little over 10% reduction in torque.

If we substitute $s = 1$ and $s = s_{fl}$ in these equations we obtain expressions for the starting current, the full-load current, the starting torque and the full-load torque. Each of these quantities is important in its own right, and they all depend on the rotor resistance and reactance. But by combining the four expressions we obtain the very far-reaching result given by equation (7.23) below.

$$ \frac{T_{st}}{T_{fl}} = \left(\frac{I_{st}}{I_{fl}}\right)^2 s_{fl} \tag{7.23} $$

The left-hand side of equation (7.23) is the ratio of starting torque to full-load torque, an important parameter for any application as it is clearly no good having a motor that can drive a load once up to speed, but either has insufficient torque to start the load from rest, or (perhaps less likely) more starting torque than is necessary leading to a too rapid acceleration.

On the right-hand side of equation (7.23), the importance of the ratio of starting to full-load current has already been emphasised: in general it is desirable to minimise this ratio in order to prevent voltage regulation at the supply terminals during a direct-on-line start. The other term is the full-load slip, which, as we have already seen should always be as low as possible in order to maximise the efficiency of the motor.

The remarkable thing about equation (7.23) is that it neither contains the rotor or stator resistances, nor the leakage reactances. This underlines the fact that this result, like those given in equations (7.20) and (7.21), reflects fundamental properties that are applicable to any induction motor.

The inescapable design trade-off faced by the designer of a constant-frequency machine is revealed by equation (7.23). We usually want to keep the full-load slip as small as possible (to maximise efficiency), and for direct starting the smaller the starting current, the better. But if both terms on the right-hand side are small, we will be left with a low starting torque, which is generally not desirable, and we must therefore seek a compromise between the starting and full-load performances, as was explained in Chapter 6.

It has to be acknowledged that the current ratio in equation (7.23) relates the load (rotor branch) currents, the magnetising current having been ignored, so there is inevitably a degree of approximation. But for the majority of machines (i.e. 2-pole and 4-pole) equation (7.23) holds good, and in view of its simplicity and value it is surprising that it does not figure in many 'machines' textbooks.

Dependence of pull out torque on motor parameters

The aim here is to quantify the dependence of the maximum or pull-out torque on the rotor parameters, for which we make use of the simplest possible (but still very useful) model shown in Figure 7.17. The magnetising branch and the stator resistance are both ignored, so that there is only one current, the referred rotor current I_2' being the same as the stator current I_1. Of the two forms shown in Figure 7.17, we will focus on the one in Figure 7.17(b), in which the actual and fictitious rotor resistances are combined.

Most of the discussion will be based on normal operation, i.e. with a constant applied voltage at a constant frequency. In practice both leakage

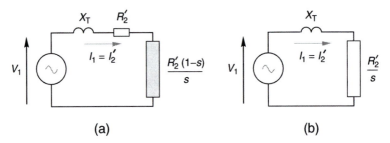

Figure 7.17   *Equivalent circuits for induction motor with magnetising branch neglected*

reactance and rotor resistance are parameters that the designer can control, but to simplify matters here we will treat the reactance as constant and regard the rotor resistance as a variable.

We will derive the algebraic relations first, then turn to a more illuminating graphical approach. To obtain the torque–slip relationship we will make use of equation (7.20), which shows that the torque is directly proportional to the total rotor power. Throughout this section our main concern will be with how motor behaviour depends on the slip, particularly over the motoring region from $s = 0$ to $s = 1$. (The treatment can easily be extended to cover the braking and generating regions, but they are not included here.)

Analysis

The total rotor power per phase is given by

$$P_2 = P_{\text{rotor}} = (I_2')^2 \frac{R_2'}{s} = \frac{V_1^2}{X_T^2 + (R_2'/s)^2} \frac{R_2'}{s} = V_1^2 \left(\frac{sR_2'}{R_2'^2 + s^2 X^2} \right) \quad (7.24)$$

The bracketed expression on the right-hand side of equation (7.24) indicates how the torque varies with slip.

At low values of slip (i.e. in the normal range of continuous operation) where sX is much less than the rotor resistance R_2', the torque becomes proportional to the slip and inversely proportional to the rotor resistance. This explains why the torque–speed curves we have seen in Chapters 5 and 6 are linear in the normal operating region, and why the curves become steeper as the rotor resistance is reduced.

At the other extreme, when the slip is 1 (i.e. at standstill), we usually find that the reactance is larger than the resistance, in which case the bracketed term in equation (7.24) reduces to R_2'/X^2. The starting torque is then proportional to the rotor resistance, a result also discussed in Chapters 5 and 6.

By differentiating the bracketed expression in equation (7.24) with respect to the slip, and equating the result to zero, we can find the slip at which the maximum torque occurs. The slip for maximum torque turns out to be given by

$$s = \frac{R_2'}{X_T} \tag{7.25}$$

(A circuit theorist could have written down this expression by inspection of Figure 7.17, provided that he knew that the condition for maximum torque was that the power in the rotor was at its peak.)

Substituting the slip for maximum torque in equation (7.24), and using equation (7.21) we obtain an expression for the maximum torque per phase as:

$$T_{max} = \frac{V_1^2}{\omega_s} \frac{1}{2X_T} \tag{7.26}$$

From these two equations we see that the slip at which maximum torque is developed is directly proportional to the rotor resistance, but that the peak torque itself is independent of the rotor resistance, and depends inversely on the leakage reactance.

Graphical interpretation via phasor diagram

We will look at the current phasor diagram as the slip is varied, for two motors, both having the same leakage reactance, X_T. One motor will be representative of the 'low-resistance' end of the scale ($R_2' = 0.1X_T$) while the other will represent the 'high-resistance' end ($R_2' = X_T$). As before, the voltage and frequency are constant throughout.

The reason for choosing total leakage reactance as the common factor linking the two motors is simply that the current loci (see below) are then very similar, and the peak torques are the same, which makes it easier to compare the shapes of the torque–slip curves.

The locus of the load current phasors as the slip varies is shown on the left-hand side of Figure 7.18, while on the right-hand side the torque–slip curves are shown.

The locus of current with slip is semicircular, the motoring condition extending only over the range from $s = 0$ to $s = 1$. The point corresponding to $s =$ infinite is important because it defines the diameter of the locus as V/X_T: it is shown by the dot on the horizontal axis. The

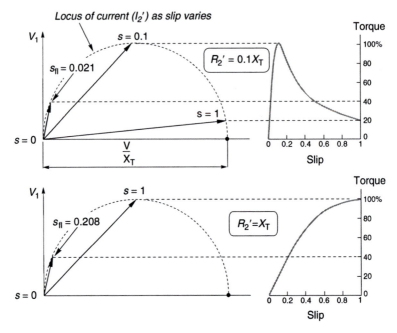

Figure 7.18   *Predicted current locus (circle diagram) and torque–speed curves for the two induction motors discussed in Section 7.8.2, based on the approximate equivalent circuit*

remainder of the circular locus (not shown) corresponds to negative slip, i.e. generator action.

The torque is proportional to total rotor power input, and the rotor power input is given by $P_{rot} = V_1 I_2' \cos \phi_r$, where ϕ_r is the rotor power-factor angle, i.e. the angle between the referred rotor current and the voltage. Since the voltage is constant and $I_2' \cos \phi_r$ is the 'in-phase' component of current, i.e. the vertical component in Figure 7.18, it follows that the torque is directly proportional to the projection of the current phasor onto the voltage, i.e. to the height of the tip of the current above the horizontal line through the point marking $s = 0$. This has been emphasised in Figure 7.18 by drawing the adjacent torque–slip curves to a scale calibrated in terms of the peak (pull-out) torque, the value of which is determined by the radius of the circle, as indicated in equation (7.26). These curves confirm the finding in equation (7.25) that the peak torque is reached when the slip is equal to the ratio of rotor resistance to reactance.

We note that although the high-resistance motor has a lower starting current, it produces a much greater starting torque because the current lags the voltage by only 45° compared with over 84° for the low-resistance motor. On the other hand the gradient of the torque–slip curve in the normal operating region is much steeper for the low-resistance motor, so its speed holding as the load changes is better and full-load efficiency can be correspondingly higher. As explained in Chapter 6, these widely differing characteristics indicate the need to match motor to application.

For the sake of completeness, full-load conditions have also been shown, based on an arbitrarily chosen specification that the full-load torque is 40% of the pull-out torque, i.e. $T_{po}/T_{fl} = 2.5$. With this constraint, the full-load slips and full-load currents can be calculated using equation (7.21), and thus we can draw up a table summarising the comparative performance of the two motors.

Rotor resistance	s_{f-l}	Full-load efficiency (%)	$\dfrac{I_{st}}{I_{f-l}}$	$\dfrac{T_{st}}{T_{f-l}}$	$\dfrac{T_{po}}{T_{f-l}}$
Low ($R_2' = 0.1X_T$)	0.021	97.9	4.85	0.5	2.5
High ($R_2' = X_T$)	0.208	79.2	3.47	2.5	2.5

The full-load points are marked on Figure 7.18: the continuous operating region (from no-load to full-load) occupies the comparatively short section of the locus betweens $s = 0$ (no-load) to the full-load slip in each case.

As already observed, the starting and low-speed performance of the high-resistance motor is superior, but its full-load rotor efficiency is very poor and rotor heating would prevent continuous operation at a slip as high as 21% (unless it was a slipring machine where most of the resistance was external). At normal running speeds the low resistance motor is superior, with its much greater rotor efficiency and steep torque–speed curve; but its starting torque is very low and it could only be used for fan-type loads, which do not require high starting torque.

As explained in Chapter 6, the double-cage rotor or the deep-bar rotor combines the merits of both high- and low-resistance rotors by having an outer cage with high resistance and relatively low reactance, in which most of the mains frequency starting current flows, and an inner cage of relatively low resistance and high reactance. The latter comes into effect as the speed rises and the rotor frequency reduces, so that at normal speed it becomes the dominant cage. The resulting torque–speed curves – shown typically in Figure 6.9 – offer good performance over the full speed range.

Equivalent circuits for double-cage motors come in a variety of guises, all using two parallel rotor branches, with a variety of methods for taking account of the significant mutual inductance between the cages.

Treatment of double-cage induction motor circuits is best left to more specialist textbooks.

MEASUREMENT OF PARAMETERS

The tests used to obtain the equivalent circuit parameters of a cage induction motor are essentially the same as those described for the transformer in Section 7.4.5.

To simulate the 'open-circuit' test we would need to run the motor with a slip of zero, so that the secondary (rotor) referred resistance (R_2'/s) became infinite and the rotor current was absolutely zero. But because the motor torque is zero at synchronous speed, the only way we could achieve zero slip would be to drive the rotor from another source, e.g. a synchronous motor. This is hardly ever necessary because when the shaft is unloaded the no-load torque is usually very small and the slip is sufficiently close to zero that the difference does not matter. From readings of input voltage, current and power (per phase) the parameters of the magnetising branch (see Figure 7.16) are derived as described earlier. The no-load friction and windage losses will be combined with the iron losses and represented in the parallel resistor – a satisfactory practice as long as the voltage and/or frequency remain constant.

The short-circuit test of the transformer becomes the locked-rotor test for the induction motor. By clamping the rotor so that the speed is zero and the slip is one, the equivalent circuit becomes the same as the short-circuited transformer. The total resistance and leakage reactance parameters are derived from voltage, current and power measurements as already described. With a cage motor the rotor resistance clearly cannot be measured directly, but the stator resistance can be obtained from a d.c. test and hence the referred rotor resistance can be obtained.

EQUIVALENT CIRCUIT UNDER VARIABLE-FREQUENCY CONDITIONS

The beauty of the conventional equivalent circuits we have explored is that everything happens at supply frequency, and there is no need for us to bother with the fact that in reality the rotor currents are at slip frequency. In addition there are well-established approximations that we can make to simplify analysis and calculations, such as moving the magnetising branch to the left-hand side, and ignoring the stator resistance in all calculations except efficiency.

But more and more induction motors now operate from variable-frequency inverters, the frequency (and voltage) being varied not only to control steady-state speed, but also to profile torque during acceleration and deceleration. Two questions that we might therefore ask are (a) does the equivalent circuit remain valid for other than mains frequencies; and (b) if so, do the approximations that have been developed for mains-frequency operation remain useful?

The answer to question (a) is that the form of the equivalent circuit is independent of the supply frequency. This is to be expected because we are simply representing in circuit form the linking of the electric and magnetic circuits that together constitute the motor, and these physical properties do not depend on the excitation frequency.

The answer to the question of the validity of approximations is less straightforward, but broadly speaking all that has so far been said is applicable except at low frequencies, say below about 10 Hz for 50 Hz or 60 Hz motors. As the frequency approaches zero, the volt-drop due to stator resistance becomes important, as the example below demonstrates.

Consider the example studied in Section 7.7, and suppose that we reduce the supply frequency from 60 to 6 Hz. All the equivalent circuit reactances reduce by a factor of 10, as shown in Figure 7.19, which assumes operation at the same (rated) torque for both cases. We can see immediately that whereas at 60 Hz the stator and rotor leakage reactances are large compared with their respective resistances, the reverse is true at 6 Hz, so we might immediately expect significant differences in

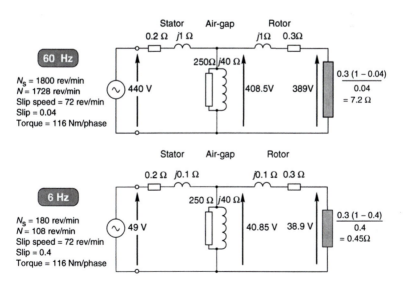

Figure 7.19 *Comparison of equivalent circuit parameters at 60 Hz and 6 Hz*

the circuit behaviour. (It should be pointed out that the calculations required to derive the voltages at 6 Hz are not trivial, as the reader may like to verify.)

We will see when we study variable-frequency speed control in Chapter 8 that to get the best out of the motor we will need the flux to be constant and the slip speed (in rev/min) to be the same when the motor is required to develop full-load torque.

To keep the air-gap flux constant we need to ensure that the magnetising current remains the same when we reduce the frequency. We know that at 6 Hz the magnetising reactance is one-tenth of its value at 60 Hz, so the voltage V_m must be reduced by a factor of 10, from 408.5 to 40.85 V, as shown in Figure 7.19. (We previously established that the condition for constant flux in an ideal transformer was that the V/f ratio must remain constant: here we see that the so-called 'air-gap voltage' (V_m) is the one that matters.)

In Section 7.7 we considered operation at 60 Hz, for which the synchronous speed was 1800 rev/min and the rated speed 1728 rev/min. This gave a slip speed of 72 rev/min and a slip of 0.04. At 6 Hz the synchronous speed is 180 rev/min, so for the same slip speed of 72 rev/min the new slip is 0.4. This explains why the 'load' resistance of 7.2 Ω at 60 Hz becomes only 0.45 Ω at 6 Hz, while the total referred rotor resistance reduces from 7.5 Ω at 60 Hz to 0.75 Ω at 6 Hz.

The 60 Hz voltages shown in the upper half of Figure 7.19 were calculated in Section 7.7. To obtain the same torque at 6 Hz as at 60 Hz, we can see from equation (7.20) that the rotor power must reduce in proportion to the frequency, i.e. by a factor of 10. The rotor resistance has reduced by a factor of 10 so we would expect the rotor voltage to also reduce by a factor of 10. This is confirmed in Figure 7.19, where the rotor voltage at 6 Hz is 38.9 V. (This means the rotor current (51.9 A) at 6 Hz is the same as it was at 60 Hz, which is what we would expect given that for the same torque we would expect the same active current.) Working backwards (from V_m) we can derive the input voltage required, i.e. 49 V.

Had we used the approximate circuit, in which the magnetising branch is moved to the left-hand side, we would have assumed that to keep the amplitude of the flux wave constant, we would have to change the voltage in proportion to the frequency, in which case we would have decided that the input voltage at 6 Hz should be 44 V, not the 49 V really needed. And if we had supplied 44 V rather than 49 V, the torque at the target speed would be almost 20% below our expectation, which is clearly significant and underlines the danger of making unjustified assumptions when operating at low frequencies.

In this example the volt-drop of 10 V across the stator resistance is much more significant at 6 Hz than at 60 Hz, for two reasons. Firstly, at 6 Hz the useful (load) voltage is 38.9 V whereas at 60 Hz the load voltage is 389 V: so a fixed drop of 10 V that might be considered negligible as compared with 389 V is certainly not negligible in comparison with 38.9 V. And secondly, when the load is predominantly resistive, as here, the reduction in the magnitude of the supply voltage due to a given series impedance is much greater if the impedance is resistive than if it is reactive. This matter was discussed in Section 1 of Chapter 6.

REVIEW QUESTIONS

1) The primary winding of an ideal transformer is rated at 240 V, 2 A. The secondary has half as many turns as the primary. Calculate the secondary-rated voltage and current.

2) An ideal transformer has 200 turns on its primary winding and 50 turns on its secondary. If a resistance of $5\,\Omega$ is connected to the secondary, what would be the apparent impedance as seen from the primary?

3) A 240 V/20 V, 50 Hz ideal mains transformer supplies a secondary load consisting of a resistance of $30\,\Omega$. Calculate the primary current and power.

4) If a small air-gap was made in the iron core of a transformer, thereby increasing the reluctance of the magnetic circuit, what effect would it have on the following:

   • The no-load current, with the primary supplied at rated voltage.
   • The magnetising reactance.
   • The secondary voltage.

5) The approximate equivalent circuit parameters of a transformer feeding a resistive load are $R_T = 0.1\,\Omega$; $X_T = 0.5\,\Omega$; $X_m = 30\,\Omega$, and the referred full-load resistance is $10\,\Omega$. Estimate the percentage fall in the secondary voltage between no-load and full-load, and the short-circuit current as a multiple of the full-load current.

6) A particular 2-pole, 60 Hz induction motor has equal stator and referred rotor resistances, and under locked rotor conditions the total power input is 12 kW. Estimate the starting torque.

7) A single-cage 5 kW, 6-pole, 950 rev/min, 50 Hz induction motor takes five times full-load current in a direct-on-line start. Estimate the starting torque.

8) What effect would a 10% increase in leakage reactance in a single-cage induction motor have on (a) the pull-out torque, (b) the pull-out slip; (c) the full-load slip, (c) the starting current.

9) An induction motor develops rated torque with a slip of 3.5% when the rotor is cold. If the rotor resistance is 20% higher when the motor is at normal running temperature, at what slip will it develop rated torque? Explain your answer using the equivalent circuit.

10) Assume that the approximate equivalent circuit is to be used in this question. An induction motor has a magnetising current of 8 A. At full-load, the referred load current is 40 A, lagging the supply voltage by 15°. Estimate the supply current and power-factor at full-load and when the slip is half of the full-load value. State any assumptions.

11) The full-load slip of a 2-pole induction motor at 50 Hz is 0.04. Estimate the speed at which the motor will develop rated torque if the frequency is reduced to (a) 25 Hz, (b) 3 Hz. Assume that in both cases the voltage is adjusted to maintain full air-gap flux. Calculate the corresponding slip in both cases, and explain why the very low-speed condition is inefficient. Explain using the equivalent circuit why the full-load currents would be the same in all the three cases.

8

INVERTER-FED INDUCTION MOTOR DRIVES

INTRODUCTION

We saw in Chapter 6 that the induction motor can only run efficiently at low slips, i.e. close to the synchronous speed of the rotating field. The best method of speed control must therefore provide for continuous smooth variation of the synchronous speed, which in turn calls for variation of the supply frequency. This is achieved using an inverter (as discussed in Chapter 2) to supply the motor. A complete speed control scheme which includes tacho (speed) feedback is shown in block diagram form in Figure 8.1.

We should recall that the function of the converter (i.e. rectifier and variable-frequency inverter) is to draw power from the fixed-frequency constant-voltage mains, and convert it to variable frequency, variable voltage for driving the induction motor. Both the rectifier and the inverter employ switching strategies (see Chapter 2), so the power conversions are accomplished efficiently and the converter can be compact.

Variable frequency inverter-fed induction motor drives are used in ratings up to hundreds of kilowatts. Standard 50 Hz or 60 Hz motors are often used (though as we will see later this limits performance), and the inverter output frequency typically covers the range from around 5–10 Hz up to perhaps 120 Hz. This is sufficient to give at least a 10:1 speed range with a top speed of twice the normal (mains frequency) operating speed. The majority of inverters are 3-phase input and 3-phase output, but single-phase input versions are available up to about 5 kW, and some very small inverters (usually less than 1 kW) are specifically intended for use with single-phase motors.

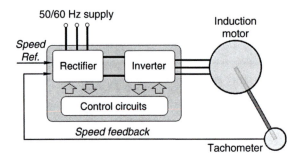

Figure 8.1   *General arrangement of inverter-fed variable-frequency induction motor speed-controlled drive*

A fundamental aspect of any converter, which is often overlooked, is the instantaneous energy balance. In principle, for any balanced three-phase load, the total load power remains constant from instant to instant, so if it was possible to build an ideal 3-phase input, 3-phase output converter, there would be no need for the converter to include any energy storage elements. In practice, all converters require some energy storage (in capacitors or inductors), but these are relatively small when the input is 3-phase because the energy balance is good. However, as mentioned above, many small and medium power converters are supplied from single-phase mains. In this case, the instantaneous input power is zero at least twice per cycle of the mains (because the voltage and current go through zero every half-cycle). If the motor is 3-phase (and thus draws power at a constant rate), it is obviously necessary to store sufficient energy in the converter to supply the motor during the brief intervals when the load power is greater than the input power. This explains why the most bulky components in many small and medium power inverters are electrolytic capacitors.

The majority of inverters used in motor drives are voltage source inverters (VSI), in which the output voltage to the motor is controlled to suit the operating conditions of the motor. Current source inverters (CSI) are still used, particularly for large applications, but will not be discussed here.

Comparison with d.c. drive

The initial success of the inverter-fed induction motor drive was due to the fact that a standard induction motor was much cheaper than a

Plate 8.1   *Inverter-fed induction motor with inverter mounted directly onto motor.*
(Alternatively the inverter can be wall-mounted, as in the upper illustration, which also
shows the user interface module.) (Photo courtesy of ABB)

comparable d.c. motor, and this saving compensated for the relatively
high cost of the inverter compared with the thyristor d.c. converter. But
whereas a d.c. drive was invariably supplied with a motor provided with
laminated field poles and through ventilation to allow it to operate
continuously at low speeds without overheating, the standard induction
motor has no such provision, having been designed primarily for fixed-
frequency full-speed operation. Thus, although the inverter is capable of
driving the induction motor with full torque at low speeds, continuous
operation is unlikely to be possible because the cooling fan will be
ineffective and the motor will overheat.

Now that inverter-fed drives dominate the market, two changes have become evident. Firstly, reputable suppliers now warn of the low-speed limitation of the standard induction motor, and encourage users to opt for a blower-cooled motor if necessary. And secondly, the fact that inverter-fed motors are not required to start direct-on-line at supply frequency means that the design need no longer be a compromise between starting and running performance. Motors can therefore be designed specifically for operation from an inverter, and have low-resistance cages giving very high steady-state efficiency and good open-loop speed holding. The majority of drives do still use standard motors, but inverter-specific motors with integral blowers are gradually gaining ground.

The steady-state performance of inverter-fed drives is broadly comparable with that of d.c. drives (except for the limitation highlighted above), with drives of the same rating having similar overall efficiencies and overall torque–speed capabilities. Speed holding is likely to be less good in the induction motor drive, though if tacho feedback is used both systems will be excellent. The induction motor is clearly more robust and better suited to hazardous environments, and can run at higher speeds than the d.c. motor, which is limited by the performance of its commutator.

Some of the early inverters did not employ pulse width modulation (PWM), and produced jerky rotation at low speed. They were also noticeably more noisy than their d.c. counterparts, but the widespread adoption of PWM has greatly improved these aspects. Most low and medium power inverters use MOSFET or IGBT devices, and may modulate at ultrasonic frequencies, which naturally result in relatively quiet operation.

The Achilles heel of the basic inverter-fed system has been the relatively poor transient performance. For fan and pump applications and high-inertia loads this is not a serious drawback, but where rapid response to changes in speed or load is called for (e.g. in machine tools or rolling mills), the d.c. drive with its fast-acting current-control loop traditionally proved superior. However, it is now possible to achieve equivalent levels of dynamic performance from induction motors, but the complexity of the control naturally reflects in a higher price. Most manufacturers now offer this so-called 'vector' or 'field-oriented' control (see Section 8.4) as an optional extra for high-performance drives.

Inverter waveforms

When we looked at the converter-fed d.c. motor we saw that the behaviour was governed primarily by the mean d.c. voltage, and that for most purposes we could safely ignore the ripple components. A similar approximation is useful when looking at how the inverter-fed induction motor performs. We make use of the fact that although the actual voltage waveform supplied by the inverter will not be sinusoidal, the motor behaviour depends principally on the fundamental (sinusoidal) component of the applied voltage. This is a somewhat surprising but extremely welcome simplification, because it allows us to make use of our knowledge of how the induction motor behaves with a sinusoidal supply to anticipate how it will behave when fed from an inverter.

In essence, the reason why the harmonic components of the applied voltage are much less significant than the fundamental is that the impedance of the motor at the harmonic frequencies is much higher than at the fundamental frequency. This causes the current to be much more sinusoidal than the voltage, as shown in Figure 8.2, and this in turn means that we can expect a sinusoidal travelling field to be set up in much the same way as discussed in Chapter 5.

It would be wrong to pretend that the harmonic components have no effects, of course. They can create unpleasant acoustic noise, and always give rise to additional iron and copper losses. As a result it is common for a standard motor to have to be de-rated (by up to perhaps 5 or 10%) for use on an inverter supply.

As with the d.c. drive the inverter-fed induction motor drive will draw non-sinusoidal currents from the utility supply. If the supply impedance is relatively high significant distortion of the mains voltage waveform is inevitable unless filters are fitted on the a.c. input side, but with normal

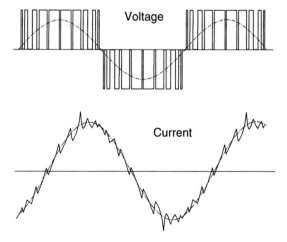

Figure 8.2 *Typical voltage and current waveforms for PWM inverter-fed induction motor. (The fundamental-frequency component is shown by the dotted line.)*

industrial supplies there is no problem for small inverters of a few kW rating.

Some inverters now include 'front-end conditioning' i.e. an extra high-frequency switching stage and filter which ensure that the current drawn from the mains is not only sinusoidal, but also at unity power factor. This feature will become widespread in medium and high power drives to meet the increasingly stringent conditions imposed by the supply authorities.

Steady-state operation – Importance of achieving full flux

Three simple relationships need to be borne in mind to simplify under-standing of how the inverter-fed induction motor behaves. Firstly, we established in Chapter 5 that for a given induction motor, the torque developed depends on the strength of the rotating flux density wave, and on the slip speed of the rotor, i.e. on the relative velocity of the rotor with respect to the flux wave. Secondly, the strength or amplitude of the flux wave depends directly on the supply voltage to the stator windings, and inversely on the supply frequency. And thirdly, the absolute speed of the flux wave depends directly on the supply frequency.

Recalling that the motor can only operate efficiently when the slip is small, we see that the basic method of speed control rests on the control of the speed of rotation of the flux wave (i.e. the synchronous speed), by control of the supply frequency. If the motor is a 4-pole one, for example, the synchronous speed will be 1500 rev/min when supplied at 50 Hz, 1200 rev/min at 40 Hz, 750 rev/min at 25 Hz and so on. The no-load speed will therefore be almost exactly proportional to the supply frequency, because the torque at no load is small and the corresponding slip is also very small.

Turning now to what happens on load, we know that when a load is applied the rotor slows down, the slip increases, more current is induced in the rotor, and more torque is produced. When the speed has reduced to the point where the motor torque equals the load torque, the speed becomes steady. We normally want the drop in speed with load to be as small as possible, not only to minimise the drop in speed with load, but also to maximise efficiency: in short, we want to minimise the slip for a given load.

We saw in Chapter 5 that the slip for a given torque depends on the amplitude of the rotating flux wave: the higher the flux, the smaller the slip needed for a given torque. It follows that having set the desired speed of rotation of the flux wave by controlling the output frequency of the inverter we must also ensure that the magnitude of the flux is adjusted so that it is at its full (rated) value, regardless of the speed of rotation. This is achieved by making the output voltage from the inverter vary in the appropriate way in relation to the frequency.

We recall that the amplitude of the flux wave is proportional to the supply voltage and inversely proportional to the frequency, so if we arrange that the voltage supplied by the inverter vary in direct proportion to the frequency, the flux wave will have a constant amplitude. This philosophy is at the heart of most inverter-fed drive systems: there are variations, as we will see, but in the majority of cases the internal control of the inverter will be designed so that the output voltage to frequency ratio (V/f) is automatically kept constant, at least up to the 'base' (50 Hz or 60 Hz) frequency.

Many inverters are designed for direct connection to the mains supply, without a transformer, and as a result the maximum inverter output voltage is limited to a value similar to that of the mains. With a 415 V supply, for example, the maximum inverter output voltage will be perhaps 450 V. Since the inverter will normally be used to supply a standard induction motor designed for say 415 V, 50 Hz operation, it is obvious that when the inverter is set to deliver 50 Hz, the voltage should be 415 V, which is within the inverter's voltage range. But when the frequency was raised to say 100 Hz, the voltage should – ideally – be increased to 830 V in order to obtain full flux. The inverter cannot supply voltages above 450 V, and it follows that in this case full flux can only be maintained up to speeds a little above base speed. (It should be noted that even if the inverter could provide higher voltages, they could not be applied to a standard motor because the winding insulation will have been designed to withstand not more than the rated voltage.)

Established practice is for the inverter to be capable of maintaining the V/f ratio constant up to the base speed (50 Hz or 60 Hz), but to accept that at all higher frequencies the voltage will be constant at its maximum value. This means that the flux is maintained constant at speeds up to base speed, but beyond that the flux reduces inversely with frequency. Needless to say the performance above base speed is adversely affected, as we will see.

Users are sometimes alarmed to discover that both voltage and frequency change when a new speed is demanded. Particular concern is expressed when the voltage is seen to reduce when a lower speed is called for. Surely, it is argued, it can't be right to operate say a 400 V induction motor at anything less than 400 V. The fallacy in this view should now be apparent: the figure of 400 V is simply the correct voltage for the motor when run directly from the mains, at say 50 Hz. If this full voltage was applied when the frequency was reduced to say 25 Hz, the implication would be that the flux would have to rise to twice its rated value.

This would greatly overload the magnetic circuit of the machine, giving rise to excessive saturation of the iron, an enormous magnetising current and wholly unacceptable iron and copper losses. To prevent this from happening, and keep the flux at its rated value, it is essential to reduce the voltage in proportion to frequency. In the case above, for example, the correct voltage at 25 Hz would be 200 V.

TORQUE–SPEED CHARACTERISTICS – CONSTANT V/F OPERATION

When the voltage at each frequency is adjusted so that the ratio *V/f* is kept constant up to base speed, and full voltage is applied thereafter, a family of torque–speed curves as shown in Figure 8.3 is obtained. These curves are typical for a standard induction motor of several kW output.

As expected, the no-load speeds are directly proportional to the frequency, and if the frequency is held constant, e.g. at 25 Hz in Figure 8.3, the speed drops only modestly from no-load (point a) to full-load (point b). These are therefore good open-loop characteristics, because the speed is held fairly well from no-load to full-load. If the application calls for the speed to be held precisely, this can clearly be achieved (with the aid of closed-loop speed control) by raising the frequency so that the full-load operating point moves to point (c).

We also note that the pull-out torque and the torque stiffness (i.e. the slope of the torque–speed curve in the normal operating region) is more or less the same at all points below base speed, except at low frequencies where the effect of stator resistance in reducing the flux becomes very pronounced. (The importance of stator resistance at low frequencies is explored quantitatively in Section 7.10.) It is clear from Figure 8.3 that

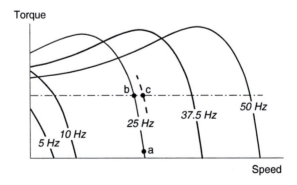

Figure 8.3   *Torque–speed curves for inverter-fed induction motor with constant voltage–frequency ratio*

the starting torque at the minimum frequency is much less than the pull-out torque at higher frequencies, and this could be a problem for loads which require a high starting torque.

The low-frequency performance can be improved by increasing the *V/f* ratio at low frequencies in order to restore full flux, a technique which is referred to as 'low-speed voltage boosting'. Most drives incorporate provision for some form of voltage boost, either by way of a single adjustment to allow the user to set the desired starting torque, or by means of more complex provision for varying the *V/f* ratio over a range of frequencies. A typical set of torque–speed curves for a drive with the improved low-speed torque characteristics obtained with voltage boost is shown in Figure 8.4.

The curves in Figure 8.4 have an obvious appeal because they indicate that the motor is capable of producing practically the same maximum torque at all speeds from zero up to the base (50 Hz or 60 Hz) speed. This region of the characteristics is known as the 'constant torque' region, which means that for frequencies up to base speed, the maximum possible torque which the motor can deliver is independent of the set speed. Continuous operation at peak torque will not be allowable because the motor will overheat, so an upper limit will be imposed by the controller, as discussed shortly. With this imposed limit, operation below base speed corresponds to the armature-voltage control region of a d.c. drive, as exemplified in Figure 3.9.

We should note that the availability of high torque at low speeds (especially at zero speed) means that we can avoid all the 'starting' problems associated with fixed-frequency operation (see Chapter 6). By starting off with a low frequency which is then gradually raised the

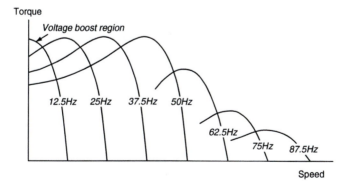

Figure 8.4 *Typical torque–speed curves for inverter-fed induction motor with low-speed voltage boost, constant voltage–frequency ratio from low speed up to base speed, and constant voltage above base speed*

slip speed of the rotor is always small, i.e. the rotor operates in the optimum condition for torque production all the time, thereby avoiding all the disadvantages of high-slip (low torque and high current) that are associated with mains-frequency starting. This means that not only can the inverter-fed motor provide rated torque at low speeds, but – perhaps more importantly – it does so without drawing any more current from the mains than under full-load conditions, which means that we can safely operate from a weak supply without causing excessive voltage dips. For some essentially fixed-speed applications, the superior starting ability of the inverter-fed system alone may justify its cost.

Beyond the base frequency, the V/f ratio reduces because V remains constant. The amplitude of the flux wave therefore reduces inversely with the frequency. Now we saw in Chapter 5 that the pull-out torque always occurs at the same absolute value of slip speed, and that the peak torque is proportional to the square of the flux density. Hence in the constant voltage region the peak torque reduces inversely with the square of the frequency and the torque–speed curve becomes less steep, as shown in Figure 8.4.

Although the curves in Figure 8.4 show what torque the motor can produce for each frequency and speed, they give no indication of whether continuous operation is possible at each point, yet this matter is of course extremely important from the user's viewpoint, and is discussed next.

Limitations imposed by the inverter – constant power and constant torque regions

The main concern in the inverter is to limit the currents to a safe value as far as the main switching devices are concerned. The current limit will be at least equal to the rated current of the motor, and the inverter control circuits will be arranged so that no matter what the user does the output current cannot exceed a safe value.

The current limit feature imposes an upper limit on the permissible torque in the region below base speed. This will normally correspond to the rated torque of the motor, which is typically about half the pull-out torque, as indicated by the shaded region in Figure 8.5.

In the region below base speed, the motor can therefore develop any torque up to rated value at any speed (but not necessarily for prolonged periods, as discussed below). This region is therefore known as the 'constant torque' region, and it corresponds to the armature voltage control region of a d.c. drive.

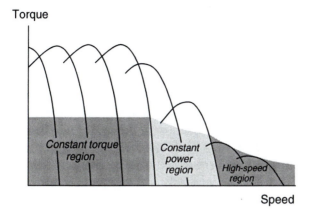

Figure 8.5   *Constant torque, constant power and high-speed motoring regions*

Above base speed of the flux is reduced inversely with the frequency; because the stator (and therefore rotor) currents are limited, the maximum permissible torque also reduces inversely with the speed, as shown in Figure 8.5. This region is therefore known as the 'constant power' region. There is of course a close parallel with the d.c. drive here, both systems operating with reduced or weak field in the constant power region. The region of constant power normally extends to somewhere around twice base speed, and because the flux is reduced the motor has to operate with higher slips than below base speed to develop the full rotor current and torque.

At the upper limit of the constant power region, the current limit coincides with the pull-out torque limit. Operation at still higher speeds is sometimes provided, but constant power is no longer available because the maximum torque is limited to the pull-out value, which reduces inversely with the square of the frequency. In this high-speed motoring region (Figure 8.5), the limiting torque–speed relationship is similar to that of a series d.c. motor.

Limitations imposed by motor

The standard practice in d.c. drives is to use a motor specifically designed for operation from a thyristor converter. The motor will have a laminated frame, will probably come complete with a tachogenerator, and – most important of all – will have been designed for through ventilation and equipped with an auxiliary air blower. Adequate ventilation is guaranteed at all speeds, and continuous operation with full torque (i.e. full current) at even the lowest speed is therefore in order.

By contrast, it is still common for inverter-fed systems to use a standard industrial induction motor. These motors are totally enclosed, with an external shaft-mounted fan, which blows air over the finned outer case. They are designed first and foremost for continuous operation from the fixed frequency mains, and running at base speed.

When such a motor is operated at a low frequency (e.g. 10 Hz), the speed is much lower than base speed and the efficiency of the cooling fan is greatly reduced. At the lower speed, the motor will be able to produce as much torque as at base speed (see Figure 8.4) but in doing so the losses in both stator and rotor will also be more or less the same as at base speed. Since the fan was only just adequate to prevent overheating at base speed, it is inevitable that the motor will overheat if operated at full torque and low speed for any length of time. Some suppliers of inverter drives do not emphasise this limitation, so users need to raise the question of whether a non-standard motor will be needed.

When through-ventilated motors with integral blowers become the accepted standard, the inverter-fed system will be freed of its low-speed limitations. Meanwhile users should note that one approach designed to combat the danger of motor overheating at low speeds is for the control circuits to be deliberately designed so that the flux and current limit are reduced at low speeds. The constant-torque facility is thus sacrificed in order to reduce copper and iron losses, but as a result the drive is only suitable for fan- or pump-type loads, which do not require high torque at low speed. These systems inevitably compare badly with d.c. drives, but manage to save face by being promoted as 'energy-saving' drives.

CONTROL ARRANGEMENTS FOR INVERTER-FED DRIVES

For speed control manufacturers offer options ranging in sophistication from a basic open-loop scheme which is adequate when precise speed holding is not essential, through closed-loop schemes with tacho or encoder feedback, up to vector control schemes which are necessary when optimum dynamic performance is called for. The variety of schemes is much greater than for the fully matured d.c. drive, so we will look briefly at some examples in the remainder of this section.

The majority of drives now provide a digital interface so that the user can input data such as maximum and minimum speeds, acceleration rates, maximum torque, etc. General purpose inverters that are not sold with a specific motor may have provision for motor parameters such as base frequency, full-load slip and current, leakage reactance and rotor resistance to be entered so that the drive can self-optimise its control routines. Perhaps the ultimate are the self-commissioning drives that apply test signals to the motor when it is first connected in order to determine the motor parameters, and then set themselves to deliver optimum performance: their detailed workings are well beyond our scope!

Open-loop speed control

In the smaller sizes the simple 'constant V/f' control is the most popular, and is shown in Figure 8.6. The output frequency, and hence the no-load speed of the motor, is set by the speed reference signal, which in an analogue scheme is either an analogue voltage (0–10 V) or current (4–20 mA). This set-speed signal may be obtained from a potentiometer on the front panel, or remotely from elsewhere. In the increasingly common digital version the speed reference will be set on the keypad. Some adjustment of the V/f ratio and low-speed voltage boost will be provided.

Typical steady-state operating torque–speed curves are shown in Figure 8.7. For each set speed (i.e. each frequency) the speed remains reasonably constant because of the stiff torque–slip characteristic of the cage motor. If the load is increased beyond rated torque, the internal current limit (not shown in Figure 8.6) comes into play to prevent the motor from reaching the unstable region beyond pull out. Instead, the frequency and speed are reduced, so that the system behaves in the same way as a d.c. drive.

Sudden changes in the speed reference are buffered by the action of an internal frequency ramp signal, which causes the frequency to be gradually increased or decreased. If the load inertia is low, the acceleration will be accomplished without the motor entering the current-limit regime. On the other hand if the inertia is large, the acceleration will take place along the torque–speed trajectory shown in Figure 8.7.

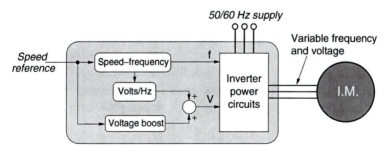

Figure 8.6   *Schematic diagram of open-loop inverter-fed induction motor speed controlled drive*

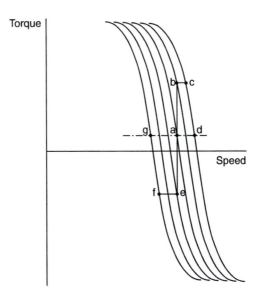

Figure 8.7   *Acceleration and deceleration trajectories in the torque–speed plane*

Suppose the motor is operating in the steady state with a constant load torque at point (a), when a new higher speed (corresponding to point (d)) is demanded. The frequency is increased, causing the motor torque to rise to point (b), where the current has reached the allowable limit. The rate of increase of frequency is then automatically reduced so that the motor accelerates under constant current conditions to point (c), where the current falls below the limit: the frequency then remains constant and the trajectory follows the curve from (c) to settle finally at point (d).

A typical deceleration trajectory is shown by the path *aefg* in Figure 8.7. The torque is negative for much of the time, the motor operating in quadrant 2 and regenerating kinetic energy to the inverter. Most small inverters do not have the capability to return power to the a.c. supply, and the excess energy therefore has to be dissipated in a resistor inside the converter. The resistor is usually connected across the d.c. link, and controlled by a chopper. When the link voltage tends to rise, because of the regenerated energy, the chopper switches the resistor on to absorb the energy. High inertia loads, which are subjected to frequent deceleration can therefore pose problems of excessive power dissipation in this 'dump' resistor.

Speed reversal poses no problem, the inverter firing sequence being reversed automatically at zero speed, thereby allowing the motor to proceed smoothly into quadrants 3 and 4.

Some schemes include slip compensation, whereby the drive senses the active component of the load current, which is a measure of the torque; deduces the slip (which at full flux is proportional to the torque); and then increases the motor frequency to compensate for the slip speed of the rotor and thereby maintain the same speed as at no load. This is similar to the 'IR' compensation used in open-loop d.c. drives and discussed in Section 4.3.5.

Closed-loop speed control

Where precision speed holding is required a closed-loop scheme must be used, with speed feedback from either a d.c. or a.c. tachogenerator, or a digital shaft encoder. Many different control strategies are employed, so we will consider the typical arrangement shown in Figure 8.8. This inverter has separate control of the a.c. output voltage (via phase-angle control in the input rectifier) and frequency (via the switching in the inverter), and this makes understanding how the control system operates easier than when voltage and frequency are controlled together, as in a PWM inverter.

Figure 8.8 has been drawn to emphasise the similarity with the closed-loop d.c. drive that was discussed at length in Section 4.3. Experience suggests that an understanding of how the d.c. drive operates is very

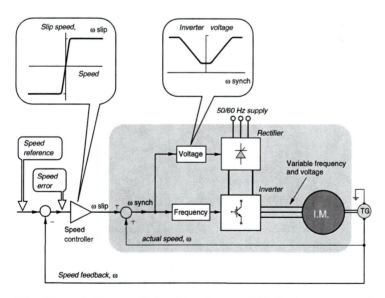

Figure 8.8   *Schematic diagram of closed-loop inverter-fed induction motor speed-controlled drive with tacho feedback*

helpful in the study of the induction motor drive, so readers may wish to refresh their ideas about the 2-loop d.c. drive by revisiting Section 4.3 before coming to grips with this section. As in the previous discussion, we will assume that the control variables are continuous analogue signals, though of course the majority of implementations will involve digital hardware.

The arrangement of the outer speed-control loop (see Figure 8.8) is identical with that of the d.c. drive (see Figure 4.11): the actual speed (represented by the voltage generated by the tachogenerator) is compared with the target or reference speed and the resulting speed error forms the input to the speed controller. The output of the speed controller provides the input or reference to the inner part of the control system, shown shaded in Figure 8.8. In both the d.c. drive and the induction motor drive, the output of the speed controller serves as a torque reference signal, and acts as the input to the inner (shaded) part of the system. We will now see that, as in the d.c. drive, the inner system of the inverter-fed drive is effectively a torque-control loop that ensures that the motor torque is directly proportional to the torque reference signal under all conditions.

We have seen that if the magnitude of the flux wave in an induction motor is kept constant, the torque in the normal operating region is directly proportional to the slip speed. (We should recall that 'normal operating region' means low values of slip, typically a few per cent of synchronous speed.) So the parameter that must be controlled in order to control torque is the slip speed. But the only variable that we can directly vary is the stator frequency (and hence the synchronous speed); and the only variable we can measure externally is the actual rotor speed. These three quantities (see Figure 8.8) are represented by the following analogue voltages:

$$\text{Slip speed} = \omega_{slip}$$
$$\text{Synchronous speed} = \omega_{synch}$$
$$\text{Rotor speed} = \omega,$$
$$\text{where } \omega_{synch} = \omega + \omega_{slip} \qquad (8.1)$$

Equation (8.1) indicates how we must vary the stator frequency (i.e. the synchronous speed) if we wish to obtain a given slip speed (and hence a given torque): we simply have to measure the rotor speed and add to it the appropriate slip speed to obtain the frequency to be supplied to the stator. This operation is performed at the summing junction at the input to the shaded inner section in Figure 8.8: the output from the summing junction directly controls the inverter output frequency (i.e. the synchronous speed), and, via a shaping function, the amplitude of the inverter output voltage.

The shaping function, shown in the call-out in Figure 8.8, provides a constant voltage–frequency ratio over the majority of the range up to base speed, with 'voltage-boost' at low frequencies. These conditions are necessary to guarantee the 'constant flux' condition that is an essential requirement for us to be able to claim that torque is proportional to slip speed. (We must also accept that as soon as the speed rises above base speed, and the voltage–frequency ratio is no longer maintained, a given slip speed reference to the inner system will yield less torque than below base speed, because the flux will be lower.)

We have noted the similarities between the structures of the induction motor and d.c. drives, but at this point we might wish to pause and reflect on the differences between the inner loops. In the d.c. drive the inner loop is a conventional (negative feedback) current control where the output (motor current) is measured directly; the torque is directly proportional to current and is therefore directly controlled by the inner loop. In contrast, the inner loop in the induction scheme provides torque control indirectly, via the regulation of slip speed, and it involves a positive feedback loop. It relies for its success on the linear relationship between torque and slip, and thus is only valid when the flux is maintained at full value and the slip speed is low; and because it involves positive feedback there is the potential for instability if the loop gain is greater than one, which means that the tachogenerator constant must be judged with care.

Returning now to the outer speed loop and assuming for the moment that the speed controller is simply a high-gain amplifier, understanding the operation of the speed-control loop is straightforward. When the speed error increases (because the load has increased a little and caused the speed to begin to fall, or the target speed has been raised modestly) the output of the speed controller increases in proportion, signalling to the inner loop that more torque is required to combat the increased load, or to accelerate to the new speed. As the target speed is approached, the speed error reduces, the torque tapers off and the target speed is reached very smoothly. If the gain of the speed error amplifier is high, the speed error under steady-state conditions will always be low, i.e. the actual speed will be very close to the reference speed.

In the discussion above, it was assumed that the speed controller remained in its linear region, i.e. the speed error was always small. But we know that in practice there are many situations where there will be very large speed errors. For instance, when the motor is at rest and the speed reference is suddenly raised to 100%, the speed error will immediately become 100%. Such a large input signal will cause the speed error amplifier output to saturate at maximum value, as shown by the sketch of the amplifier characteristic in Figure 8.8. In this case the slip reference will be at maximum value and the torque and acceleration will also be at maximum, which is what we want in order to reach the target speed in the minimum time. As the speed increases the motor terminal voltage and frequency will both rise in order to maintain maximum slip until the speed error falls to a low value and the speed error amplifier comes out of saturation. Control then enters the linear regime in which the torque becomes proportional to the speed error, giving a smooth approach to the final steady-state speed.

In relatively long-term transients of the type just discussed, where changes in motor frequency occur relatively slowly (e.g. the frequency increases at perhaps a few per cent per cycle) the behaviour of the standard inverter-fed drive is very similar to that of the two-loop d.c. drive, which as we have already seen has long been regarded as the yardstick by which others are judged. Analogue control using a proportional and integral speed error amplifier (see Appendix) can give a good transient response and steady-state speed holding of better than 1% for a speed range of 20:1 or more. For higher precision, a shaft encoder together with a phase-locked loop is used. The need to fit a tacho or encoder can be a problem if a standard induction motor is used, because there is normally no shaft extension at the non-drive end. The user then faces the prospect of paying a great deal more for what amounts to a relatively minor modification, simply because the motor then ceases to be standard.

VECTOR (FIELD-ORIENTED) CONTROL

Where very rapid changes in speed are called for, however, the standard inverter-fed drive compares unfavourably with d.c. drive. The superiority of the d.c. drive stems firstly from the relatively good transient response of the d.c. motor, and secondly from the fact that the torque can be directly controlled even under transient conditions by controlling the armature current. In contrast, the induction motor has inherently poor transient performance.

For example, when we start an unloaded induction motor direct-on-line we know that it runs up to speed, but if we were to look in detail at what happens immediately after switching on we might be very surprised. We would see that the instantaneous torque fluctuates wildly for the first few cycles of the supply, until the flux wave has built up and all three phases have settled into a quasi-steady-state condition while the motor completes its run-up. (The torque–speed curves found in this and most other textbooks ignore this phenomenon, and present only the average steady-state curve.) We might also find that the speed oscillated around synchronous before finally settling with a small slip.

For the majority of applications the standard inverter-fed induction motor is perfectly adequate, but for some very demanding tasks, such as high-speed machine tool spindle drives, the dynamic performance is extremely important and 'vector' or 'field-oriented' control is warranted. Understanding all the ins and outs of vector control is well beyond our scope, but it is worthwhile outlining how it works, if only to dispel some of the mystique surrounding the matter. Some recent textbooks on electrical machines now cover the theory of vector control (which is still considered difficult to understand, even for experts) but the majority concentrate on the control theory and very few explain what actually happens inside a motor when operated under vector control.

Transient torque control

We have seen previously that in both the induction motor and the d.c. motor, torque is produced by the interaction of currents on the rotor with the radial flux density produced by the stator. Thus to change the torque, we must either change the magnitude of the flux, or the rotor current, or both; and if we want a sudden (step) increase in torque, we must make the change (or changes) instantaneously.

Since every magnetic field has stored energy associated with it, it should be clear that it is not possible to change a magnetic field instantaneously, as this would require the energy to change in zero time, which calls for a pulse of infinite power. In the case of the main field of a motor, we could not hope to make changes fast enough even to approximate the step change in torque we are seeking, so the only alternative is to make the rotor current change as quickly as possible.

In the d.c. motor it is relatively easy to make very rapid changes in the armature (rotor) current because we have direct access to the armature current via the brushes. The armature circuit inductance is relatively low, so as long as we have plenty of voltage available, we can apply a large voltage (for a very short time) whenever we want to make a sudden change in the armature current and torque. This is done automatically by the inner (current-control) loop in the d.c. drive (see Chapter 4).

In the induction motor, matters are less straightforward because we have no direct access to the rotor currents, which have to be induced from the stator side. Nevertheless, because the stator and rotor windings are tightly coupled via the air-gap field (see Chapter 5), it is possible to make more or less instantaneous changes to the induced currents in the rotor, by making instantaneous changes to the stator currents. Any sudden change in the stator MMF pattern (resulting from a change in the stator currents) is immediately countered by an opposing rotor MMF set up by the additional rotor currents which suddenly spring up. All tightly coupled circuits behave in this way, the classic example being the transformer, in which any sudden change in say the secondary current is immediately accompanied by a corresponding change in the primary current. Organising these sudden step changes in the rotor currents represents both the essence and the challenge of the vector-control method.

We have already said that we have to make sudden step changes in the stator currents, and this is achieved by providing each phase with a fast-acting closed-loop current controller. Fortunately, under transient conditions the effective inductance looking in at the stator is quite small (it is equal to the leakage inductance), so it is possible to obtain very rapid changes in the stator currents by applying high, short-duration impulsive voltages to the stator windings. In this respect each stator current controller closely resembles the armature current controller used in the d.c. drive.

When a step change in torque is required, the magnitude, frequency and phase of the three stator currents are changed (almost) instantaneously in such a way that the frequency, magnitude and phase of the rotor current wave (see Chapter 5) jump suddenly from one steady state to another. This change is done without altering the amplitude or position of the resultant rotor flux linkage relative to the rotor, i.e. without altering the stored energy significantly. The flux density term (B) in equation (5.8) therefore remains the same while the terms I_r and φ_r change instantaneously to their new steady-state values, corresponding to the new steady-state slip and torque.

We can picture what happens by asking what we would see if we were able to observe the stator MMF wave at the instant that a step increase in torque was demanded. For the sake of simplicity, we will assume that the rotor speed remains constant, in which case we would find that:

(a) the stator MMF wave suddenly increases its amplitude;
(b) it suddenly accelerates to a new synchronous speed;
(c) it jumps forward to retain its correct relative phase with respect to the rotor flux and current waves.

Thereafter the stator MMF retains its new amplitude, and rotates at its new speed. The rotor experiences a sudden increase in its current and torque, the new current being maintained by the new (higher) stator currents and slip frequency.

We should note that both before and after the sudden changes, the motor operates in the normal fashion, as discussed earlier. The 'vector control' is merely the means by which we are able to make a sudden stepwise transition from one steady state operating condition to another, and it has no effect whatsoever once we have reached the steady state.

The unique feature of the vector drive which differentiates it from the ordinary or scalar drive (in which only the magnitude and frequency of the stator MMF wave changes when more torque is required) is that by making the right sudden change to the instantaneous *position* of the stator MMF wave, the transition from one steady state to the other is achieved instantaneously, without the variables hunting around before settling to their new values. In particular, the vector approach allows us to overcome the long electrical time-constant of the rotor cage, which is responsible for the inherently sluggish transient response of the induction motor. It should also be pointed out that, in practice, the speed of the rotor will not remain constant when the torque changes (as assumed in the discussion above) so that, in order to keep track of the exact position of the rotor flux wave, it will be necessary to have a rotor position feedback signal.

Because the induction motor is a multi-variable non-linear system, an elaborate mathematical model of the motor is required, and implementation of the complex control algorithms calls for a large number of fast computations to be continually carried out so that the right instantaneous voltages are applied to each stator winding. This has only recently been made possible by using sophisticated and powerful signal processing in the drive control.

No industry standard approach to vector control has yet emerged, but systems fall into two broad categories, depending on whether or not they employ feedback from a shaft-mounted encoder to track the instantaneous position of the rotor. Those that do are known as 'direct' methods, whereas those which rely entirely on a mathematical model of the motor are known as 'indirect' methods. Both systems use current feedback as an integral part of each stator current controllers, so at least two stator current sensors are required. Direct systems are inherently more robust and less sensitive to changes in machine parameters, but call for a non-standard (i.e. more expensive) motor and encoder.

The dynamic performance of direct vector drives is now so good that they are found in demanding roles that were previously the exclusive preserve of the d.c. drive, such as reversing drives and positioning applications. The achievement of such outstandingly impressive performance from a motor whose inherent transient behaviour is poor represents a major milestone in the already impressive history of the induction motor.

CYCLOCONVERTER DRIVES

We conclude this chapter with a discussion of the cycloconverter variable-frequency drive, which has never become very widespread but is sometimes used in very large low-speed induction motor or synchronous motor drives. Cycloconverters are only capable of producing acceptable output waveforms at frequencies well below the mains frequency, but this, coupled with the fact that it is feasible to make large induction or synchronous motors with high-pole numbers (e.g. 20) means that a very low-speed direct (gearless) drive becomes practicable. A 20-pole motor, for example, will have a synchronous speed of only 30 rev/min at 5 Hz, making it suitable for mine winders, kilns, crushers, etc.

Most of the variable-frequency sources discussed in this book have been described as inverters because they convert power from d.c. to a.c. The power usually comes from a fixed-frequency mains supply, which is first rectified to give an intermediate stage – the 'd.c. link' – which is then chopped up to form a variable-frequency output. In contrast, the cycloconverter is a 'direct' converter, i.e. it does not have a d.c. stage (see also Section 2.4.6). Instead, the output voltage is synthesised by switching the load directly across whichever phase of the mains gives the best approximation to the desired load voltage at each instant of time. The principal advantage of the cycloconverter is that naturally commutated devices (thyristors) can be used instead of self-commutating devices, which means that the cost of each device is lower and higher powers can be achieved. In principle cycloconverters can have any combination of input and output phase numbers, but in practice the 3-phase input, 3-phase output version is used for drives, mainly in small numbers at the highest powers (e.g. 1 MW and above).

Textbooks often include bewilderingly complex circuit diagrams of the cycloconverter, which are not much help to the user who is seeking to understand how such systems work. Our understanding can be eased by recognising that the power conversion circuit for each of the three output phases is the same, so we can consider the simpler question of how to obtain a variable frequency, variable voltage supply, suitable for one phase of an induction motor, from a 3-phase supply of fixed frequency and constant voltage. It will also assist us to bear in mind that cycloconverters are only used to synthesise output frequencies that are low in comparison with the mains frequency. With a 50 Hz supply for example, we can expect to be able to achieve reasonably satisfactory approximations to a sinusoidal output voltage for frequencies from zero (d.c.) up to about 15 Hz; but at higher frequencies the harmonic distortion of the waveform will be so awful that even the normally tolerant induction motor will balk at the prospect.

To understand why we use any particular power electronic circuit configuration, we first need to address the question of what combination(s) of voltages and currents will be required in the load. Here the load is an induction motor, and we know that under sinusoidal supply conditions the power factor varies with load but never reaches unity. In other words, the stator current is never in phase with the stator voltage. So during the positive half-cycle of the voltage waveform the current will be positive for some of the time, but negative for the remainder; while during the negative voltage half-cycle the current will be negative for some of the time and positive for the rest of the time. This means that the supply to the motor has to be able to handle any combination of both positive and negative voltage and current.

We have already explored how to achieve a variable-voltage d.c. supply, which can handle both positive and negative currents, in Chapter 4. We saw that what was needed was two fully controlled 3-phase converters (as shown in Figure 2.11), connected back to back, as shown in Figure 4.8. We also saw in Chapter 4 that by varying the firing delay angle of the thyristors, the positive-current bridge could produce a range of mean (d.c.) output voltages from a positive maximum, through zero, to a negative maximum; and likewise the negative-current bridge could give a similar range of mean output voltages from negative maximum to positive maximum. Typical 'd.c.' output voltages over the range of firing angles from $\alpha = 0°$ (maximum d.c. voltage) to $\alpha = 90°$ (zero d.c. voltage) are shown in Figure 2.12.

In Chapter 4, the discussion focused on the mean or d.c. level of the output voltages, because we were concerned with the d.c. motor drive. But here we want to provide a low-frequency (preferably sinusoidal) output voltage for an induction motor, and the means for doing this should now be becoming clear. Once we have a double thyristor converter, and assuming for the moment that the load is resistive, we can generate a low-frequency sinusoidal output voltage simply by varying the firing angle of the positive-current bridge so that its output voltage increases from zero in a sinusoidal manner with respect to time. Then, when we have completed the positive half-cycle and arrived back at zero voltage, we bring the negative bridge into play and use it to generate the negative half-cycle, and so on.

In practice, as we have seen above, the load (induction motor) is not purely resistive, so matters are rather more complicated, because as we saw earlier for some part of the positive half-cycle of the output voltage wave the motor current will be negative. This negative current can only be supplied by the negative-current bridge (see Chapter 4) so as soon as the current reverses the negative bridge will have to be brought into action, initially to provide positive voltage (i.e. in the inverting mode) but later – when the current goes negative – providing negative voltage (i.e. rectifying). As long as the current remains continuous (see Chapter 2) the synthesised output voltage waveform will be typically as shown in Figure 8.9.

We see that the output voltage wave consists of chunks of the incoming mains voltage, and that it offers a reasonable approximation to the fundamental frequency sine wave shown by the dotted line in Figure 8.9. The output voltage waveform is no worse than the voltage waveform from a d.c. link inverter (see Figure 8.2), and as we saw in that context, the current waveform in the motor will be a good deal smoother than the

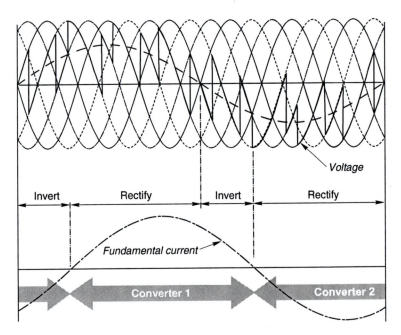

Figure 8.9 *Typical output voltage waveform for one phase of six-pulse cycloconverter supplying an inductive (motor) load.* (The output frequency shown in the figure is one third of the mains frequency, and the amplitude of the fundamental component of the output voltage (shown by the dotted line) is about 75% of the maximum that which could be obtained. The fundamental component of the load current is shown in order to define the modes of operation of the converters.)

voltage, because of the filtering action of the stator leakage inductance. The motor performance will therefore be acceptable, despite the extra losses that arise from the unwanted harmonic components. We should note that because each phase is supplied from a double converter, the motor can regenerate when required (e.g. to restrain an overhauling load, or to return kinetic energy to the supply when the frequency is lowered to reduce speed). This is one of the major advantages of the cycloconverter.

It is not necessary to go into the detail of how the firing angle scheme is implemented, but it should be clear that by varying the amplitude and frequency of the reference signal to the firing angle control, we can expect the output voltage to vary in sympathy. We then have the ability to keep the voltage–frequency ratio constant, so that the flux in the induction motor remains constant and we obtain a constant torque characteristic. It should also be evident from Figure 8.9 that as the output frequency is raised it becomes increasingly difficult to achieve a reasonable approximation to a sine wave, because there are too few 'samples' available in each half-cycle of the output. Cycloconverters are therefore seldom operated at more than one third of the mains frequency.

With the configuration described above, each phase of the motor requires its own double-bridge converter, consisting of 12 thyristors, so the complete cycloconverter requires 36 thyristors. To avoid short-circuits between the incoming mains lines, the three motor phase-windings must be isolated from each other (i.e. the motor cannot be connected in the conventional star or delta fashion, but must have both ends of each winding brought out), or each double converter can be supplied from separate transformer secondaries.

In practice there are several power-circuit configurations that can be used with star-connected motors, and which differ in detail from the set-up described above, but all require the same number of thyristors to achieve the waveform shown in Figure 8.9. This waveform is referred to as 6-pulse (see Chapter 2), because the output has six pulses per cycle of the mains. A worse (3-pulse) waveform is obtained with 18 thyristors, while a much better (12-pulse) waveform can be obtained by using 72 thyristors.

REVIEW QUESTIONS

1) A 2-pole, 440 V, 50 Hz induction motor develops rated torque at a speed of 2960 rev/min; the corresponding stator and rotor currents are 60 A and 150 A, respectively. If the stator voltage and frequency are adjusted so that the flux remains constant, calculate the speed at

which full torque is developed when the supply frequency is (a) 30 Hz, (b) 3 Hz.

2) Estimate the stator and rotor currents and the rotor frequency for the motor in question 1 at 30 Hz and at 3 Hz.

3) What is 'voltage boosting' in a voltage-source inverter, and why is it necessary?

4) An induction motor with a synchronous speed of N_s is driving a constant torque load at base frequency, and the slip is 5%. If the frequency of the supply is then doubled, but the voltage remains the same, estimate the new slip speed and the new percentage slip.

5) Approximately how would the efficiency of an inverter-fed motor be expected to vary between full (base) speed, 50% speed and 10% speed, assuming that the load torque was constant at 100% at all speeds and that the efficiency at base speed was 80%.

6) Why is it unwise to expect a standard induction motor driving a high-torque load to run continuously at low speed?

7) What problems might there be in using a single inverter to supply more than one induction motor with the intention of controlling the speeds of all of them simultaneously?

8) Explain briefly why an inverter-fed induction motor will probably be able to produce more starting torque per ampere of supply current than the same motor would if connected directly to the mains supply. Why is this likely to be particularly important if the supply impedance is high?

9) Why is the harmonic content of inverter-fed induction motor current waveform less than the harmonic content of the voltage waveforms?

10) An inverter-fed induction motor drive has closed-loop control with tacho feedback. The motor is initially at rest, and unloaded. Sketch graphs showing how you would expect the stator voltage and frequency and the slip speed to vary following a step demand for 150% of base speed if (a) the drive was programmed to run up to speed slowly (say in 10 s); (b) the drive was programmed to run up to speed as quickly as possible.

9

STEPPING MOTORS

INTRODUCTION

Stepping motors are attractive because they can be controlled directly by computers or microcontrollers. Their unique feature is that the output shaft rotates in a series of discrete angular intervals, or steps, one step being taken each time a command pulse is received. When a definite number of pulses has been supplied, the shaft will have turned through a known angle, and this makes the motor ideally suited for open-loop position control.

The idea of a shaft progressing in a series of steps might conjure up visions of a ponderous device laboriously indexing until the target number of steps has been reached, but this would be quite wrong. Each step is completed very quickly, usually in a few milliseconds; and when a large number of steps is called for the step command pulses can be delivered rapidly, sometimes as fast as several thousand steps per second. At these high stepping rates the shaft rotation becomes smooth, and the behaviour resembles that of an ordinary motor. Typical applications include disc head drives, and small numerically controlled machine tool slides, where the motor would drive a lead screw; and print feeds, where the motor might drive directly, or via a belt.

Most stepping motors look very much like conventional motors, and as a general guide we can assume that the torque and power of a stepping motor will be similar to the torque and power of a conventional totally enclosed motor of the same dimensions and speed range. Step angles are mostly in the range 1.8°–90°, with torques ranging from 1 μNm (in a tiny wristwatch motor of 3 mm diameter) up to perhaps 40 Nm in a motor of 15 cm diameter suitable for a machine tool application where speeds of 500 rev/min might be called for. The majority of applications fall between these limits, and use motors that can comfortably be held in the hand.

Open-loop position control

A basic stepping motor system is shown in Figure 9.1. The drive contains the electronic switching circuits, which supply the motor, and is discussed later. The output is the angular position of the motor shaft, while the input consists of two low-power digital signals. Every time a

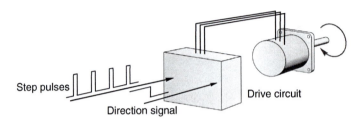

Figure 9.1   *Open-loop position control using a stepping motor*

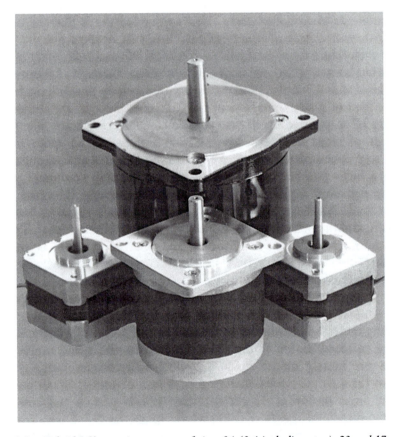

Plate 9.1   *Hybrid 1.8° stepping motors, of sizes 34 (3.4 inch diameter), 23 and 17. (Photo courtesy of Astrosyn)*

pulse occurs on the step input line, the motor takes one step, the shaft remaining at its new position until the next step pulse is supplied. The state of the direction line ('high' or 'low') determines whether the motor steps clockwise or anticlockwise. A given number of step pulses will therefore cause the output shaft to rotate through a definite angle.

This one to one correspondence between pulses and steps is the great attraction of the stepping motor: it provides *position* control, because the output is the angular position of the output shaft. It is a *digital* system, because the total angle turned through is determined by the *number* of pulses supplied; and it is *open-loop* because no feedback need be taken from the output shaft.

Generation of step pulses and motor response

The step pulses may be produced by an oscillator circuit, which itself is controlled by an analogue voltage, digital controller or microprocessor. When a given number of steps is to be taken, the oscillator pulses are gated to the drive and the pulses are counted, until the required number of steps is reached, when the oscillator is gated off. This is illustrated in Figure 9.2, for a six-step sequence. There are six-step command pulses, equally spaced in time, and the motor takes one step following each pulse.

Three important general features can be identified with reference to Figure 9.2. Firstly, although the total angle turned through (six steps) is governed only by the number of pulses, the average speed of the shaft (which is shown by the slope of the broken line in Figure 9.2) depends on the oscillator frequency. The higher the frequency, the shorter the time taken to complete the six steps.

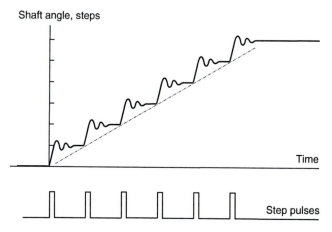

Figure 9.2 *Typical step response to low-frequency train of step command pulses*

Secondly, the stepping action is not perfect. The rotor takes a finite time to move from one position to the other, and then overshoots and oscillates before finally coming to rest at the new position. Overall single-step times vary with motor size, step angle and the nature of the load, but are commonly within the range 5–100 ms. This is often fast enough not to be seen by the unwary newcomer, though individual steps can usually be heard; small motors 'tick' when they step, and larger ones make a satisfying 'click' or 'clunk'.

Thirdly, in order to be sure of the absolute position at the end of a stepping sequence, we must know the absolute position at the beginning. This is because a stepping motor is an incremental device. As long as it is not abused, it will always take one step when a drive pulse is supplied, but in order to keep track of absolute position simply by counting the number of drive pulses (and this is after all the main virtue of the system) we must always start the count from a known datum position. Normally the step counter will be 'zeroed' with the motor shaft at the datum position, and will then count up for clockwise direction, and down for anticlockwise rotation. Provided no steps are lost (see later) the number in the step counter will then always indicate the absolute position.

High-speed running and ramping

The discussion so far has been restricted to operation when the step command pulses are supplied at a constant rate, and with sufficiently long intervals between the pulses to allow the rotor to come to rest between steps. Very large numbers of small stepping motors in watches and clocks do operate continuously in this way, stepping perhaps once every second, but most commercial and industrial applications call for a more exacting and varied performance.

To illustrate the variety of operations which might be involved, and to introduce high-speed running, we can look briefly at a typical industrial application. A stepping motor-driven table feed on a numerically controlled milling machine nicely illustrates both of the key operational features discussed earlier. These are the ability to control position (by supplying the desired number of steps) and velocity (by controlling the stepping rate).

The arrangement is shown diagrammatically in Figure 9.3. The motor turns a leadscrew connected to the worktable, so that each motor step causes a precise incremental movement of the workpiece relative to the cutting tool. By making the increment small enough, the fact that the motion is discrete rather than continuous will not cause any difficulties in the machining process. We will assume that we have selected the step

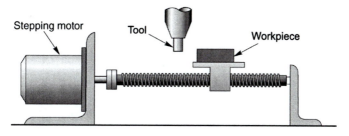

Figure 9.3 *Application of stepping motor for open-loop position control*

angle, the pitch of the leadscrew, and any necessary gearing so as to give a table movement of 0.01 mm per motor step. We will also assume that the necessary step command pulses will be generated by a digital controller or computer, programmed to supply the right number of pulses, at the right speed for the work in hand.

If the machine is a general-purpose one, many different operations will be required. When taking heavy cuts, or working with hard material, the work will have to be offered to the cutting tool slowly, at say, 0.02 mm/s. The stepping rate will then have to be set to 2 steps/s. If we wish to mill out a slot 1-cm long, we will therefore programme the controller to put out 1000 steps, at a uniform rate of 2 steps per second, and then stop. On the other hand, the cutting speed in softer material could be much higher, with stepping rates in the range 10–100 steps per second being in order. At the completion of a cut, it will be necessary to traverse the work back to its original position, before starting another cut. This operation needs to be done as quickly as possible, to minimise unproductive time, and a stepping rate of perhaps 2000 steps per second (or even higher), may be called for.

It was mentioned earlier that a single step (from rest) takes upwards of several milliseconds. It should therefore be clear that if the motor is to run at 2000 steps per second (i.e. 0.5 ms/step), it cannot possibly come to rest between successive steps, as it does at low stepping rates. Instead, we find in practice that at these high stepping rates, the rotor velocity becomes quite smooth, with hardly any outward hint of its stepwise origins. Nevertheless, the vital one-to-one correspondence between step command pulses and steps taken by the motor is maintained throughout, and the open-loop position control feature is preserved. This extraordinary ability to operate at very high stepping rates (up to 20 000 steps per second in some motors), and yet to remain fully in synchronism with the command pulses, is the most striking feature of stepping motor systems.

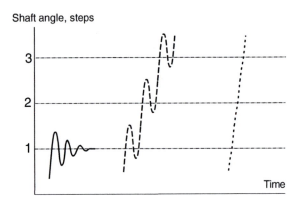

Figure 9.4   *Position-time responses at low, medium and high stepping rates*

Operation at high speeds is referred to as 'slewing'. The transition from single stepping (as shown in Figure 9.2) to high-speed slewing is a gradual one and is indicated by the sketches in Figure 9.4. Roughly speaking, the motor will 'slew' if its stepping rate is above the frequency of its single-step oscillations. When motors are in the slewing range, they generally emit an audible whine, with a fundamental frequency equal to the stepping rate.

It will come as no surprise to learn that a motor cannot be started from rest and expected to 'lock on' directly to a train of command pulses, at say, 2000 steps per second, which is well into the slewing range. Instead, it has to be started at a more modest stepping rate, before being accelerated (or 'ramped') up to speed: this is discussed more fully later in Section 9.5. In undemanding applications, the ramping can be done slowly, and spread over a large number of steps; but if the high stepping rate has to be reached quickly, the timings of individual step pulses must be very precise.

We may wonder what will happen if the stepping rate is increased too quickly. The answer is simply that the motor will not be able to remain 'in step' and will stall. The step command pulses will still be delivered, and the step counter will be accumulating what it believes are motor steps, but, by then, the system will have failed completely. A similar failure mode will occur if, when the motor is slewing, the train of step pulse is suddenly stopped, instead of being progressively slowed. The stored kinetic energy of the motor (and load) will cause it to overrun, so that the number of motor steps will be greater than the number of command pulses. Failures of this sort are prevented by the use of closed-loop control, as discussed later.

Finally, it is worth mentioning that stepping motors are designed to operate for long periods with their rotor held in a fixed (step) position, and with rated current in the winding (or windings). We can therefore anticipate that stalling is generally not a problem for a stepping motor, whereas for most other types of motor, stalling results in a collapse of back e.m.f. and a very high current which can rapidly lead to burnout.

PRINCIPLE OF MOTOR OPERATION

The principle on which stepping motors are based is very simple. When a bar of iron or steel is suspended so that it is free to rotate in a magnetic field, it will align itself with the field. If the direction of the field is changed, the bar will turn until it is again aligned, by the action of the so-called reluctance torque. (The mechanism is similar to that of a compass needle, except that if a compass had an iron needle instead of a permanent magnet it would settle along the earth's magnetic field but it might be rather slow and there would be ambiguity between N and S!)

Before exploring constructional details, it is worth saying a little more about reluctance torque, and its relationship with the torque-producing mechanism we have encountered so far in this book. The alert reader will be aware that, until this chapter, there has been no mention of reluctance torque, and might therefore wonder if it is entirely different from what we have considered so far.

The answer is that in the vast majority of electrical machines, from generators in power stations down to induction and d.c. motors, torque is produced by the interaction of a magnetic field (produced by the stator windings) with current-carrying conductors on the rotor. We based our understanding of how d.c. and induction motors produce torque on the simple formula $F = BIl$ for the force on a conductor of length l carrying a current I perpendicular to a magnetic flux density B (see Chapter 1). There was no mention of reluctance torque because (with very few exceptions) machines which exploit the 'BIl' mechanism do not have reluctance torque.

As mentioned above, reluctance torque originates in the tendency of an iron bar to align itself with magnetic field: if the bar is displaced from its alignment position it experiences a restoring torque. The rotors of machines that produce torque by reluctance action are therefore designed so that the rotor iron has projections or 'poles' (see Figure 9.5) that align with the magnetic field produced by the stator windings. All the torque is then produced by reluctance action, because with no conductors on the rotor to carry current, there is obviously no 'BIl' torque. In contrast, the iron in the rotors of d.c. and induction motors is (ideally) cylindrical, in which case there is no 'preferred' orientation of the rotor iron, i.e. no reluctance torque.

Because the two torque-producing mechanisms appear to be radically different, the approaches taken to develop theoretical models have also diverged. As we have seen, simple equivalent circuits are available to allow us to understand and predict the behaviour of mainstream 'BIl' machines such as d.c. and induction motors, and this is fortunate because of the overwhelming importance of these machines. Unfortunately, no such simple treatments are available for stepping and other reluctance-based machines. Circuit-based numerical models for performance prediction are widely used by manufacturers but they are not really of much use for illuminating behaviour, so we will content ourselves with building up a picture of behaviour from a study of typical operating characteristics.

The two most important types of stepping motor are the variable reluctance (VR) type and the hybrid type. Both types utilise the reluctance principle, the difference between them lying in the method by which the magnetic fields are produced. In the VR type the fields are produced solely by sets of stationary current-carrying windings. The hybrid type also has sets of windings, but the addition of a permanent magnet (on the rotor) gives rise to the description 'hybrid' for this type of motor. Although both types of motor work on the same basic principle, it turns out in practice that the VR type is attractive for the larger step angles (e.g. 15°, 30°, 45°), while the hybrid tends to be best suited when small angles (e.g. 1.8°, 2.5°) are required.

Variable reluctance motor

A simplified diagram of a 30° per step VR stepping motor is shown in Figure 9.5. The stator is made from a stack of steel laminations, and has six equally spaced projecting poles, or teeth, each carrying a separate coil. The rotor, which may be solid or laminated, has four projecting teeth, of the same width as the stator teeth. There is a very small air-gap – typically between 0.02 and 0.2 mm – between rotor and stator teeth. When no current is flowing in any of the stator coils, the rotor will therefore be completely free to rotate.

Diametrically opposite pairs of stator coils are connected in series, such that when one of them acts as a N pole, the other acts as a S pole. There are thus three independent stator circuits, or phases, and each one can be supplied with direct current from the drive circuit (not shown in

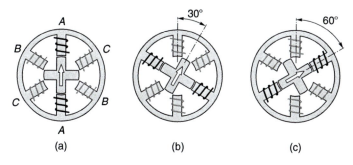

Figure 9.5   *Principle of operation of 30° per step variable reluctance stepping motor*

Figure 9.5). When phase A is energised (as indicated by the thick lines in Figure 9.5(a)), a magnetic field with its axis along the stator poles of phase A is created. The rotor is therefore attracted into a position where the pair of rotor poles distinguished by the marker arrow line up with the field, i.e. in line with the phase A pole, as shown in Figure 9.5(a). When phase A is switched-off, and phase B is switched-on instead, the second pair of rotor poles will be pulled into alignment with the stator poles of phase B, the rotor moving through 30° clockwise to its new step position, as shown in Figure 9.5(b). A further clockwise step of 30° will occur when phase B is switched-off and phase C is switched-on. At this stage the original pair of rotor poles come into play again, but this time they are attracted to stator poles C, as shown in Figure 9.5(c). By repetitively switching on the stator phases in the sequence ABCA, etc. the rotor will rotate clockwise in 30° steps, while if the sequence is ACBA, etc. it will rotate anticlockwise. This mode of operation is known as 'one-phase-on', and is the simplest way of making the motor step. Note that the polarity of the energising current is not significant: the motor will be aligned equally well regardless of the direction of current.

An alternative form of VR motor is the multi-stack type, consisting of several (typically three) magnetically independent sections or 'stacks' within a single housing. In a three-stack motor the rotor will consist of three separate toothed sections on a common shaft, each having the same number of equispaced teeth, but with the teeth on each section displaced by one-third of a tooth pitch from its neighbour. The stator also has three separate stacks each of which looks rather like the stator of a hybrid motor (see Figure 9.6), with the teeth on each stator pole having the same pitch as the rotor teeth. The three stator stacks have their teeth aligned, and each stator has a winding, which excites all of its poles.

For one-phase-on operation each stator is energised sequentially, so that its respective rotor teeth are pulled into alignment with the stator teeth. Stepping occurs because successive sets of rotor teeth are mis-aligned by one step. The principal difference as compared with the single-stack VR motor (see Figure 9.5) is that all the rotor teeth on any one stack contribute to torque when that particular stack is ener-gised. The overall utilisation of materials is no better than the single-stack type, however, because only one-third of the material is energised at a time.

Hybrid motor

A cross-sectional view of a typical 1.8° hybrid motor is shown in Figure 9.6. The stator has eight main poles, each with five teeth, and each main pole carries a simple coil. The rotor has two steel end-caps, each with 50

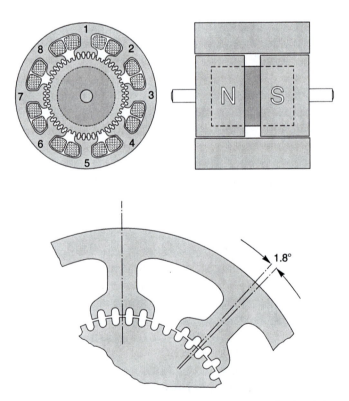

Figure 9.6   *Hybrid (200 steps per revolution) stepping motor. The detail shows the rotor and stator tooth alignments, and indicates the step angle of 1.8°*

Plate 9.2 *Rotor of size 34 (3.4 inch or 8 cm diameter) 3-stack hybrid 1.8° stepping motor. The dimensions of the rotor end-caps and the associated axially-magnetised permanent magnet are optimised for the single-stack version. Extra torque is obtained by adding a second or third stack, the stator simply being stretched to accommodate the longer rotor. (Photo by courtesy of Astrosyn)*

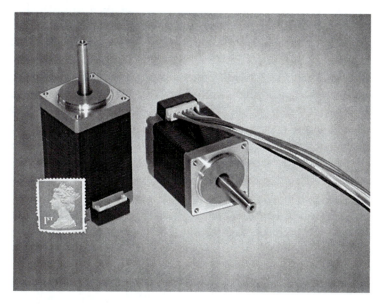

Plate 9.3 *Size 11 (1.1 inch) hybrid motors. (Photo by courtesy of Astrosyn)*

teeth, and separated by a permanent magnet. The rotor teeth have the same pitch as the teeth on the stator poles, and are offset so that the centreline of a tooth at one end-cap coincides with a slot at the other end-cap. The permanent magnet is axially magnetised, so that one set of rotor teeth is given a N polarity, and the other a S polarity.

When no current is flowing in the windings, the only source of magnetic flux across the air-gap is the permanent magnet. The magnetic flux crosses the air-gap from the N end-cap into the stator poles, flows axially along the body of the stator, and returns to the magnet by crossing the air-gap to the S end-cap. If there were no offset between the two sets of rotor teeth, there would be a strong periodic alignment torque when the rotor was turned, and every time a set of stator teeth was in line with the rotor teeth we would obtain a stable equilibrium position. However, there is an offset, and this causes the alignment torque due to the magnet to be almost eliminated. In practice a small 'detent' torque remains, and this can be felt if the shaft is turned when the motor is de-energised: the motor tends to be held in its step positions by the detent torque. This is sometimes very useful: for example, it is usually enough to hold the rotor stationary when the power is switched-off, so the motor can be left overnight without fear of it being accidentally nudged into to a new position.

The eight coils are connected to form two phase-windings. The coils on poles 1, 3, 5 and 7 form phase A, while those on 2, 4, 6 and 8 form phase B. When phase A carries positive current stator poles 1 and 5 are magnetised as S, and poles 3 and 7 become N. The offset teeth on the N end of the rotor are attracted to poles 1 and 5 while the offset teeth at the S end of the rotor are attracted into line with the teeth on poles 3 and 7. To make the rotor step, phase A is switched-off, and phase B is energised with either positive current or negative current, depending on the sense of rotation required. This will cause the rotor to move by one-quarter of a tooth pitch (1.8°) to a new equilibrium (step) position.

The motor is continuously stepped by energising the phases in the sequence +A, − B, − A, + B, + A (clockwise) or +A, + B, − A, − B, +A (anticlockwise). It will be clear from this that a bipolar supply is needed (i.e. one which can furnish +ve or −ve current). When the motor is operated in this way it is referred to as 'two-phase, with bipolar supply'.

If a bipolar supply is not available, the same pattern of pole energisation may be achieved in a different way, as long as the motor windings consist of two identical ('bifilar wound') coils. To magnetise pole 1 north, a positive current is fed into one set of phase A coils. But to magnetise pole 1 south, the same positive current is fed into the other set of phase A coils, which have the opposite winding sense. In total, there are then four separate windings, and when the motor is operated in this way it is referred to as '4-phase, with unipolar supply'. Since each winding only occupies half of the space, the MMF of each winding is only half of that of the full coil, so the thermally rated output is clearly reduced as compared with bipolar operation (for which the whole winding is used).

We round off this section on hybrid motors with a comment on identifying windings, and a warning. If the motor details are not known, it is usually possible to identify bifilar windings by measuring the resistance from the common to the two ends. If the motor is intended for unipolar drive only, one end of each winding may be commoned inside the casing; for example, a 4-phase unipolar motor may have only five leads, one for each phase and one common. Wires are also usually colour-coded to indicate the location of the windings; for example, a bifilar winding on one set of poles will have one end red, the other end red and white and the common white. Finally, it is not advisable to remove the rotor of a hybrid motor because they are magnetised in situ: removal typically causes a 5–10% reduction in magnet flux, with a corresponding reduction in static torque at rated current.

Summary

The construction of stepping motors is simple, the only moving part being the rotor, which has no windings, commutator or brushes: they are therefore robust and reliable. The rotor is held at its step position solely by the action of the magnetic flux between stator and rotor. The step angle is a property of the tooth geometry and the arrangement of the stator windings, and accurate punching and assembly of the stator and rotor laminations is therefore necessary to ensure that adjacent step positions are exactly equally spaced. Any errors due to inaccurate punching will be non-cumulative, however.

The step angle is obtained from the expression

$$\text{Step angle} = \frac{360°}{(\text{rotor teeth}) \times (\text{stator phases})}$$

The VR motor in Figure 9.5 has four rotor teeth, three stator phase windings and the step angle is therefore 30°, as already shown. It should also be clear from the equation why small angle motors always have to have a large number of rotor teeth. Probably the most widely used motor is the 200 steps per revolution hybrid type (see Figure 9.6). This has a 50 tooth rotor, 4-phase stator, and hence a step angle of 1.8° ($= 360°/(50 \times 4)$).

The magnitude of the aligning torque clearly depends on the magnitude of the current in the phase winding. However, the equilibrium positions itself does not depend on the magnitude of the current, because it is simply the position where the rotor and stator teeth are in line. This property underlines the digital nature of the stepping motor.

MOTOR CHARACTERISTICS

Static torque–displacement curves

From the previous discussion, it should be clear that the shape of the torque–displacement curve, and in particular the peak static torque, will depend on the internal electromagnetic design of the rotor. In particular the shapes of the rotor and stator teeth, and the disposition of the stator windings (and permanent magnet(s)) all have to be optimised to obtain the maximum static torque.

We now turn to a typical static torque–displacement curve, and look at how it determines motor behaviour. Several aspects will be discussed, including the explanation of basic stepping (which has already been looked at in a qualitative way); the influence of load torque on step position accuracy; the effect of the amplitude of the winding current; and half-step and mini-stepping operation. For the sake of simplicity, the discussion will be based on the 30° per step 3-phase VR motor introduced earlier, but the conclusions reached apply to any stepping motor.

Typical static torque–displacement curves for a 3-phase 30° per step VR motor are shown in Figure 9.7. These show the torque that has to be

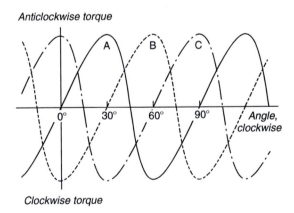

Figure 9.7 *Static torque–displacement curves for 30° per step variable reluctance stepping motor*

applied to move the rotor away from its aligned position. Because of the rotor–stator symmetry, the magnitude of the restoring torque when the rotor is displaced by a given angle in one direction is the same as the magnitude of the restoring torque when it is displaced by the same angle in the other direction, but of opposite sign.

There are three curves in Figure 9.7, one for each of the three phases, and for each curve we assume that the relevant phase winding carries its full (rated) current. If the current is less than rated, the peak torque will be reduced, and the shape of the curve is likely to be somewhat different. The convention used in Figure 9.7 is that a clockwise displacement of the rotor corresponds to a movement to the right, while a positive torque tends to move the rotor anticlockwise.

When only one phase, say A, is energised, the other two phases exert no torque, so their curves can be ignored and we can focus attention on the solid line in Figure 9.7. Stable equilibrium positions (for phase A excited) exist at $\theta = 0°$, $90°$, $180°$ and $270°$. They are stable (step) positions because any attempt to move the rotor away from them is resisted by a counteracting or restoring torque. These points correspond to positions where successive rotor poles (which are $90°$ apart) are aligned with the stator poles of phase A, as shown in Figure 9.5(a). There are also four unstable equilibrium positions, (at $\theta = 45°$, $135°$, $225°$ and $315°$) at which the torque is also zero. These correspond to rotor positions where the stator poles are midway between two rotor poles, and they are unstable because if the rotor is deflected slightly in either direction, it will be accelerated in the same direction until it reaches the next stable position. If the rotor is free to turn, it will therefore always settle in one of the four stable positions.

Single-stepping

If we assume that phase A is energised, and the rotor is at rest in the position $\theta = 0°$ (see Figure 9.7) we know that if we want to step in a clockwise direction, the phases must be energised in the sequence ABCA, etc., so we can now imagine that phase A is switched-off, and phase B is energised instead. We will also assume that the decay of current in phase A and the build-up in phase B take place very rapidly, before the rotor moves significantly.

The rotor will find itself at $\theta = 0°$, but it will now experience a clockwise torque (see Figure 9.7) produced by phase B. The rotor will therefore accelerate clockwise, and will continue to experience clockwise torque, until it has turned through $30°$. The rotor will be accelerating all the time, and it will therefore overshoot the $30°$ position, which is of course its target (step) position for phase B. As soon as it overshoots, however, the torque reverses, and the rotor experiences a braking torque, which brings it to rest before accelerating it back towards the 30° position. If there was no friction or other cause of damping, the rotor would continue to oscillate; but in practice it comes to rest at its new position quite quickly in much the same way as a damped second-order system. The next 30° step is achieved in the same way, by switching-off the current in phase B, and switching-on phase C.

In the discussion above, we have recognised that the rotor is acted on sequentially by each of the three separate torque curves shown in Figure 9.7. Alternatively, since the three curves have the same shape, we can think of the rotor being influenced by a single torque curve, which 'jumps' by one step (30° in this case) each time the current is switched from one phase to the next. This is often the most convenient way of visualising what is happening in the motor.

Step position error and holding torque

In the previous discussion the load torque was assumed to be zero, and the rotor was therefore able to come to rest with its poles exactly in line with the excited stator poles. When load torque is present, however, the rotor will not be able to pull fully into alignment, and a 'step position error' will be unavoidable.

The origin and extent of the step position error can be appreciated with the aid of the typical torque–displacement curve shown in Figure 9.8. The true step position is at the origin in the figure, and this is where the rotor would come to rest in the absence of load torque. If we imagine the rotor is

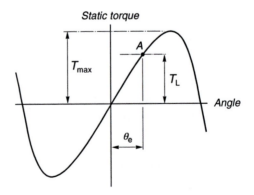

Figure 9.8   *Static torque–angle curve showing step position error (θ_e) resulting from load torque T_L*

initially at this position, and then consider that a clockwise load (T_L) is applied, the rotor will move clockwise, and as it does so it will develop progressively more anticlockwise torque. The equilibrium position will be reached when the motor torque is equal and opposite to the load torque, i.e. at point A in Figure 9.8. The corresponding angular displacement from the step position (θ_e in Figure 9.8) is the step position error.

The existence of a step position error is one of the drawbacks of the stepping motor. The motor designer attempts to combat the problem by aiming to produce a steep torque–angle curve around the step position, and the user has to be aware of the problem and choose a motor with a sufficiently steep curve to keep the error within acceptable limits. In some cases this may mean selecting a motor with a higher peak torque than would otherwise be necessary, simply to obtain a steep enough torque–angle curve around the step position.

As long as the load torque is less than T_{max} (see Figure 9.8), a stable rest position is obtained, but if the load torque exceeds T_{max}, the rotor will be unable to hold its step position. T_{max} is therefore known as the 'holding' torque. The value of the holding torque immediately conveys an idea of the overall capability of any motor, and it is – after step angle – the most important single parameter, which is looked for in selecting a motor. Often, the adjective 'holding' is dropped altogether: for example 'a 1-Nm motor' is understood to be one with a peak static torque (holding torque) of 1 Nm.

Half stepping

We have already seen how to step the motor in 30° increments by energising the phases one at a time in the sequence ABCA, etc. Although this 'one-phase-on' mode is the simplest and most widely used, there are two other modes, which are also frequently employed. These are referred to as the 'two-phase-on' mode and the 'half-stepping' mode. The two-phase-on can provide greater holding torque and a much better damped single-step response than the one-phase-on mode; and the half-stepping mode permits the effective step angle to be halved – thereby doubling the resolution – and produces a smoother shaft rotation.

In the two-phase-on mode, two phases are excited simultaneously. When phases A and B are energised, for example, the rotor experiences torques from both phases, and comes to rest at a point midway between the two adjacent full step positions. If the phases are switched in the sequence AB, BC, CA, AB, etc., the motor will take full (30°) steps, as in the one-phase-on mode, but its equilibrium positions will be interleaved between the full step positions.

To obtain 'half stepping' the phases are excited in the sequence A, AB, B, BC, etc., i.e. alternately in the one-phase-on and two-phase-on modes. This is sometimes known as 'wave' excitation, and it causes the rotor to advance in steps of 15°, or half the full step angle. As might be expected, continuous half stepping usually produces a smoother shaft rotation than full stepping, and it also doubles the resolution.

We can see what the static torque curve looks like when two phases are excited by superposition of the individual phase curves. An example is shown in Figure 9.9, from which it can be seen that for this machine, the holding torque (i.e. the peak static torque) is higher with two phases excited than with only one excited. The stable equilibrium (half-step) position is at 15°, as expected. The price to be paid for the increased holding torque is the increased power dissipation in the windings, which is doubled as compared with the one-phase-on mode. The holding torque increases by a factor less than two, so the torque per watt (which is a useful figure of merit) is reduced.

A word of caution is needed in regard to the addition of the two separate one-phase-on torque curves to obtain the two-phase-on curve. Strictly, such a procedure is only valid where the two phases are magnetically independent, or the common parts of the magnetic circuits are unsaturated. This is not the case in most motors, in which the phases share a common magnetic circuit, which operates under highly saturated conditions. Direct addition of the one-phase-on curves cannot therefore be expected to give an accurate result for the two-phase-on curve, but it is easy to do, and provides a reasonable estimate.

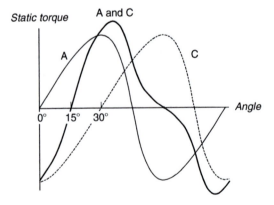

Figure 9.9   *Static torque–angle curve (thick line) corresponding to two-phase-on excitation*

Apart from the higher holding torque in the two-phase-on mode, there is another important difference which distinguishes the static behaviour from that of the one-phase-on mode. In the one-phase-on mode, the equilibrium or step positions are determined solely by the geometry of the rotor and stator: they are the positions where the rotor and stator are in line. In the two-phase-on mode, however, the rotor is intended to come to rest at points where the rotor poles are lined-up midway between the stator poles. This position is not sharply defined by the 'edges' of opposing poles, as in the one-phase-on case; and the rest position will only be exactly midway if (a) there is exact geometrical symmetry and, more importantly (b) the two currents are identical. If one of the phase currents is larger than the other, the rotor will come to rest closer to the phase with the higher current, instead of halfway between the two. The need to balance the currents to obtain precise half stepping is clearly a drawback to this scheme. Paradoxically, however, the properties of the machine with unequal phase currents can sometimes be turned to good effect, as we now see.

Step division – mini-stepping

There are some applications (e.g. in printing and phototypesetting) where very fine resolution is called for, and a motor with a very small step angle – perhaps only a fraction of a degree – is required. We have already seen that the step angle can only be made small by increasing the number of rotor teeth and/or the number of phases, but in practice it is inconvenient to have more than four or five phases, and it is difficult to manufacture rotors with more than 50–100 teeth. This means it is rare for motors to have step angles below about 1°. When a smaller step angle is required a technique known as mini-stepping (or step division) is used.

Mini-stepping is a technique based on two-phase-on operation which provides for the subdivision of each full motor step into a number of 'substeps' of equal size. In contrast with half stepping, where the two currents have to be kept equal, the currents are deliberately made unequal. By correctly choosing and controlling the relative amplitudes of the currents, the rotor equilibrium position can be made to lie anywhere between the step positions for each of the two separate phases.

Closed-loop current control is needed to prevent the current from changing as a result of temperature changes in the windings, or variations in the supply voltage; and if it is necessary to ensure that the holding torque stays constant for each ministep both currents must be changed according to a prescribed algorithm. Despite the difficulties referred to above, mini-stepping is used extensively, especially in photographic and printing applications where a high resolution is needed. Schemes involving between 3 and 10 ministeps for a 1.8° step motor are numerous, and there are instances where up to 100 ministeps (20 000 ministeps/rev) have been successfully achieved.

So far, we have concentrated on those aspects of behaviour, which depend only on the motor itself, i.e. the static performance. The shape of the static torque curve, the holding torque and the slope of the torque curve about the step position have all been shown to be important pointers to the way the motor can be expected to perform. All of these characteristics depend on the current(s) in the windings, however, and when the motor is running the instantaneous currents will depend on the type of drive circuit employed, as discussed in the next two sections.

STEADY-STATE CHARACTERISTICS – IDEAL (CONSTANT-CURRENT) DRIVE

In this section, we will look at how the motor would perform if it were supplied by an ideal drive circuit, which turns out to be one that is capable of supplying rectangular pulses of current to each winding when required, and regardless of the stepping rate. Because of the inductance of the windings, no real drive circuit will be able to achieve this, but the most sophisticated (and expensive) ones achieve near-ideal operation up to very high stepping rates.

Requirements of drive

The basic function of the complete drive is to convert the step command input signals into appropriate patterns of currents in the motor windings. This is achieved in two distinct stages, as shown in Figure 9.10, which relates to a 3-phase motor.

The 'translator' stage converts the incoming train of step command pulses into a sequence of on/off commands to each of the three power stages. In the one-phase-on mode, for example, the first step command pulse will be routed to turn on phase A, the second will turn on phase B and so on. In a very simple drive, the translator will probably provide for only one mode of operation (e.g. one-phase-on), but most commercial drives provide the option of one-phase-on, two-phase-on and half stepping. Single-chip ICs with these three operating modes and with both three-phase and four-phase outputs are readily available.

The power stages (one per phase) supply the current to the windings. An enormous diversity of types are in use, ranging from simple ones with one switching transistor per phase, to elaborate chopper-type circuits with four transistors per phase, and some of these are discussed in Section 9.5. At this point, however, it is helpful to list the functions required of the 'ideal' power stage. These are firstly that when the translator calls for a phase to be energised, the full (rated) current should be established immediately; secondly, the current should be maintained constant (at its rated value) for the duration of the 'on' period and finally, when the translator calls for the current to be turned off, it should be reduced to zero immediately.

The ideal current waveforms for continuous stepping with one-phase-on operation are shown in the lower part of Figure 9.10. The currents have a square profile because this leads to the optimum value of running torque from the motor. But because of the inductance of the windings, no real drive will achieve the ideal current waveforms, though many drives come close to the ideal, even at quite high stepping rates. Drives which produce such rectangular current waveforms are (not surprisingly) called constant-current drives. We now look at the running torque produced by a motor when operated from an ideal constant-current drive. This will act as a yardstick for assessing the performance of other drives, all of which will be seen to have inferior performance.

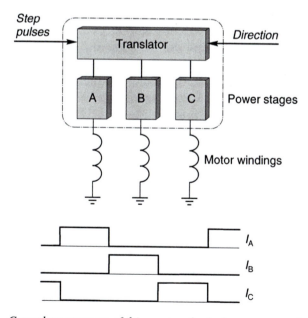

Figure 9.10   *General arrangement of drive system for 3-phase motor, and winding currents corresponding to an 'ideal' drive*

Pull-out torque under constant-current conditions

If the phase currents are taken to be ideal, i.e. they are switched-on and switched-off instantaneously, and remain at their full-rated value during each 'on' period, we can picture the axis of the magnetic field to be advancing around the machine in a series of steps, the rotor being urged to follow it by the reluctance torque. If we assume that the inertia is high enough for fluctuations in rotor velocity to be very small, the rotor will be rotating at a constant rate, which corresponds exactly to the stepping rate.

Now if we consider a situation where the position of the rotor axis is, on average, lagging behind the advancing field axis, it should be clear that, on average, the rotor will experience a driving torque. The more it lags behind, the higher will be the average forward torque acting on it, but only up to a point. We already know that if the rotor axis is displaced too far from the field axis, the torque will begin to diminish, and eventually reverse, so we conclude that although more torque will be developed by increasing the rotor lag angle, there will be a limit to how far this can be taken.

Turning now to a quantitative examination of the torque on the rotor, we will make use of the static torque–displacement curves discussed earlier, and look at what happens when the load on the shaft is varied, the stepping rate being kept constant. Clockwise rotation will be studied, so the phases will be energised in the sequence ABC. The instantaneous torque on the rotor can be arrived at by recognising (a) that the rotor speed is constant, and it covers one-step angle (30°) between step command pulses, and (b) the rotor will be 'acted on' sequentially by each of the set of torque curves.

When the load torque is zero, the net torque developed by the rotor must be zero (apart from a very small torque required to overcome friction). This condition is shown in Figure 9.11(a) . The instantaneous torque is shown by the thick line, and it is clear that each phase in turn exerts first a clockwise torque, then an anticlockwise torque while the rotor angle turns through 30°. The average torque is zero, the same as the load torque, because the average rotor lag angle is zero.

When the load torque on the shaft is increased, the immediate effect is to cause the rotor to fall back in relation to the field. This causes the clockwise torque to increase, and the anticlockwise torque to decrease. Equilibrium is reached when the lag angle has increased sufficiently for the motor torque to equal the load torque. The torque developed at an intermediate load condition like this is shown by the thick line in Figure 9.11(b). The highest average torque that can possibly be developed is shown by the thick line in Figure 9.11(c): if the load torque exceeds this value (which is

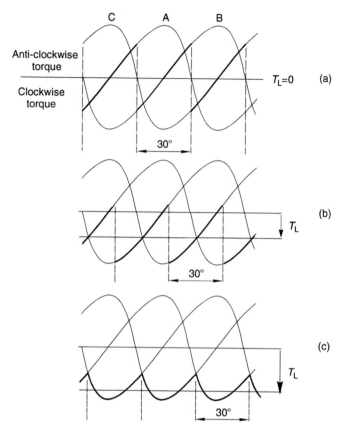

Figure 9.11 *Static torque curves indicating how the average steady-state torque (T_L) is developed during constant-frequency operation*

known as the pull-out torque) the motor loses synchronism and stalls, and the vital one-to-one correspondences between pulses and steps are lost.

Since we have assumed an ideal constant-current drive, the pull-out torque will be independent of the stepping rate, and the pull-out torque–speed curve under ideal conditions is therefore as shown in Figure 9.12. The shaded area represents the permissible operating region: at any particular speed (stepping rate) the load torque can have any value up to the pull-out torque, and the motor will continue to run at the same speed. But if the load torque exceeds the pull-out torque, the motor will suddenly pull out of synchronism and stall.

As mentioned earlier, no real drive will be able to provide the ideal current waveforms, so we now turn to look briefly at the types of drives in common use, and at their pull-out torque–speed characteristics.

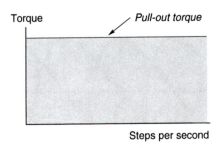

Figure 9.12 *Steady-state operating region with ideal constant-current drive.* (In such idealised circumstances there would be no limit to the stepping rate, but as shown in Figure 9.14 a real drive circuit imposes an upper limit.)

DRIVE CIRCUITS AND PULL-OUT TORQUE–SPEED CURVES

Users often find difficulty in coming to terms with the fact that the running performance of a stepping motor depends so heavily on the type of drive circuit being used. It is therefore important to emphasise that in order to meet a specification, it will always be necessary to consider the motor and drive together, as a package.

There are three commonly used types of drive. All use transistors, which are operated as switches, i.e. they are either turned fully on, or they are cut-off. A brief description of each is given below, and the pros and cons of each type are indicated. In order to simplify the discussion, we will consider one phase of a 3-phase VR motor and assume that it can be represented by a simple series R–L circuit in which R and L are the resistance and self-inductance of the winding, respectively. (In practice the inductance will vary with rotor position, giving rise to motional e.m.f. in the windings, which, as we have seen previously in this book, is an inescapable manifestation of an electromechanical energy-conversion process. If we needed to analyse stepping motor behaviour fully we would have to include the motional e.m.f. terms. Fortunately, we can gain a pretty good appreciation of how the motor behaves if we model each winding simply in terms of its resistance and self-inductance.)

Constant-voltage drive

This is the simplest possible drive: the circuit for one of the three phases is shown in the upper part of Figure 9.13, and the current waveforms at low and high stepping rates are shown in the lower part of the figure. The d.c. voltage V is chosen so that when the transistor is on, the steady

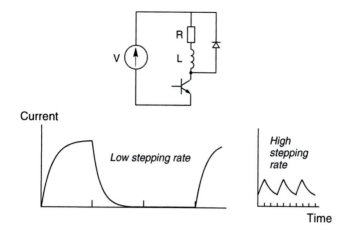

Figure 9.13 *Basic constant-voltage drive circuit and typical current waveforms*

current (equal to V/R if we neglect the on-state voltage drop across the transistor) is the rated current as specified by the motor manufacturer.

The current waveforms display the familiar rising exponential shape that characterises a first-order system: the time-constant is L/R, the current reaching its steady state after several time-constants. When the transistor switches off, the stored energy in the inductance cannot instantaneously reduce to zero, so although the current through the transistor suddenly becomes zero, the current in the winding is diverted into the closed path formed by the winding and the freewheel diode, and it then decays exponentially to zero, again with time-constant L/R. In this phase the stored energy in the magnetic field is dissipated as heat in the resistance of the winding and diode.

At low stepping rates (low speed), the drive provides a reasonably good approximation to the ideal rectangular current waveform. (We are considering a 3-phase motor, so ideally one phase should be on for one step pulse and off for the next two, as in Figure 9.10.) But at higher frequencies (right-hand waveform in Figure 9.13), where the 'on' period is short compared with the winding time-constant, the current waveform degenerates, and is nothing like the ideal rectangular shape. In particular the current never gets anywhere near its full value during the on pulse, so the torque over this period is reduced; and even worse, a substantial current persists when the phase is supposed to be off, so during this period the phase will contribute a negative torque to the rotor. Not surprisingly all this results in a very rapid fall-off of pull-out torque with speed, as shown in Figure 9.16(a).

Curve (a) in Figure 9.16 should be compared with the pull-out torque under ideal constant-current conditions shown in Figure 9.12 in order to appreciate the severely limited performance of the simple constant-voltage drive.

Current-forced drive

The initial rate of rise of current in a series R–L circuit is directly proportional to the applied voltage, so in order to establish the current more quickly at switch-on, a higher supply voltage (V_f) is needed. But if we simply increased the voltage, the steady-state current (V_f/R) would exceed the rated current and the winding would overheat.

To prevent the current from exceeding the rated value, an additional 'forcing' resistor has to be added in series with the winding. The value of this resistance (R_f) must be chosen so that $V_f/(R + R_f) = I$, where I is the rated current. This is shown in the upper part of Figure 9.14, together with the current waveforms at low and high stepping rates. Because the rates of rise and fall of current are higher, the current waveforms approximate more closely to the ideal rectangular shape, especially at low stepping rates, though at higher rates they are still far from ideal, as shown in Figure 9.14. The low-frequency pull-out torque is therefore maintained to a higher stepping rate, as shown in Figure 9.16(b). Values of R_f from 2 to 10 times the motor resistance (R) are common. Broadly speaking, if $R_f = 10R$, for example, a given pull-out torque will be available at 10 times the stepping rate, compared with an unforced constant-voltage drive.

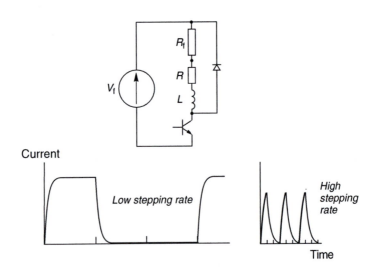

Figure 9.14   *Current-forced (L/R) drive and typical current waveforms*

Some manufacturers call this type of drive an 'R/L' drive, while others call it an 'L/R' drive, or even simply a 'resistor drive'. Often, sets of pull-out torque–speed curves in catalogues are labelled with values R/L (or L/R) = 5, 10, etc. This means that the curves apply to drives where the forcing resistor is five (or ten) times the winding resistance, the implication being that the drive voltage has also been adjusted to keep the static current at its rated value. Obviously, it follows that the higher R_f is made, the higher the power rating of the supply; and it is the higher power rating which is the principal reason for the improved torque–speed performance.

The major disadvantage of this drive is its inefficiency, and the consequent need for a high power-supply rating. Large amounts of heat are dissipated in the forcing resistors, especially when the motor is at rest and the phase current is continuous, and disposing of the heat can lead to awkward problems in the siting of the forcing resistors.

It was mentioned earlier that the influence of the motional e.m.f. in the winding would be ignored. In practice, however, the motional e.m.f. always has a pronounced influence on the current, especially at high stepping rates, so it must be borne in mind that the waveforms shown in Figures 9.13 and 9.14 are only approximate. Not surprisingly, it turns out that the motional e.m.f. tends to make the current waveforms worse (and the torque less) than the discussion above suggests. Ideally therefore, we need a drive, which will keep the current constant throughout the on period, regardless of the motional e.m.f. The closed-loop chopper-type drive (below) provides the closest approximation to this, and also avoids the waste of power, which is a feature of R/L drives.

Chopper drive

The basic circuit for one phase of a VR motor is shown in the upper part of Figure 9.15 together with the current waveforms. A high-voltage power supply is used in order to obtain very rapid changes in current when the phase is switched-on or off.

The lower transistor is turned on for the whole period during which current is required. The upper-transistor turns on whenever the actual current falls below the lower threshold (shown dotted in Figure 9.15) and it turns off when the current exceeds the upper threshold. The chopping action leads to a current waveform that is a good approximation to the ideal (see Figure 9.10). At the end of the on period both transistors switch off and the current freewheels through both diodes and back to the supply. During this period the stored energy in the

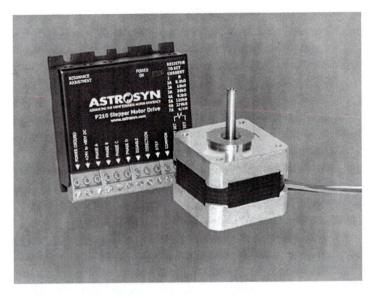

Plate 9.4 *Bipolar constant-current chopper drive with size 17 hybrid motor. This versatile drive draws its power from a d.c. supply (between 24 V and 80 V) and the output phase current can be set (using programming resistors) to any value in the range 0.3 A to 7 A. A resonance adjustment is provided. (Photo courtesy of Astrosyn)*

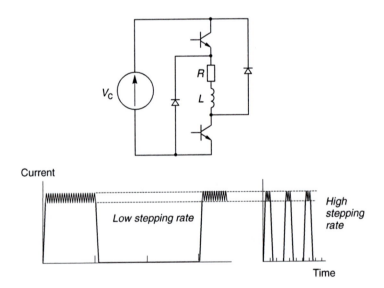

Figure 9.15 *Constant-current chopper drive and typical current waveforms*

Pull-out torque

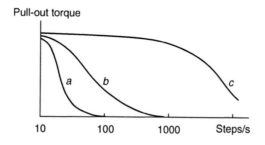

Figure 9.16 *Typical pull-out torque–speed curves for a given motor with different types of drive circuit.* (a) *constant voltage drive;* (b) *current-forced drive;* (c) *chopper drive*

inductance is returned to the supply, and because the winding terminal voltage is then $-V_c$, the current decays as rapidly as it built up.

Because the current-control system is a closed-loop one, distortion of the current waveform by the motional e.m.f. is minimised, and this means that the ideal (constant-current) torque–speed curve is closely followed up to high stepping rates. Eventually, however, the 'on' period reduces to the point where it is less than the current rise time, and the full current is never reached. Chopping action then ceases, the drive reverts essentially to a constant-voltage one, and the torque falls rapidly as the stepping rate is raised even higher, as in Figure 9.16(c). There is no doubt about the overall superiority of the chopper-type drive, and it is gradually becoming the standard drive. Single-chip chopper modules can be bought for small (say 1–2 A) motors, and complete plug-in chopper cards, rated up to 10 A or more are available for larger motors.

The discussion in this section relates to a VR motor, for which unipolar current pulses are sufficient. If we have a hybrid or other permanent magnet motor we will need a bipolar current source (i.e. one that can provide positive or negative current), and for this we will find that each phase is supplied from a four-transistor H-bridge, as discussed in Chapter 2, Section 2.4.1.

Resonances and instability

In practice, measured torque–speed curves frequently display severe dips at or around certain stepping rates. Manufacturers are not keen to stress this feature, and sometimes omit the dips from their curves, so it is doubly important for the user to be on the look out for them. A typical measured characteristic for a hybrid motor with a voltage-forced drive is shown as (a) in Figure 9.17.

The magnitude and location of the torque dips depend in a complex way on the characteristics of the motor, the drive, the operating mode and the load. We will not go into detail here, apart from mentioning the underlying causes and remedies.

There are two distinct mechanisms that cause the dips. The first is a straightforward 'resonance-type' problem, which manifests itself at low stepping rates, and originates from the oscillatory nature of the single-step response. Whenever the stepping rate coincides with the natural frequency of rotor oscillations, the oscillations can be enhanced, and this in turn makes it more likely that the rotor will fail to keep in step with the advancing field.

The second phenomenon occurs because at certain stepping rates it is possible for the complete motor/drive system to exhibit positive feedback, and become unstable. This instability usually occurs at relatively high stepping rates, well above the 'resonance' regions discussed above. The resulting dips in the torque–speed curve are extremely sensitive to the degree of viscous damping present (mainly in the bearings), and it is not uncommon to find that a severe dip, which is apparent on a warm day (such as that shown at around 1000 steps per second in Figure 9.17) will disappear on a cold day.

The dips are most pronounced during steady-state operation, and it may be that their presence is not serious provided that continuous operation at the relevant speeds is not required. In this case, it is often possible to accelerate through the dips without adverse effect. Various special-drive techniques exist for eliminating resonances by smoothing out the stepwise nature of the stator field, or by modulating the supply frequency to damp out the instability, but the simplest remedy in open-loop operation is to fit a damper to the motor shaft. Dampers of the

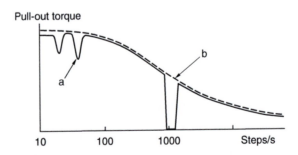

Figure 9.17   *Pull-out torque–speed curves for a hybrid stepping motor showing (curve a) low-speed resonance dips and mid-frequency instability at around 1000 steps per second; and improvement brought about by adding an inertial damper (curve b)*

Lanchester type or of the viscously coupled inertia (VCID-type) are used. These consist of a lightweight housing, which is fixed rigidly to the motor shaft, and an inertia, which can rotate relative to the housing. The inertia and the housing are separated either by a viscous fluid (VCID-type) or by a friction disc (Lanchester type). Whenever the motor speed is changing, the assembly exerts a damping torque, but once the motor speed is steady, there is no drag torque from the damper. By selecting the appropriate damper, the dips in the torque–speed curve can be eliminated, as shown in Figure 9.17(b). Dampers are also often essential to damp the single-step response, particularly with VR motors, many of which have a highly oscillatory step response. Their only real drawback is that they increase the effective inertia of the system, and thus reduce the maximum acceleration.

TRANSIENT PERFORMANCE

Step response

It was pointed out earlier that the single-step response is similar to that of a damped second-order system. We can easily estimate the natural frequency ω_n in rad/s from the equation

$$\omega_n^2 = \frac{\text{slope of torque} - \text{angle curve}}{\text{total inertia}}$$

Knowing ω_n, we can judge what the oscillatory part of the response will look like, by assuming the system is undamped. To refine the estimate, and to obtain the settling time, however, we need to estimate the damping ratio, which is much more difficult to determine as it depends on the type of drive circuit and mode of operation as well as on the mechanical friction. In VR motors the damping ratio can be as low as 0.1, but in hybrid types it is typically 0.3–0.4. These values are too low for many applications where rapid settling is called for.

Two remedies are available, the simplest being to fit a mechanical damper of the type mentioned above. Alternatively, a special sequence of timed command pulses can be used to brake the rotor so that it reaches its new step position with zero velocity and does not overshoot. This procedure is variously referred to as 'electronic damping', 'electronic braking' or 'back phasing'. It involves re-energising the previous phase for a precise period before the rotor has reached the next step position, in order to exert just the right degree of braking. It can only be used successfully when the load torque and inertia are predictable and not subject to change. Because it is an open-loop scheme it is extremely sensitive to apparently minor changes such as day-to-day variation in friction, which can make it unworkable in many instances.

Starting from rest

The rate at which the motor can be started from rest without losing steps is known as the 'starting' or 'pull-in' rate. The starting rate for a given motor depends on the type of drive, and the parameters of the load. This is entirely as expected since the starting rate is a measure of the motor's ability to accelerate its rotor and load and pull into synchronism with the field. The starting rate thus reduces if either the load torque, or the load inertia are increased. Typical pull-in torque–speed curves, for various inertias, are shown in Figure 9.18. The pull-out torque–speed curve is also shown, and it can be seen that for a given load torque, the maximum steady (slewing) speed at which the motor can run is much higher than the corresponding starting rate. (Note that only one pull-out torque is usually shown, and is taken to apply for all inertia values. This is because the inertia is not significant when the speed is constant.)

It will normally be necessary to consult the manufacturer's data to obtain the pull-in rate, which will apply only to a particular drive. However, a rough assessment is easily made: we simply assume that the motor is producing its pull-out torque, and calculate the acceleration that this would produce, making due allowance for the load torque and inertia. If, with the acceleration as calculated, the motor is able to reach the steady speed in one step or less, it will be able to pull-in; if not, a lower pull-in rate is indicated.

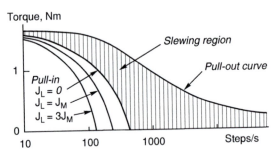

Figure 9.18 *Typical pull-in and pull-out curves showing effect of load inertia on the pull-in torque.* (J_M = motor inertia; J_L = load inertia)

Optimum acceleration and closed-loop control

There are some applications where the maximum possible accelerations and decelerations are demanded, in order to minimise point-to-point times. If the load parameters are stable and well defined, an open-loop approach is feasible, and this is discussed first. Where the load is unpredictable, however, a closed-loop strategy is essential, and this is dealt with later.

To achieve maximum possible acceleration calls for every step command pulse to be delivered at precisely optimised intervals during the acceleration period. For maximum torque, each phase must be on whenever it can produce positive torque, and off when its torque would be negative. Since the torque depends on the rotor position, the optimum switching times have to be calculated from a full dynamic analysis. This can usually be accomplished by making use of the static torque–angle curves (provided appropriate allowance is made for the rise and fall times of the stator currents), together with the torque–speed characteristic and inertia of the load. A series of computations is required to predict the rotor angle–time relationship, from which the switchover points from one phase to the other are deduced. The train of accelerating pulses is then pre-programmed into the controller, for subsequent feeding to the drive in an open-loop fashion. It is obvious that this approach is only practicable if the load parameters do not vary, since any change will invalidate the computed optimum stepping intervals.

When the load is unpredictable, a much more satisfactory arrangement is obtained by employing a closed-loop scheme, using position feedback from a shaft-mounted encoder. The feedback signals indicate the instantaneous position of the rotor, and are used to ensure that the phase-windings are switched at precisely the right rotor position for maximising the developed torque. Motion is initiated by a single command pulse, and subsequent step command pulses are effectively self-generated by the encoder. The motor continues to accelerate until its load torque equals the load torque, and then runs at this (maximum) speed until the deceleration sequence is initiated. During all this time, the step counter continues to record the number of steps taken.

Closed-loop operation ensures that the optimum acceleration is achieved, but at the expense of more complex control circuitry, and the need to fit a shaft encoder. Relatively cheap encoders are, however, now available for direct fitting to some ranges of motors, and single chip microcontrollers are available which provide all the necessary facilities for closed-loop control.

An appealing approach aimed at eliminating an encoder is to detect the position of the rotor by online analysis of the signals (principally the rates of change of currents) in the motor windings themselves: in other words, to use the motor as its own encoder. A variety of approaches have been tried, including the addition of high-frequency alternating voltages superimposed on the excited phase, so that as the rotor moves the variation of inductance results in a modulation of the alternating current component. Some success has been achieved with particular motors, but the approach has not yet achieved widespread commercial exploitation.

To return finally to encoders, we should note that they are also used in open-loop schemes when an absolute check on the number of steps taken is required. In this context the encoder simply provides a tally of the total steps taken, and normally plays no part in the generation of the step pulses. At some stage, however, the actual number of steps taken will be compared with the number of step command pulses issued by the controller. When a disparity is detected, indicating a loss (or gain) of steps), the appropriate additional forward or backward pulses can be added.

REVIEW QUESTIONS

1) Why are the step positions likely to be less well defined when a motor is operated in 'two-phase-on' mode as compared with one-phase-on mode?

2) What is meant by detent torque, and in what type of motors does detent torque occur?

3) What is meant by the 'holding torque' of a stepping motor?

4) The static torque curve of a 3-phase VR stepper is approximately sinusoidal, the peak torque at rated current being 0.8 Nm. Find the step position error when a steady load torque of 0.25 Nm is present.

5) For the motor in question 4, estimate the low-speed pull-out torque when the motor is driven by a constant-current drive.

6) The static torque curve of a particular $1.8°$ hybrid step motor can be approximated by a straight line with a slope of 2 Nm per degree, and the total inertia (motor plus load) is 1.8×10^{-3} kg/m^2. Estimate the frequency of oscillation of the rotor following a single step.

7) Find the step angle of the following stepping motors: (a) 3-phase, VR, 12 stator teeth and 8 rotor teeth; (b) 3-phase, VR, three-stack, 16 rotor teeth; (c) 4-phase unipolar, hybrid, 50 rotor teeth.

8) What simple tests could be done on an unmarked stepping motor to decide whether it was a VR motor or a hybrid motor?

9) At what speed would a 1.8° hybrid steeping motor run if its two phases were each supplied from the 60 Hz mains supply, the current in one of the phases being phase shifted by 90° with respect to the other?

10) The rated current of a 4-phase unipolar stepping motor is 3 A per phase and its winding resistance is 1.5 Ω. When supplied from a simple constant-voltage drive without additional forcing resistance the pull-out torque at a speed of 50 steps per second is 0.9 Nm. Estimate the voltage and forcing resistance that will allow the same pull-out torque to be achieved at a speed of 250 steps per second.

11) An experimental scientist read that stepping motors typically complete each single step in a few milliseconds. He decided to use one for a display aid, so he mounted a size 18 (approximately 4 cm diameter) 15° per step VR motor so that its shaft was vertical, and fixed a lightweight (30 cm) aluminium pointer about 40 cm long onto the shaft. When he operated the motor he was very disappointed to discover that after every step the pointer oscillated wildly and took almost 2 s before coming to rest. Why should he not have been surprised?

10

SYNCHRONOUS, BRUSHLESS D.C. AND SWITCHED RELUCTANCE DRIVES

INTRODUCTION

In this chapter the common feature which links the motors is that they are all a.c. motors in which the electrical power that is converted to mechanical power is fed into the stator, so, as with the induction motor, there are no sliding contacts in the main power circuits. All except the switched reluctance motor also have stators that are identical (or very similar) to the induction motor.

We begin by looking at motors that are intended to be operated directly off the mains supply, usually at either 50 or 60 Hz. These motors are known as 'synchronous' or 'reluctance' motors, and they provide a precise, specific and constant speed for a wide range of loads, and are therefore used in preference to induction motors when constant speed operation is essential. Such machines are available over a very wide range from tiny single-phase versions in domestic timers to multi-megawatt machines in large industrial applications such as gas compressors. Their principal disadvantage is that if the load torque becomes too high, the motor will suddenly lose synchronism and stall.

To overcome the fixed-speed limitation that results from the constant frequency of the mains, controlled-speed synchronous motor drives simply use a variable-frequency inverter to provide for variation of the synchronous speed. These 'open-loop' drives are dealt with next.

We then look at what are perhaps best referred to as 'self-synchronous' drives, which potentially offer competition for d.c. and induction motor drives. In these drives the motor is basically a synchronous motor with the stator fed from a variable-frequency inverter; but the frequency is determined by a speed signal from a transducer mounted on the rotor.

This closed-loop arrangement ensures that the motor can never lose synchronism, hence the name 'self-synchronous'. Amongst this category is the so-called 'brushless d.c.' drive, where the motor is specifically designed to operate from its own converter, and cannot be supplied directly from conventional mains supplies.

Finally, the most recent addition to the family of industrial drives – the switched reluctance drive – is briefly discussed. The switched reluctance motor is perhaps the simplest of all electrical machines, but it was only with the advent of power-electronic switching and sophisticated digital control that its potential could be fully demonstrated.

SYNCHRONOUS MOTORS

In Chapter 5 we saw that the 3-phase stator windings of an induction motor produce a sinusoidal rotating magnetic field in the air-gap. The speed of rotation of the field (the synchronous speed) was shown to be directly proportional to the supply frequency, and inversely proportional to the pole number of the winding. We also saw that in the induction motor the rotor is dragged along by the field, but that the higher the load on the shaft, the more the rotor has to slip with respect to the field in order to induce the rotor currents required to produce the torque. Thus although at no-load the speed of the rotor can be close to the synchronous speed, it must always be less; and as the load increases, the speed has to fall.

In the synchronous motor, the stator windings are exactly the same as in the induction motor, so when connected to the 3-phase supply, a rotating magnetic field is produced. But instead of having a cylindrical rotor with a cage winding, the synchronous motor has a rotor with either a d.c. excited winding (supplied via sliprings), or permanent magnets, designed to cause the rotor to 'lock-on' or 'synchronise with' the rotating magnetic field produced by the stator. Once the rotor is synchronised, it will run at exactly the same speed as the rotating field despite load variation, so under constant-frequency operation the speed will remain constant as long as the supply frequency is stable.

As previously shown, the synchronous speed (in rev/min) is given by the expression

$$N_s = \frac{120f}{p} \qquad (10.1)$$

where f is the supply frequency and p is the pole number of the winding. Hence for two-, four- and six-pole industrial motors the running speeds on a 50 Hz supply are 3000, 1500 and 1000 rev/min, while on a 60 Hz supply they become 3600, 1800 and 1200 rev/min, respectively. At the other extreme, the little motor in a central heating timer with its cup-shaped rotor with 20 axially projecting fingers and a circular coil in the middle is a 20-pole reluctance synchronous motor that will run at 300 rev/min when fed from 50 Hz mains. Users who want speeds different from these will be disappointed, unless they are prepared to invest in a variable-frequency inverter.

We discussed a similar mechanism whereby the rotor locked onto a magnetic field in connection with the stepping motor (see Chapter 9), but there the field proceeds in a stepwise fashion, rather than smoothly. With the synchronous machine we again find that, as with the stepper, there is a limit to the maximum (pull-out) torque which can be developed before the rotor is forced out of synchronism with the rotating field. This 'pull-out' torque will typically be 1.5 times the continuous rated torque, but for all torques below pull-out the steady running speed will be absolutely constant. The torque–speed curve is therefore simply a vertical line at the synchronous speed, as shown in Figure 10.1. We can see from Figure 10.1 that the vertical line extends into quadrant 2, which indicates that if we try to force the speed above the synchronous speed the machine will act as a generator.

The mains-fed synchronous motor is clearly ideal where a constant speed is essential, and also where several motors must run at precisely the same speed. Examples where 3-phase motors are used include artificial fibre spinning lines, and film and tape transports. Small single-phase reluctance versions are used in clocks and timers for washing

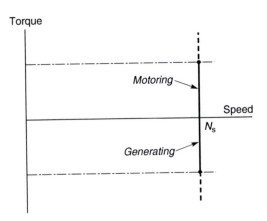

Figure 10.1 *Steady-state torque–speed curve for a synchronous motor supplied at constant frequency*

machines, heating systems, etc. They are also used where precise integral speed ratios are to be maintained: for example, a 3:1 speed ratio can be guaranteed by using a 2-pole and a 6-pole motor, fed from the same supply.

We will now look briefly at the various types of synchronous motor, mentioning the advantages and disadvantages of each. The excited-rotor type is given most weight not only because of its importance in large sizes but also because its behaviour can be analysed, and its mechanism of operation illuminated, by means of a relatively simple equivalent circuit. Parallels are drawn with both the d.c. motor and the induction motor to emphasise that despite their obvious differences, most electrical machines also have striking similarities.

Excited-rotor motors

The rotor carries a 'field' winding which is supplied with direct current via a pair of sliprings on the shaft, and is designed to produce an air-gap field of the same pole number and spatial distribution (usually sinus-oidal) as that produced by the stator winding. The rotor may be more or less cylindrical, with the field winding distributed in slots (see Figure 10.2(a)), or it may have projecting ('salient') poles around which the winding is concentrated (see Figure 10.2(b)). As the discussion in Section 10.2 revealed, a cylindrical-rotor motor has little or no reluctance torque, so it can only produce torque when current is fed into the rotor. On the other hand, the salient-pole type also produces some reluctance torque even when the rotor winding has no current. In both cases, however, the rotor 'excitation' power is relatively small, since all the mechanical output power is supplied from the stator side.

Excited-rotor motors are used in sizes ranging from a few kW up to several MW. The large ones are effective alternators (as used for power

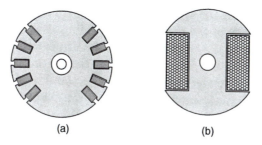

(a)   (b)

Figure 10.2   *Rotors for synchronous motors. 2-pole cylindrical (a) with field coils distributed in slots, and 2-pole salient pole (b) with concentrated field winding*

generation) but used as motors. Wound-rotor induction motors (see Chapter 6) can also be made to operate synchronously by supplying the rotor with d.c. through the sliprings.

The simplest way to visualise the mechanism of torque production is to focus on a static picture, and consider the alignment force between the stator and rotor field patterns. When the two are aligned with N facing S, the torque is zero and the system is in stable equilibrium, with any displacement to right or left causing a restoring torque to come into play. If the fields are distributed sinusoidally in space, the restoring torque will reach a maximum when the poles are misaligned by half a pole pitch, or 90°. Beyond 90° the torque reduces with angle, giving an unstable region, zero torque being reached again when N is opposite to N.

When the motor is running synchronously, we can use much the same mental picture because the field produced by the 3-phase alternating currents in the stator windings rotates at precisely the same speed as the field produced by the d.c. current in the rotor. At no-load there is little or no angular displacement between the field patterns, because the torque required to overcome friction is small. But each time the load increases, the rotor slows momentarily before settling at the original speed but with a displacement between the two field patterns that is sufficient to furnish the torque needed for steady-state running. This angle is known as the 'load-angle', and we can actually see it when we illuminate the shaft of the motor with a mains-frequency stroboscope: a reference mark on the shaft is seen to drop back by a few degrees each time the load is increased.

Equivalent circuit of excited-rotor synchronous motor

Predicting the current and power-factor drawn from the mains by a cylindrical-rotor synchronous motor is possible by means of the very simple per-phase a.c. equivalent circuit shown in Figure 10.3. In this circuit X_s (known as the synchronous reactance) represents the effective inductive reactance of the stator phase winding; R is the stator winding resistance; V the applied voltage and E the e.m.f. induced in the stator winding by the rotating field produced by the d.c. current on the rotor. (For the benefit of readers who tackled Chapter 7, it should be pointed out that X_s is effectively equal to the sum of the magnetising and leakage reactances, i.e. $X_s = X_m + X_1$: but because the air-gap in synchronous machines is usually larger than in induction motors, their per unit synchronous reactance is usually lower than that of an induction machine with the same stator winding.)

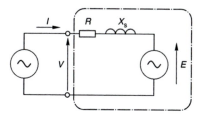

Figure 10.3   *Equivalent circuit for synchronous machine*

The similarity between this circuit and that of the d.c. machine (see Figure 3.6) and the induction motor (see Figure 5.8) is clear, and it stems from the fact that these machines produce torque by the interaction of a magnetic field and current-carrying conductors (the so-called '*BIl*' effect). In the case of the d.c. machine the inductance was seen not to be important under steady-state conditions, because the current was steady (i.e. d.c.), and the resistance emerged as the dominant parameter. In the case of the mains-fed synchronous motor the current is alternating at mains frequency, so not surprisingly we find that the synchronous reactance is the dominant impedance and resistance plays only a minor role.

At this point readers who are not familiar with a.c. circuit theory can be reassured that they will not be seriously disadvantaged by skipping the rest of this and the following sections. But although no seminal truths are to follow, discussion of the equivalent circuit and the associated phasor diagram greatly assists the understanding of motor behaviour, especially its ability to operate over a range of power factors.

Our aim is to find the current drawn from the mains, which from Figure 10.3 clearly depends on all the parameters therein. But for a given machine operating from a constant-voltage, constant-frequency supply, the only variables are the load on the shaft and the d.c. current (the excitation) fed into the rotor, so we will look at the influence of both, beginning with the effect of the load on the shaft.

The speed is constant and therefore the mechanical output power (torque times speed) is proportional to the torque being produced, which in the steady state is equal and opposite to the load torque. Hence if we neglect the losses in the motor, the electrical input power is also determined by the load on the shaft. The input power per phase is given by $VI \cos\phi$, where I is the current and the power-factor angle is ϕ. But V is fixed, so the in-phase (or real) component of input current ($I \cos\phi$) is determined by the mechanical load on the shaft. We recall that, in the same way, the current in the d.c. motor (see Figure 3.6) was determined by the load. This discussion reminds us that although the equivalent circuits in Figures 10.3 and 3.6 are very informative, they should perhaps carry a 'health warning' to the effect that the single most important determinant of the current (the load torque) does not actually appear explicitly on the diagrams.

Turning now to the influence of the d.c. excitation current, at a given supply frequency (i.e. speed) the mains-frequency e.m.f., (E) induced in the stator is proportional to the d.c. field current fed into the rotor. (If we want to measure this e.m.f. we could disconnect the stator windings from the supply, drive the rotor at synchronous speed by an external means, and measure the voltage at the stator terminals, performing the so-called 'open-circuit' test. If we were to vary the speed at which we drove the rotor, keeping the field current constant, we would of course find that E was proportional to the speed.) We discovered a very similar state of affairs when we studied the d.c. machine (see Chapter 3): its induced ('back') e.m.f. (E) turned out to be proportional to the field current, and to the speed of rotation of the armature. The main difference between the d.c. machine and the synchronous machine is that in the d.c. machine the field is stationary and the armature rotates, whereas in the synchronous machine the field system rotates while the stator windings are at rest: in other words, one could describe the synchronous machine, loosely, as an 'inside-out' d.c. machine.

We also saw in Chapter 3 that when the unloaded d.c. machine was connected to a constant voltage d.c. supply, it ran at a speed such that the induced e.m.f. was (almost) equal to the supply voltage, so that the no-load current was almost zero. When a load was applied to the shaft, the speed fell, thereby reducing E and increasing the current drawn from the supply until the motoring torque produced was equal to the load torque. Conversely if we applied a driving torque to the shaft of the machine, the speed rose, E became greater than V, current flowed out to the supply and the machine acted as a generator. These findings are based on the assumption that the field current remains constant, so that changes in E are a reflection of changes in speed. Our overall conclusion was the simple statement that if E is less than V, the d.c. machine acts as a motor, while if E is greater than V, it acts as a generator.

The situation with the synchronous motor is similar, but now the speed is constant and we can control E independently via the control of the d.c. excitation current fed to the rotor. We might again expect that if E was less than V the machine would draw in current and act as a motor, and vice versa if E was greater than V. But we are no longer dealing with simple d.c. circuits in which phrases such as 'draw in current' have a clear meaning in terms of what it tells us about power flow. In the synchronous motor equivalent circuit the voltages and currents are a.c., so we have to be more careful with our language and pay due respect to the phase of the current, as well as its magnitude. Things turn out to be rather different from what we found in the d.c. motor, but there are also similarities.

Phasor diagram and Power-factor control

To see how the magnitude of the e.m.f. influences behaviour we can examine the phasor diagrams of a synchronous machine operating as a motor, as shown in Figure 10.4. The first point to clarify is that our sign convention is that motoring corresponds to positive input power to the machine. The power is given by $VI\cos\phi$, so when the machine is motoring (positive power) the angle ϕ lies in the range $\pm90°$. If the current lags or leads the voltage by more than $90°$ the machine will be generating.

Figure 10.4 shows three phasor diagrams corresponding to low, medium and high values of the induced e.m.f., (E), the shaft load (i.e. mechanical power) being constant. As discussed above, if the mechanical power is constant, so is $I\cos\phi$, and the locus of the current is therefore shown by the horizontal dotted line. The load angle (δ), discussed in Section 10.2.1, is the angle between V and E in the phasor diagram. In Figure 10.4, the voltage phasor diagram embodies Kirchhoff's law as applied to the equivalent circuit in Figure 10.3, i.e. $V = E + IR + jIX_s$, but for the sake of simplicity R is neglected so the phasor diagram simply consists of the volt-drop IX_s (which leads the current I by $90°$) added to E to yield V.

Figure 10.4(a) represents a condition where the field current has been set so that the magnitude of the induced e.m.f., (E) is less than V. This is called an 'underexcited' condition, and as can be seen the current is lagging the terminal voltage and the power-factor is $\cos\phi_a$ lagging. When the field current is increased (increasing the magnitude of E) the magnitude of the input current reduces and it moves more into phase with V: the special case shown in Figure 10.4(b) shows that the motor can be operated at unity power-factor if the field current is suitably chosen. Finally, in Figure 10.4(c), the field current is considerably higher (the 'overexcited' case) which causes the current to increase again but this time the current leads the voltage and the power-factor is $\cos\phi_c$ leading. We see that we can obtain any desired power-factor by appropriate choice of rotor excitation, and in particular we can operate with a leading power-factor. This is a freedom not afforded to users of

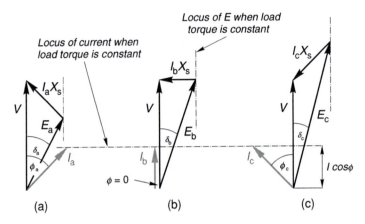

Figure 10.4   *Phasor diagrams for synchronous motor operating with constant load torque, for three different values of the rotor (excitation) current*

induction motors, and arises because in the synchronous machine there is an additional mechanism for providing excitation, as we will now see.

When we studied the induction motor we discovered that the magnitude and frequency of the supply voltage V governed the magnitude of the resultant flux density wave in the machine, and that the current drawn by the motor could be considered to consist of two components. The real (in-phase) component represented the real power being converted from electrical to mechanical form, so this component varied with the load. On the other hand the lagging reactive (quadrature) component represented the 'magnetising' current that was responsible for producing the flux, and it remained constant regardless of load.

The stator winding of the synchronous motor is the same as the induction motor, so it is to be expected that the resultant flux will be determined by the magnitude and frequency of the applied voltage. This flux will therefore remain constant regardless of the load, and there will be an associated requirement for magnetising MMF. But now we have two possible means of providing the excitation MMF, namely the d.c. current fed into the rotor and the lagging component of a.c. current in the stator.

When the rotor is underexcited, i.e. the induced e.m.f. E is less than V (Figure 10.4(a)), the stator current has a lagging component to make up for the shortfall in excitation needed to yield the resultant field that must be present as determined by the terminal voltage, V. With more field current (Figure 10.4(b)), however, the rotor excitation alone is sufficient and no lagging current is drawn by the stator. And in the overexcited case (Figure 10.4(c)), there is so much rotor excitation that there is effectively some reactive power to spare and the leading power-factor represents the export of lagging reactive power that could be used to provide excitation for induction motors elsewhere on the same system.

To conclude our look at the synchronous motor we can now quantify the qualitative picture of torque production introduced in Section 10.2.1 by noting from the phasor diagrams that if the mechanical power (i.e. load torque) is constant, the variation of the load-angle (δ) with E is such that $E \sin \delta$ remains constant. As the rotor excitation is reduced, and E becomes smaller, the load angle increases until it eventually reaches its maximum of $90°$, at which point the rotor will lose synchronism and stall. This means that there will always be a lower limit to the excitation required for the machine to be able to transmit the specified torque. This is just what our simple mental picture of torque being developed between two magnetic fields, one of which becomes very weak, would lead us to expect.

Starting

It should be clear from the discussion of how torque is produced that unless the rotor is running at the same speed as the rotating field, no steady torque can be produced. If the rotor is running at a different speed, the two fields will be sliding past each other, giving rise to a pulsating torque with an average value of zero. Hence a basic synchronous machine is not self-starting, and some alternative method of producing a run-up torque is required.

Most synchronous motors are therefore equipped with some form of rotor cage, similar to that of an induction motor, in addition to the main field winding. When the motor is switched onto the mains supply, it operates as an induction motor during the run-up phase, until the speed is just below synchronous. The excitation is then switched on so that the rotor is able to make the final acceleration and 'pull-in' to synchronism with the rotating field. Because the cage is only required during starting, it can be short time rated, and therefore comparatively small. Once the rotor is synchronised, and the load is steady, no currents are induced in the cage, because the slip is zero. The cage does, however, come into play when the load changes, when it provides an effective method for damping out the oscillations of the rotor as it settles at its new steady-state load angle.

Large motors will tend to draw a very heavy current during run-up, so some form of reduced voltage starter is often required (see Chapter 6). Sometimes, a separate small induction motor is used simply to run-up the main motor before synchronisation, but this is only feasible where the load is not applied until after the main motor has been synchronised.

No special starter is required for the wound-rotor induction motor of course, which runs up in the usual way (see Chapter 6) before the d.c. excitation is applied. Motors operated like this are sometimes known as 'Inductosyns'.

Permanent magnet synchronous motors

Permanent magnets are used on the rotor instead of a wound field: typical 2-pole and 4-pole surface-mounted versions are shown in Figure 10.5, the direction in which the magnets have been magnetised being represented by the arrows. Motors of this sort have outputs ranging from about 100 W up to perhaps 100 kW.

For starting from a fixed-frequency supply a rotor cage is required, as discussed above. The advantages of the permanent magnet type are that no supply is needed for the rotor and the rotor construction can be robust and reliable. The disadvantage is that the excitation is fixed, so the designer must either choose the shape and disposition of the magnets to match the requirements of one specific load, or seek a general-purpose compromise. Control of power-factor via excitation is no longer possible.

Early permanent magnet motors suffered from the tendency for the magnets to be demagnetised by the high stator currents during starting, and from a restricted maximum allowable temperature. Much improved versions using high coercivity rare-earth magnets were developed during the 1970s to overcome these problems. They are usually referred to as 'Line-Start' motors, to indicate that they are designed for direct-on-line

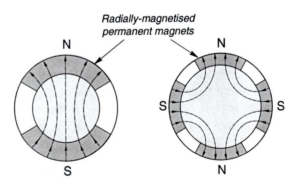

Figure 10.5   *Permanent magnet synchronous motor rotors. 2-pole (left); 2-pole (right)*

starting. The steady-state efficiency and power-factor at full load are in most cases better than the equivalent induction motor, and they can pull-in to synchronism with inertia loads of many times rotor inertia.

Hysteresis motors

Whereas most motors can be readily identified by inspection when they are dismantled, the hysteresis motor is likely to baffle anyone who has not come across it before. The rotor consists simply of a thin-walled cylinder of what looks like steel, while the stator has a conventional single-phase or 3-phase winding. Evidence of very weak magnetism may just be detectable on the rotor, but there is no hint of any hidden magnets as such, and certainly no sign of a cage. Yet the motor runs up to speed very sweetly and settles at exactly synchronous speed with no sign of a sudden transition from induction to synchronous operation.

These motors (the operation of which is quite complex) rely mainly on the special properties of the rotor sleeve, which is made from a hard steel which exhibits pronounced magnetic hysteresis. Normally in machines we aim to minimise hysteresis in the magnetic materials, but in these motors the effect (which arises from the fact that the magnetic flux density B depends on the previous 'history' of the MMF) is deliberately accentuated to produce torque. There is actually also some induction motor action during the run-up phase, and the net result is that the torque remains roughly constant at all speeds.

Small hysteresis motors are used extensively in tape recorders, office equipment, fans, etc. The near constant torque during run-up and the very modest starting current (of perhaps 1.5 times rated current) means that they are also suited to high inertia loads such as gyrocompasses and small centrifuges.

Reluctance motors

The reluctance motor is arguably the simplest synchronous motor of all, the rotor consisting simply of a set of laminations shaped so that it tends to align itself with the field produced by the stator. This 'reluctance torque' action was discussed when we looked at the variable reluctance stepping motor (in Chapter 9) and again briefly in Section 10.2.

Here we are concerned with mains-frequency reluctance motors, which differ from steppers in that they only have saliency on the rotor, the stator being identical with that of a 3-phase induction motor. In fact, since induction motor action is required in order to get the rotor up to synchronous speed, a reluctance-type rotor resembles a cage induction

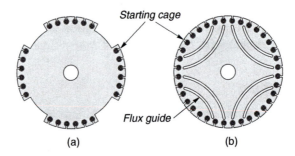

Figure 10.6   *Reluctance motor rotors (4-pole):* (a) *salient type,* (b) *flux-guided type*

motor, with parts of the periphery cut away in order to force the flux from the stator to enter the rotor in the remaining regions where the air-gap is small, as shown in Figure 10.6(a). Alternatively, the 'preferred flux paths' can be imposed by removing iron inside the rotor so that the flux is guided along the desired path, as shown in Figure 10.6(b).

The rotor will tend to align itself with the field, and hence is able to remain synchronised with the travelling field set up by the 3-phase winding on the stator in much the same way as a permanent magnet rotor. Early reluctance motors were invariably one or two frame sizes bigger than an induction motor for a given power and speed, and had low power-factor and poor pull-in performance. As a result they fell from favour except for some special applications such as textile machinery where cheap constant speed motors were required. Understanding of reluctance motors is now much more advanced, and they can compete on almost equal terms with the induction motor as regards power-output, power-factor and efficiency. They are nevertheless relative expensive because they are not produced in large numbers.

CONTROLLED-SPEED SYNCHRONOUS MOTOR DRIVES

As soon as variable-frequency inverters became a practicable proposition it was natural to use them to supply synchronous motors, thereby freeing the latter from the fixed-speed constraint imposed by mains-frequency operation and opening up the possibility of a simple open-loop controlled speed drive. The obvious advantage over the inverter-fed induction motor is that the speed of the synchronous motor is exactly determined by the frequency, whereas the induction motor always has to run with a finite slip. A precision frequency source (oscillator) controlling the inverter switching is all that is necessary to give an accurate speed control with a synchronous motor, while speed feedback is essential to achieve accuracy with an induction motor.

In practice, open-loop operation of inverter-fed synchronous motors is not as widespread as might be expected, though it is commonly used in multi-motor drives (see below). Closed-loop or self-synchronous operation is however rapidly gaining momentum, and is already well established in two distinct guises at opposite ends of the size range. At one extreme, large excited-rotor synchronous motors are used in place of d.c. drives, particularly where high speeds are required or when the motor must operate in a hazardous atmosphere (e.g. in a large gas compressor). At the other end of the scale, small permanent magnet synchronous motors are used in brushless d.c. drives. We will look at these closed-loop applications after a brief discussion of open-loop operation.

Open-loop inverter-fed synchronous motor drives

This simple method is attractive in multi-motor installations where all the motors must run at exactly the same speed. Individually the motors (permanent magnet or reluctance) are more expensive than the equivalent mass-produced induction motor, but this is offset by the fact that speed feedback is not required, and the motors can all be supplied from a single inverter, as shown in Figure 10.7.

The inverter voltage–frequency ratio will usually be kept constant (see Chapter 8) to ensure that the motors operate at full flux at all speeds, and therefore have a 'constant-torque' capability. If prolonged low-speed operation is called for, improved cooling of the motors may be necessary. Speed is precisely determined by the inverter frequency, but speed changes (including run-up from rest) must be made slowly, under ramp control, to avoid the possibility of exceeding the pull-out load angle, which would result in stalling.

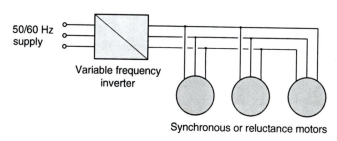

Figure 10.7   *Open-loop operation of a group of several synchronous or reluctance motors supplied from a single variable-frequency inverter*

A problem which can sometimes occur with this sort of open-loop operation is that the speed of the motor exhibits apparently spontaneous oscillation or 'hunting'. The supply frequency may be absolutely constant but the rotor speed is seen to fluctuate about its expected (synchronous) value, sometimes with an appreciable amplitude, and usually at a low frequency of perhaps 1 Hz. The origin of this unstable behaviour lies in the fact that the motor and load constitute at least a fourth-order system, and can therefore become very poorly damped or even unstable for certain combinations of the system parameters. Factors that influence stability are terminal voltage, supply frequency, motor time-constants and load inertia and damping. Unstable behaviour in the strict sense of the term (i.e. where the oscillations build-up without limit) is rare, but bounded instability is not uncommon, especially at speeds well below the base (50 Hz or 60 Hz) level, and under light-load conditions. It is very difficult to predict exactly when unstable behaviour might be encountered, and provision must be made to combat it. Some inverters therefore include circuitry that detects any tendency for the currents to fluctuate (indicating hunting) and to modulate the voltage and/or frequency to suppress the unwanted oscillations.

Self-synchronous (closed-loop) operation

In the open-loop scheme outlined above, the frequency of the supply to the motor is under the independent control of the oscillator driving the switching devices in the inverter. The inverter has no way of knowing whether the rotor is correctly locked-on to the rotating field produced by the stator, and if the pull-out torque is exceeded, the motor will simply stall.

In the self-synchronous mode, however, the inverter output frequency is determined by the speed of the rotor. More precisely, the instants at which the switching devices operate to turn the stator windings on and off are determined by rotor position-dependent signals obtained from a rotor position transducer (RPT) mounted on the rotor shaft. In this way, the stator currents are always switched on at the right time to produce the desired torque on the rotor, because the inverter effectively knows where the rotor is at every instant of time. The use of rotor position feedback signals to control the inverter accounts for the description 'closed-loop' used above. If the rotor slows down (as a result of an increase in load, for example), the stator supply frequency automatically reduces so that the rotor remains synchronised with the rotating field, and the motor therefore cannot 'pull-out' in the way it does under open-loop operation.

An analogy with the internal combustion engine may help to clarify the difference between closed-loop and open-loop operations. An engine invariably operates as a closed-loop system in the sense that the opening and closing of the inlet and exhaust valves is automatically synchronised with the position of the pistons by means of the camshaft and timing belt. The self-synchronous machine is much the same in that the switching devices in the inverter turn the current on and off according to the position of the rotor. By contrast, open-loop operation of the engine would imply that we had removed the timing belt and chosen to operate the valves by driving the camshaft independently, in which case it should be clear that the engine would only be capable of producing power at one speed at which the up and down motion of the pistons corresponded exactly with the opening and closing of the valves.

It turns out that the overall operating characteristics of a self-synchronous a.c. motor are very similar to those of a conventional d.c. motor. This is really not surprising when we recall that in a d.c. motor, the mechanical commutator reverses the direction of the current in each (rotating) armature coil at the appropriate point such that, regardless of speed, the current under each (stationary) field pole is always in the right direction to produce the desired torque. In the self-synchronous motor the roles of stator and rotor are reversed compared with the d.c. motor. The field is rotating and the 'armature' winding (consisting of three discrete groups of coils or phases) is stationary. The timing and direction of the current in each phase is governed by the inverter switching, which in turn is determined by the rotor position sensor. Hence, regardless of speed, the torque is always in the right direction.

The combination of the rotor position sensor and inverter performs effectively the same function as the commutator in a conventional d.c. motor. There are of course usually only three windings to be switched by the inverter, as compared with many more coils and commutator segments to be switched by the brushes in the d.c. motor, but otherwise the comparison is valid. Not surprisingly the combination of position sensor and inverter is sometimes referred to as an 'electronic commutator', while the overall similarity of behaviour gives rise to the rather clumsy term 'electronically commutated motor' (ECM) or the even worse 'commutatorless d.c. motor' (CLDCM) to describe self-synchronous machines.

Operating characteristics and control

If the d.c. input voltage to the inverter is kept constant and the motor starts from rest, the motor current will be large at first, but will decrease with speed until the motional e.m.f. generated inside the motor is almost equal to the applied voltage. When the load on the shaft is increased, the speed begins to fall, the motional e.m.f. reduces and the current increases until a new equilibrium is reached where the extra motor torque is equal to the load torque. This behaviour parallels that of the conventional d.c. motor, where the no-load speed depends on the applied armature voltage. The speed of the self-synchronous motor can therefore be controlled by controlling the d.c. link voltage to the inverter. The d.c. link will usually be provided by a controlled rectifier, so the motor speed can be controlled by varying the input converter firing angle, as shown in Figure 10.8.

The overall similarity with the d.c. drive (see Chapter 4) is deliberately emphasised in Figure 10.8. The dotted line enclosing the a.c. motor together with its rotor position detector and inverter is in effect the replacement for the conventional d.c. motor. We note, however, that a tachogenerator is not necessary for closed-loop speed control because the speed feedback signal can be derived from the frequency of the rotor position signal. And, as with the d.c. drive, current control, as distinct from voltage control can be used where the output torque rather than speed is to be controlled. Full four-quadrant operation is possible, as long as the inverter is supplied from a fully controlled converter.

In simple cost terms the self-synchronous system looks attractive when the combined cost of the inverter and synchronous motor is lower than the equivalent d.c. motor. When such schemes were first introduced (in the 1970s) they were only cost-effective in very large sizes (say over 1 MW), but the break-even point for wound-field motors is falling and drives with ratings in the hundreds of kW are now common. They may utilise converter-grade (relatively cheap) thyristors in the inverter bridge because the thyristors will commutate naturally with the aid of the motor's generated e.m.f. At very low speeds, however, the generated

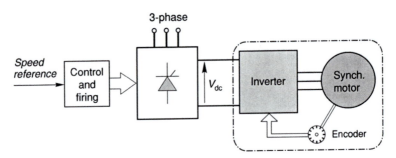

Figure 10.8 *Self-synchronous motor–inverter system. In large sizes this arrangement is sometimes referred to as a 'synchdrive'; in smaller sizes it would be known as a brushless d.c. motor drive*

e.m.f. is insufficient, so the motor is started under open-loop current-fed operation, in the manner of a stepping motor.

As inverter costs have fallen, lower power drives using permanent magnet motors have become attractive, especially where very high speeds are required and the conventional brushed d.c. motor is unsuitable because of commutator limitations.

BRUSHLESS D.C. MOTORS

Much of the impetus for the development of brushless d.c. motors came from the computer peripheral and aerospace industries, where high performance coupled with reliability and low maintenance are essential. Very large numbers of brushless d.c. motors are now used, particularly in sizes up to a few hundred watts. The small versions (less than 100 W) are increasingly made with all the control and power electronic circuits integrated at one end of the motor, so that they can be directly retrofitted as a replacement for a conventional d.c. motor. Because all the heat-dissipating circuits are on the stator, cooling is much better than in a conventional motor, so higher specific outputs can be achieved. The rotor inertia can also be less than that of a conventional armature, which means that the torque–inertia ratio is better, giving a higher acceleration. Higher speeds are practicable because there is no mechanical commutator.

In principle, there is no difference between a brushless d.c. motor and the self-synchronous permanent magnet motor discussed earlier in this chapter. The reader may therefore be puzzled as to why some motors are described as brushless d.c. while others are not. In fact, there is no logical reason at all, nor indeed is there any universal definition or agreed terminology.

Broadly speaking, however, the accepted practice is to restrict the term 'brushless d.c. motor' to a particular type of self-synchronous permanent magnet motor in which the rotor magnets and stator windings are arranged to produce an air-gap flux density wave which has a trapezoidal shape. Such motors are fed from inverters that produce rectangular current waveforms, the switch-on being initiated by digital signals from a relatively simple rotor position sensor. This combination permits the motor to develop a more or less smooth torque, regardless of speed, but does not require an elaborate position sensor. (In contrast, many self-synchronous machines have sinusoidal air-gap fields, and therefore require more sophisticated position sensing and current profiling if they are to develop continuous smooth torque.)

The brushless d.c. motor is essentially an inside out electronically commutated d.c. motor, and can therefore be controlled in the same way as a conventional d.c. motor (see Chapter 4). Many brushless motors are used in demanding servo-type applications, where they need to be integrated with digitally controlled systems. For this sort of application, complete digital control systems, which provide for torque, speed and position control are available.

SWITCHED RELUCTANCE MOTOR DRIVES

The switched reluctance drive was developed in the 1980s to offer advantages in terms of efficiency, power per unit weight and volume, robustness and operational flexibility. The motor and its associated power-electronic drive must be designed as an integrated package, and optimised for a particular specification, e.g. for maximum overall efficiency with a

Plate 10.1   *Switched reluctance motors. The motors with finned casings are TEFV for use in general-purpose industrial controlled speed drives, the largest being rated at 75 kW (100 h.p.) at 1500 rev/min. The force-ventilated motor (centre rear) is for high-performance ('d.c. equivalent') applications. Most of the other motors are designed for specific OEM applications, including domestic white goods (left front) automotive (centre front), food processor and vacuum cleaner (right front). Very high speeds can be used because the rotor is very robust and there are no brushes, and thus very high specific outputs are obtained; some of the small motors run at up to 30 000 rev/min. (Photograph by courtesy of Switched Reluctance Drives Ltd)*

specific load, or maximum speed range, or peak short-term torque. Despite being relatively new, the technology has been applied to a wide range of applications including general-purpose industrial drives, compressors, domestic appliances and office and business equipment.

Principle of operation

The switched reluctance motor differs from the conventional reluctance motor in that both the rotor and the stator have salient poles. This doubly salient arrangement (as shown in Figure 10.9) proves to be very effective as far as electromagnetic energy conversion is concerned.

The stator carries coils on each pole, while the rotor, which is made from laminations in the usual way, has no windings or magnets and is therefore cheap to manufacture and extremely robust. The particular example shown in Figure 10.9 has 12 stator poles and 8 rotor poles, and represents a widely used arrangement, but other pole combinations are used to suit different applications. In Figure 10.9, the 12 coils are grouped to form three phases, which are independently energised from a 3-phase converter.

The motor rotates by exciting the phases sequentially in the sequence A, B, C for anticlockwise rotation or A, C, B for clockwise rotation, the 'nearest' pair of rotor poles being pulled into alignment with the appropriate stator poles by reluctance torque action. In Figure 10.9 the four coils forming phase A are shown by thick line, the polarities of the coil MMFs being indicated by the letters N and S on the back of the core. Each time a new phase is excited the equilibrium position of the rotor advances by 15°, so after one complete cycle (i.e. each of the three phases has been excited once) the angle turned through is 45°. The machine therefore rotates once for eight fundamental cycles of supply to the stator windings, so in terms of the relationship between the

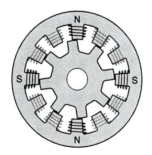

Figure 10.9   *Typical switched reluctance (SR) motor. Each of the 12-stator poles carries a concentrated winding, while the 8-pole rotor has no windings or magnets*

fundamental supply frequency and the speed of rotation, the machine in Figure 10.9 behaves as a 16-pole conventional machine.

Readers familiar with stepping motors (see Chapter 9) will correctly identify the SR motor as a variable reluctance stepping motor. There are of course important design differences which reflect the different objectives (continuous rotation for the SR, stepwise progression for the stepper), but otherwise the mechanisms of torque production are identical. However, while the stepper is designed first and foremost for open-loop operation, the SR motor is designed for self-synchronous operation, the phases being switched by signals derived from a shaft-mounted rotor position detector (RPT). In terms of performance, at all speeds below the base speed continuous operation at full torque is possible. Above the base speed, the flux can no longer be maintained at full amplitude and the available torque reduces with speed. The operating characteristics are thus very similar to those of the other most important controlled-speed drives, but with the added advantage that overall efficiencies are generally a per cent or two higher.

Given that the mechanism of torque production in the switched reluctance motor appears to be very different from that in d.c. machines, induction motors and synchronous machines (all of which exploit the 'BIl' force on a conductor in a magnetic field) it might have been expected that one or other type would offer such clear advantages that the other would fade away. In fact, despite claims and counter claims, there appears to be little to choose between them overall, and one wonders whether the mechanisms of operation are really so fundamentally different as our ingrained ways of looking at things lead us to believe. Perhaps a visitor from another planet would note the similarity in terms of volume, quantities and disposition of iron and copper, and overall performance, and bring some fresh enlightenment to bear so that we emerge recognising some underlying truth that hitherto has escaped us.

Torque prediction and control

If the iron in the magnetic circuit is treated as ideal, analytical expressions can be derived to express the torque of a reluctance motor in terms of the rotor position and the current in the windings. In practice, however, this analysis is of little real use, not only because switched reluctance motors are designed to operate with high levels of magnetic saturation in parts of the magnetic circuit, but also because, except at low speeds, it is not practicable to achieve specified current profiles.

The fact that high levels of saturation are involved makes the problem of predicting torque at the design stage challenging, but despite the highly non-linear relationships it is possible to compute the flux, current and torque as functions of rotor position, so that optimum control strategies can be devised to meet particular performance specifications. Unfortunately this complexity means that there is no simple equivalent circuit available to illuminate behaviour.

As we saw when we discussed the stepping motor, to maximise the average torque it would (in principle) be desirable to establish the full current in each phase instantaneously, and to remove it instantaneously at the end of each positive torque period. But, as illustrated in Figure 9.14, this is not possible even with a small stepping motor, and certainly out of the question for switched reluctance motors (which have much higher inductance) except at low speeds where current chopping (see Figure 9.14) is employed. For most of the speed range, the best that can be done is to apply the full voltage available from the converter at the start of the 'on' period, and (using a circuit such as that shown in Figure 9.15) apply full negative voltage at the end of the pulse by opening both of the switches.

Operation using full positive voltage at the beginning and full negative voltage at the end of the 'on' period is referred to as 'single-pulse' operation. For all but small motors (of less than say 1 kW) the phase resistance is negligible and consequently the magnitude of the phase flux-linkage is determined by the applied voltage and frequency, as we have seen many times previously with other types of motor.

The relationship between the flux-linkage (ψ) and the voltage is embodied in Faraday's law, i.e. $v = d\psi/dt$, so with the rectangular voltage waveform of single-pulse operation the phase flux linkage waveforms have a very simple triangular shape, as in Figure 10.10 which shows the waveforms for phase A of a 3-phase motor. (The waveforms for phases B and C are identical, but are not shown: they are displaced by one third and two thirds of a cycle, as indicated by the arrows.) The upper half of the diagram represents the situation at speed N, while the lower half corresponds to a speed of 2N. As can be seen, at the higher speed (high frequency) the 'on' period halves, so the amplitude of the flux halves, leading to a reduction in available torque. The same limitation was seen in the case of the inverter-fed induction motor drive, the only difference being that the waveforms in that case were sinusoidal rather than triangular.

It is important to note that these flux waveforms do not depend on the rotor position, but the corresponding current waveforms do because the MMF needed for a given flux depends on the effective reluctance of the magnetic circuit, and this of course varies with the position of the rotor.

To get the most motoring torque for any given phase flux waveform, it is obvious that the rise and fall of the flux must be timed to coincide

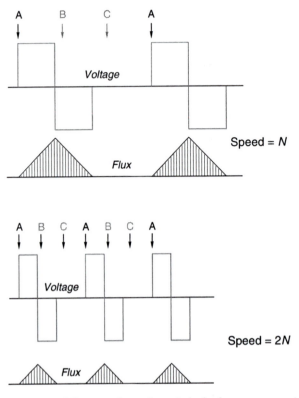

Figure 10.10   *Voltage and flux waveforms for switched-reluctance motor in 'single-pulse' mode*

with the rotor position: ideally, the flux should only be present when it produces positive torque, and be zero whenever it would produce negative torque, but given the delay in build-up of the flux it may be better to switch on early so that the flux reaches a decent level at the point when it can produce the most torque, even if this does lead to some negative torque at the start and finish of the cycle.

The job of the torque control system is to switch each phase on and off at the optimum rotor position in relation to the torque being demanded at the time, and this is done by keeping track of the rotor position using a RPT. Just what angles constitute the optimum depends on what is to be optimised (e.g. average torque, overall efficiency), and this in turn is decided by reference to the data stored digitally in the controller 'memory map' that relates current, flux, rotor position and torque for the particular machine. Torque control is thus considerably less straightforward than in d.c. drives, where torque is directly proportional to armature current, or induction motor drives where torque is proportional to slip.

Power converter and overall drive characteristics

An important difference between the SR motor and all other self-synchronous motors is that its full torque capability can be achieved without having to provide for both positive and negative currents in the phases. This is because the torque does not depend on the direction of current in the phase-winding. The advantage of such 'unipolar' drives is that because each of the main switching devices is permanently con-nected in series with one of the motor windings (as in Figure 9.15), there is no possibility of the 'shoot-through' fault (see Chapter 2) which is a major headache in the conventional inverter.

Overall closed-loop speed control is obtained in the conventional way with the speed error acting as a torque demand to the torque control system described above. However, in most cases it is not necessary to fit a tacho as the speed feedback signal can be derived from the RPT.

In common with other self-synchronous drives, a wide range of oper-ating characteristics is available. If the input converter is fully controlled, continuous regeneration and full four-quadrant operation is possible, and the usual constant torque, constant power and series type character-istic is regarded as standard. Low speed torque can be uneven unless special measures are taken to profile the current pulses, but continuous low speed operation is usually better than for most competing systems in terms of overall efficiency.

REVIEW QUESTIONS

1) What pole number would be needed for a synchronous motor to run at a speed of 300 rev/min from a 60 Hz supply?

2) What voltage should be used to allow a 420 V, 60 Hz, 4-pole syn-chronous motor to be used on a 50 Hz supply?

3) What purpose might be served by a pair of 3-phase synchronous machines (one of which has 10 polar projections on its rotor and the other 12) mounted on a bedplate with their shafts coupled together, but with no shaft projections at their outer ends?

4) In this chapter it is claimed that the speed of a synchronous motor supplied at constant frequency is absolutely constant, regardless of load: it also talks of the rotor falling back with respect to the rotating

field as the load on the shaft increases. Since the field is rotating at a constant speed, how can the rotor fall back unless its speed is less than that of the field?

5) The book explains that in excited-rotor synchronous machines the field winding is supplied with d.c. current via sliprings. Given that the field winding rotates, why is there no mention of any motional e.m.f. in the rotor circuit?

6) A large synchronous motor is running without any load on its shaft, and it is found that when the d.c. excitation on the rotor is set to either maximum or minimum, the a.c. current in the stator is large, but that at an intermediate level the stator current becomes almost zero. The stator power seems to remain low regardless of the rotor current. Explain these observations by reference to the equivalent circuit and phasor diagram. Under what conditions does the motor look like a capacitor when viewed from the supply side?

7) What effect would doubling the total effective inertia have on (a) the run-up time and (b) the pull-out torque of a mains-fed synchronous motor?

8) A large synchronous motor is running with a load angle of 40°. If the rotor excitation is adjusted so that the induced e.m.f. is increased by 50%, estimate the new load angle. How would the input power be expected to change when the excitation was increased?

9) In a synchronous motor the magnetic field in the rotor is steady (apart from the brief periods when the load or excitation changes), so there will be no danger of eddy currents. Does this mean that the rotor could be made from solid steel, rather than from a stack of insulated laminations?

10) Why do the majority of self-synchronous motors have three stator phases, rather than say four or five?

11) What is the principal difference between a brushless d.c. motor and a self-synchronous motor?

12) Why are the drive circuits for switched-reluctance motors referred to as 'unipolar', and what advantage does a unipolar circuit have over the more common bipolar?

13) What is the purpose of the substantial 'dump' resistor found in the drive converter of many of the low-power versions of the drives described in this chapter? What does the presence of a dump resistor

imply about the capability of the drive to operate continuously outside quadrant 1 of the torque–speed plane?

14) Which of the drive types discussed in this chapter are theoretically capable of operating in generating mode? What factors determine whether or not generation is possible in practice?

11

MOTOR/DRIVE SELECTION

INTRODUCTION

The selection process often highlights difficulties in three areas. Firstly, as we have discovered in the preceding chapters, there is a good deal of overlap between the major types of motor and drive. This makes it impossible to lay down a set of hard and fast rules to guide the user straight to the best solution for a particular application. Secondly, users tend to underestimate the importance of starting with a comprehensive specification of what they really want, and they seldom realise how much weight attaches to such things as the steady-state torque–speed curve, the inertia of the load, the pattern of operation (continuous or intermittent) and the question of whether or not the drive needs to be capable of regeneration. And thirdly, they may be unaware of the existence of standards and legislation, and hence can be baffled by questions from any potential supplier.

The aim in this chapter is to assist the user by giving these matters an airing. We begin by drawing together broad guidelines relating to power and speed ranges for the various types of motor, then move on to the questions which need to be asked about the load and pattern of operation, and finally look briefly at the matter of standards. The whole business of selection is so broad that it really warrants a book to itself, but the cursory treatment here should at least help the user to specify the drive rating and arrive at a shortlist of possibilities.

POWER RANGE FOR MOTORS AND DRIVES

The diagrams (Figures 11.1 and 11.2) give a broad indication of the power range for the most common types of motor and drive. Because

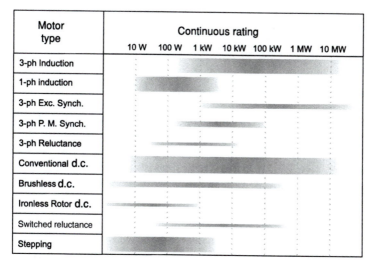

Motor type	Continuous rating						
	10 W	100 W	1 kW	10 kW	100 kW	1 MW	10 MW
3-ph Induction							
1-ph induction							
3-ph Exc. Synch.							
3-ph P. M. Synch.							
3-ph Reluctance							
Conventional d.c.							
Brushless d.c.							
Ironless Rotor d.c.							
Switched reluctance							
Stepping							

Figure 11.1   *Continuous power rating for various types of motor*

the power scales are logarithmic it would be easy to miss the exception-ally wide power range of some types of motor: induction and d.c. motors, for example, extend from watts to megawatts, an astonishing range that few other inventions can match.

The width of the bands is intended to give some idea of relative importance, while the shading reflects the fact that there is no sharp cut-off at the extremities of the range. We should also bear in mind that we are talking here about the continuously rated maximum power at the

Type of drive	Power range					
	100 W	1 kW	10 kW	100 kW	1 MW	10 MW
Inverter-fed cage induction						
Cycloconverter-fed cage induction						
Slipring induction						
Inverter-fed synchronous						
Closed-loop synchronous						
Conventional d.c.						
Brushless d.c.						
Switched reluctance						

Figure 11.2   *Power range for various types of drive*

normal base speed, and as we have seen most motors will be able to exceed this for short periods, and also to run faster than base speed, provided that reduced torque is acceptable.

Maximum speed and speed range

We saw in Chapter 1 that as a general rule, for a given power the higher the base speed the smaller the motor. In practice, there are only a few applications where motors with base speeds below a few hundred rev/min are attractive, and it is usually best to obtain low speeds by means of the appropriate mechanical speed reduction.

Speeds over 10 000 rev/min are also unusual except in small universal motors and special-purpose inverter-fed motors. The majority of medium-size motors have base speeds between 1500 and 3000 rev/min. Base speeds in this range are attractive as far as motor design is concerned, because good power/weight ratios are obtained, and are also satisfactory as far as any mechanical transmission is concerned.

In controlled-speed applications, the range over which the steady-state speed must be controlled, and the accuracy of the speed holding, are significant factors in the selection process. In general, the wider the speed range, the more expensive the drive: a range of 10:1 would be unexceptional, whereas 100:1 would be demanding. Figures for accuracy of speed holding can sometimes cause confusion, as they are usually given as a percentage of the base speed. Hence with a drive claiming a decent speed holding accuracy of 0.2% and a base speed of 2000 rev/min, the user can expect the actual speed to be between 1996 and 2004 rev/min when the speed reference is 2000 rev/min. But if the speed reference is set for 100 rev/min, the actual speed can be anywhere between 96 and 104 rev/min, and still be within the specification.

For constant torque loads, which require operation at all speeds, the inverter-fed induction motor, the d.c. drive and any of the self-synchronous drives are possibilities, but only the d.c. drive would automatically come with a force-ventilated motor capable of continuous operation with full torque at low speeds.

Fan-type loads (see below) with a wide operating speed range are a somewhat easier proposition because the torque is low at low speeds. In the medium and low power ranges the inverter-fed induction motor (using a standard motor) is satisfactory, and will probably be cheaper than the d.c. drive. For restricted speed ranges (say from base speed down to 75%) and particularly with fan-type loads where precision speed control is unnecessary, the simple voltage-controlled induction motor is likely to be the cheapest solution.

LOAD REQUIREMENTS – TORQUE–SPEED CHARACTERISTICS

The most important things we need to know about the load are the steady-state torque–speed characteristic, and the effective inertia as seen by the motor. In addition, we clearly need to know what performance is required. At one extreme, for example, in a steel-rolling mill, it may be necessary for the speed to be set at any value over a wide range, and for the mill to react very quickly when a new target speed is demanded. Having reached the set speed, it may be essential that it is held very precisely even when subjected to sudden load changes. At the other extreme, for example, a large ventilating fan, the range of set speed may be quite limited (perhaps from 80% to 100%); it may not be important to hold the set speed very precisely; and the time taken to change speeds, or to run-up from rest, are unlikely to be critical.

At full speed both of these examples may demand the same power, and at first sight might therefore be satisfied by the same drive system. But the ventilating fan is obviously an easier proposition, and it would be overkill to use the same system for both. The rolling mill would call for a regenerative d.c. or a.c. drive with tacho or encoder feedback, while the fan could quite happily manage with a cheaper open-loop inverter-fed induction motor drive, or even perhaps a simple voltage-controlled induction motor.

Although loads can vary enormously, it is customary to classify them into two major categories, referred to as 'constant-torque' or 'fan or pump' types. We will use the example of a constant-torque load to illustrate in detail what needs to be done to arrive at a specification for the torque–speed curve. An extensive treatment is warranted because this is often the stage at which users come unstuck.

Constant-torque load

A constant torque load implies that the torque required to keep the load running is the same at all speeds. A good example is a drum-type hoist, where the torque required varies with the load on the hook, but not with the speed of hoisting. An example is shown in Figure 11.3.

The drum diameter is 0.5 m, so if the maximum load (including the cable) is say 1000 kg, the tension in the cable (mg) will be 9810 N, and the torque applied by the load at the drum will be given by force × radius = $9810 \times 0.25 \approx 2500$ Nm. When the speed is constant (i.e. the load is not accelerating), the torque provided by the motor at

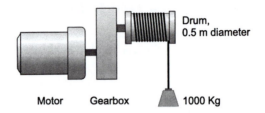

Drum,
0.5 m diameter

Motor   Gearbox   1000 Kg

Figure 11.3   *Motor-driven hoist – a constant-torque load*

the drum must be equal and opposite to that exerted at the drum by the
load. (The word 'opposite' in the last sentence is often omitted, it being
understood that steady-state motor and load torque must necessarily act
in opposition.)

Suppose that the hoisting speed is to be controllable at any value up to
a maximum of 0.5 m/s, and that we want this to correspond with a
maximum motor speed of around 1500 rev/min, which is a reasonable
speed for a wide range of motors. A hoisting speed of 0.5 m/s corresponds to a drum speed of 19 rev/min, so a suitable gear ratio would be
say 80:1, giving a maximum motor speed of 1520 rev/min.

The load torque, as seen at the motor side of the gearbox, will be
reduced by a factor of 80, from 2500 to 31 Nm at the motor. We must
also allow for friction in the gearbox, equivalent to perhaps 20% of the
full-load torque, so the maximum motor torque required for hoisting
will be 37 Nm, and this torque must be available at all speeds up to the
maximum of 1520 rev/min.

We can now draw the steady-state torque–speed curve of the load as
seen by the motor, as shown in Figure 11.4.

The steady-state motor power is obtained from the product of torque
(Nm) and angular velocity (rad/s). The maximum continuous motor
power for hoisting is therefore given by

$$P_{\max} = 37 \times 1520 \times \frac{2\pi}{60} = 5.9\,\text{kW} \qquad (11.1)$$

At this stage it is always a good idea to check that we would obtain
roughly the same answer for the power by considering the work done
per second at the load. The force (F) on the load is 9810 N, the velocity
(v) is 0.5 m/s so the power (Fv) is 4.9 kW. This is 20% less than we
obtained above, because here we have ignored the power lost in the
gearbox.

So far we have established that we need a motor capable of continuously delivering 5.9 kW at 1520 rev/min in order to lift the heaviest load

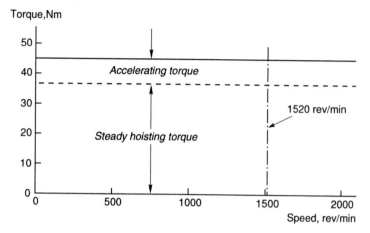

Figure 11.4   *Torque requirements for motor in hoist application (Figure 11.3)*

at the maximum required speed. However, we have not yet addressed the question of how the load is accelerated from rest and brought up to the maximum speed. During the acceleration phase the motor must produce a torque greater than the load torque, or else the load will descend as soon as the brake is lifted. The greater the difference between the motor torque and the load torque, the higher the acceleration. Suppose we want the heaviest load to reach full speed from rest in say 1 s, and suppose we decide that the acceleration is to be constant. We can calculate the required accelerating torque from the equation of motion, i.e.

$$\text{Torque (Nm)} = \text{Inertia (kg m}^2) \times \text{Angular acceleration (rad/s}^2)$$

$$(11.2)$$

We usually find it best to work in terms of the variables as seen by the motor, and therefore we first need to find the effective total inertia as seen at the motor shaft, then calculate the motor acceleration, and finally use equation (11.2) to obtain the accelerating torque.

The effective inertia consists of the inertia of the motor itself, the referred inertia of the drum and gearbox, and the referred inertia of the load on the hook. The term 'referred inertia' means the apparent inertia, viewed from the motor side of the gearbox. If the gearbox has a ratio of $n{:}1$ (where n is greater than 1), an inertia of J on the low-speed side appears to be an inertia of J/n^2 at the high-speed side. (This formula is the same as that for finding the referred impedance as seen at the primary of an ideal transformer, as discussed in Chapter 7.)

In this example the load actually moves in a straight line, so we need to ask what the effective inertia of the load is, as 'seen' at the drum. The geometry here is simple, and it is not difficult to see that as far as the inertia seen by the drum is concerned the load appears to be fixed to the surface of the drum. The load inertia at the drum is then obtained by using the formula for the inertia of a mass m located at radius r, i.e. $J = mr^2$, yielding the effective load inertia at the drum as $1000\,\text{kg} \times (0.25\,\text{m})^2 = 62.5\,\text{kg m}^2$.

The effective inertia of the load as seen by the motor is $1/(80)^2 \times 62.5 \approx 0.01\,\text{kg m}^2$. To this must be added firstly the motor inertia (which we can only estimate by consulting the manufacturer's catalogue for a 5.9 kW, 1520 rev/min motor, which yields a figure of $0.02\,\text{kg m}^2$), and secondly the referred inertia of the drum and gearbox, which again we have to look up. Suppose this yields a further $0.02\,\text{kg m}^2$. The total effective inertia is thus $0.05\,\text{kg m}^2$, of which 40% is due to the motor itself.

The acceleration is easy to obtain, since we know the motor speed is required to rise from 0 to 1520 rev/min in 1 s. The angular acceleration is given by the increase in speed divided by the time taken, i.e.

$$\left(1520 \times \frac{2\pi}{60}\right) \div 1 = 160\,\text{rad/s}^2$$

We can now calculate the accelerating torque from equation (11.2) as

$$T = 0.05 \times 160 = 8\,\text{Nm}$$

Hence in order to meet both the steady-state and dynamic torque requirements, a drive capable of delivering a torque of 45 Nm ($= 37 + 8$) at all speeds up to 1520 rev/min is required, as indicated in Figure 11.4.

In the case of a hoist, the anticipated pattern of operation may not be known, but it is likely that the motor will spend most of its time hoisting rather than accelerating. Hence although the peak torque of 45 Nm must be available at all speeds, this will not be a continuous demand, and will probably be within the short-term overload capability of a drive which is continuously rated at 5.9 kW.

We should also consider what happens if it is necessary to lower the fully loaded hook. We allowed for friction of 20% of the load torque (31 Nm), so during descent we can expect the friction to exert a braking torque equivalent to 6.2 Nm. But in order to prevent the hook from running away, we will need a total torque of 31 Nm, so to restrain the load the motor will have to produce a torque of 24.8 Nm. We would naturally refer to this as a braking torque because it is necessary to prevent the load on the hook from running away, but in fact the torque remains in the same direction as when hoisting. The speed is however negative, and in terms of a 'four-quadrant' diagram (e.g. Figure 3.17) we have moved from quadrant 1 to quadrant 4, and thus the power flow is reversed and the motor is regenerating, the loss of potential energy of the descending load being converted back into electrical form. Hence, if we wish to cater for this situation we must go for a drive that is capable of continuous regeneration: such a drive would also have the facility for operating in quadrant 3 to produce negative torque to drive down the empty hook if its weight was insufficient to lower itself.

In this example the torque is dominated by the steady-state requirement, and the inertia-dependent accelerating torque is comparatively modest. Of course if we had specified that the load was to be accelerated in one-fifth of a second rather than 1 s, we would require an accelerating torque of 40 Nm rather than 8 Nm, and as far as torque requirements are concerned the acceleration torque would be more or less the same as the steady-state running torque. In this case it would be necessary to consult the drive manufacturer to determine the drive rating, which would depend on the frequency of the start–stop sequence.

The question of how to rate the motor when the loading is intermittent is explored more fully in Section 11.4.2, but it is worth noting that if the inertia is appreciable the stored rotational kinetic energy $(1/2J\omega^2)$ may become very significant, especially when the drive is required to bring the load to rest. Any stored energy either has to be dissipated in the motor and drive itself, or returned to the supply. All motors are inherently capable of regenerating, so the arrangement whereby the kinetic energy is recovered and dumped as heat in a resistor within the drive enclosure is the cheaper option, but is only practicable when the energy to be absorbed is modest. If the stored kinetic energy is large, the drive must be capable of returning energy to the supply, and this inevitably pushes up the cost of the converter.

In the case of our hoist, the stored kinetic energy is only $1/2\times 0.05 (1520 \times 2\pi/60)^2 = 633\,J$, or about 1% of the energy needed to heat up a mug of water for a cup of coffee. Such modest energies could easily be absorbed by a resistor, but given that in this instance we are providing a regenerative drive, this energy would also be returned to the supply.

Inertia matching

There are some applications where the inertia dominates the torque requirement, and the question of selecting the right gearbox ratio has to be addressed. In this context the term 'inertia matching' often causes confusion, so it worth explaining what it means.

Suppose we have a motor with a given torque, and we want to drive an inertial load via a gearbox. As discussed previously, the gear ratio determines the effective inertia as 'seen' by the motor: a high step-down ratio (i.e. load speed much less than motor speed) leads to a very low referred inertia, and vice-versa.

If the specification calls for the acceleration of the load to be maximised, it turns out that the optimum gear ratio is that which causes the referred inertia of the load to be equal to the inertia of the motor. Applications in which load acceleration is important include all types of positioning drives, e.g. in machine tools and phototypesetting. (There is another electrical parallel here – to get the most power into a load from a source with internal resistance R, the load resistance must be made equal to R.)

It is important to note, however, that inertia matching only maximises the *acceleration* of the load. Frequently it turns out that some other aspect of the specification (e.g. the maximum required load speed) cannot be met if the gearing is chosen to satisfy the inertia matching criterion, and it then becomes necessary to accept reduced acceleration of the load in favour of higher speed.

Fan and pump loads

Fans and pumps have steady-state torque–speed characteristics which generally have the shapes shown in Figure 11.5.

These characteristics are often approximately represented by assuming that the torque required is proportional to the square or the cube of the speed, giving rise to the terms 'square-law' or 'cube-law' load. We should note, however, that the approximation is seldom valid at low speeds because most real fans or pumps have a significant static friction or breakaway torque (as shown in Figure 11.5), which must be overcome when starting.

When we consider the power–speed relationships the striking difference between the constant-torque and fan-type load is underlined. If the motor is rated for continuous operation at the full speed, it will be very lightly loaded (typically around 20%) at half speed, whereas with the constant torque load the power rating will be 50% at half speed.

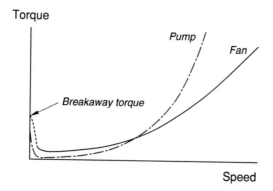

Figure 11.5 *Torque–speed characteristics for fan- and pump-type loads*

Fan-type loads which require speed control can therefore be handled by drives which can only allow reduced power at such low speeds, such as the inverter-fed cage induction motor without additional cooling, or the voltage-controlled cage motor. If we assume that the rate of acceleration required is modest, the motor will require a torque–speed characteristic, which is just a little greater than the load torque at all speeds. This defines the operating region in the torque–speed plane, from which the drive can be selected.

Many fans do not require speed control of course, and are well served by mains-frequency induction motors. We looked at a typical example in Chapter 6, the run-up behaviour being contrasted with that of a constant-torque load in Figure 6.6.

GENERAL APPLICATION CONSIDERATIONS

Regenerative operation and braking

All motors are inherently capable of regenerative operation, but in drives the basic power converter as used for the 'bottom of the range' version will not normally be capable of continuous regenerative operation. The cost of providing for fully regenerative operation is usually considerable, and users should always ask the question 'do I really need it?'

In most cases it is not the recovery of energy for its own sake, which is of prime concern, but rather the need to achieve a specified dynamic performance. Where rapid reversal is called for, for example, kinetic energy has to be removed quickly, and, as discussed in the previous section, this implies that the energy is either returned to the supply (regenerative operation) or dissipated (usually in a braking resistor).

An important point to bear in mind is that a non-regenerative drive will have an asymmetrical transient speed response, so that when a higher speed is demanded, the extra kinetic energy can be provided quickly, but if a lower speed is demanded, the drive can do no better than reduce the torque to zero and allow the speed to coast down.

Duty cycle and rating

This is a complex matter, which in essence reflects the fact that whereas all motors are governed by a thermal (temperature rise) limitation, there are different patterns of operation which can lead to the same ultimate temperature rise.

Broadly speaking the procedure is to choose the motor on the basis of the r.m.s. of the power cycle, on the assumption that the losses (and therefore the temperature rise) vary with the square of the load. This is a reasonable approximation for most motors, especially if the variation in power is due to variations in load torque at an essentially constant speed, as is often the case, and the thermal time-constant of the motor is long compared with the period of the loading cycle. (The thermal time-constant has the same significance as it does in relation to any first-order linear system, e.g. an R/C circuit. If the motor is started from ambient temperature and run at a constant load, it takes typically four or five time-constants to reach its steady operating temperature.) Thermal time-constants vary from more than an hour for the largest motors (e.g. in a steel mill) through tens of minutes for medium power machines

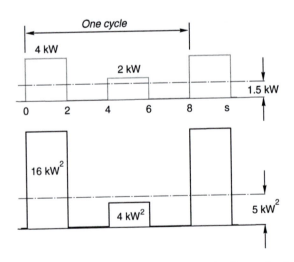

Figure 11.6   *Calculation of r.m.s. power rating for periodically varying load*

down to minutes for fractional horsepower motors and seconds for small stepping motors.

To illustrate the estimation of rating when the load varies periodically, suppose a mains-fed cage induction motor is required to run at a power of 4 kW for 2 min, followed by 2 min running light, then 2 min at 2 kW, then 2 min running light, this 8-min pattern is repeated continuously. To choose an appropriate power rating we need to find the r.m.s. power, which means exactly what it says, i.e. it is the square root of the mean (average) of the square of the power. The variation of power is shown in the upper part of Figure 11.6, which has been drawn on the basis that when running light the power is negligible. The 'power squared' is shown in the lower part of the figure.

The average power is 1.5 kW, the average of the power squared is $5\,kW^2$, and the r.m.s. power is therefore $\sqrt{5}\,kW$, i.e. 2.24 kW. A motor that is continuously rated at 2.24 kW would therefore be suitable for this application, provided of course that it is capable of meeting the overload torque associated with the 4 kW period. The motor must therefore be able to deliver a torque that is greater than the continuous rated torque by a factor of 4/2.25, i.e. 178%: this would be within the capability of most general-purpose induction motors.

Motor suppliers are accustomed to recommending the best type of motor for a given pattern of operation, and they will typically classify the duty type in one of eight standard categories, which cover the most commonly encountered modes of operation. As far as rating is concerned the most common classifications are maximum continuous rating, where the motor is capable of operating for an unlimited period, and short time rating, where the motor can only be operated for a limited time (typically 10, 30 or 60 min) starting from ambient temperature.

Enclosures and cooling

There is clearly a world of difference between the harsh environment faced by a winch motor on the deck of an ocean-going ship, and the comparative comfort enjoyed by a motor driving the drum of an office photocopier. The former must be protected against the ingress of rain and seawater, while the latter can rely on a dry and largely dust-free atmosphere.

Classifying the extremely diverse range of environments poses a potential problem, but fortunately this is one area where international standards have been agreed and are widely used. The International Electrotechnical Committee (IEC) standards for motor enclosures are now almost universal and take the form of a classification number prefixed by the letters IP, and followed by two digits. The first digit indicates the protection level against ingress of solid particles ranging from 1 (solid bodies greater than 50-mm diameter) to 5 (dust), while the second relates to the level of protection against ingress of water ranging from 1 (dripping water) through 5 (jets of water) to 8 (submersible). A zero in either the first or second digit indicates no protection.

Methods of motor cooling have also been classified and the more common arrangements are indicated by the letters IC followed by two digits, the first of which indicates the cooling arrangement (e.g. 4 indicates cooling through the surface of the frame of the motor) while the second shows how the cooling circuit power is provided (e.g. 1 indicates motor-driven fan).

Dimensional standards

Standardisation is improving in this area, though it remains far from universal. Such matters as shaft diameter, centre height, mounting arrangements, terminal box position and overall dimensions are fairly closely defined for the mainstream motors (induction, d.c.) over a wide size range, but standardisation is relatively poor at the low-power end because so many motors are tailor-made for specific applications.

Supply interaction and harmonics

Most converter-fed drives cause distortion of the mains voltage which can upset other sensitive equipment, particularly in the immediate vicinity of the installation. There are some drives that are equipped with 'front-end' conditioning (whereby the current drawn from the mains is forced to approximate closely to a sinewave at unity power-factor), but this increases the cost of the power-electronics and is limited to small- and medium-power drives. With more and larger drives being installed the problem of mains distortion is increasing, and supply authorities therefore react by imposing increasingly stringent statutory limits governing what is allowable.

The usual pattern is for the supply authority to specify the maximum amplitude and spectrum of the harmonic currents at various levels in the power system. If the proposed installation exceeds these limits, appropriate filter circuits must be connected in parallel with the installation. These can be costly, and their design is far from simple because the electrical characteristics of the supply system need to be known in advance in order to avoid unwanted resonance phenomena. Users need to be alert to the potential problem, and to ensure that the drive supplier takes responsibility for handling it.

REVIEW QUESTIONS

1) The speed-holding accuracy of a 1500 rev/min drive is specified as 0.5% at all speeds below base speed. If the speed reference is set at 75 rev/min, what are the maximum and minimum speeds between which the drive can claim to meet the specification?

2) A servo motor drives an inertial load via a toothed belt. The motor carries a 12-tooth pulley, and the total inertia of motor and pulley is $0.001 \, kg \, m^2$. The load inertia (including the load pulley) is $0.009 \, kg \, m^2$. Find the number of teeth on the load pulley that will maximise the acceleration of the load.

3) Assuming that the temperature rise of a motor follows an exponential curve, that the final temperature rise is proportional to the total losses and that the losses are proportional to the square of the load power, for how long could a motor with a 30-min thermal time-constant be started from cold and overloaded by 60%?

4) A special-purpose numerically controlled machine-tool spindle has a maximum speed of 10 000 rev/min, and requires full torque at all speeds. Peak steady-state power is of the order of 1200 W. The manufacturer wishes to use a direct-drive motor. What options are available and what would you recommend? What additional information is required in order to seek the best solution?

5) When planning to purchase a variable-frequency inverter to provide speed control of an erstwhile fixed-frequency induction motor driving a hoist, what problems should be anticipated if low-speed operation is envisaged?

6) List three controlled-speed applications for which conventional d.c. motors are not well-suited, and suggest an alternative for each.

7) An induction motor was specifically chosen for a duty that required it to run day and night delivering 2 kW for 1 min, followed by 1-min running light. What continuous power rating would be appropriate for the motor if it was redeployed to drive a constant-torque load?

8) A 50 Hz pump drive requires a torque of 60 Nm at approximately 1400 rev/min for 1 min, followed by 5 min during which the motor

runs unloaded. This cyclic pattern is repeated continuously. The motor is to be selected from a range of general-purpose cage motors with continuous ratings of 2.2, 3, 4, 5.5, 7.5, 11 and 15 kW. The motors all have full-load slips of approximately 5% and pull-out torques of 200% at slips of approximately 15%. Select the pole number and power rating, and estimate the running speed when the motor drives the pump.

9) A speed-controlled drive rated 50 kW at its base speed of 1200 rev/ min drives a large circular stonecutting saw. When the drive is started from rest with the speed reference set to base speed, it accelerates to 1180 rev/min in 4 s, during which time the acceleration is more or less uniform. It takes a further second to settle at full speed, after which time the saw engages with the workpiece. Estimate the total effective inertia of motor and saw. Make clear what assumptions you have had to make. Estimate the stored kinetic energy at full speed, and compare it with the energy supplied during the first 4 s.

10) When the saw in question 9 is running light at base speed, and the power is switched-off, the speed falls approximately linearly, taking 20 s to reach 90% of base speed. Estimate the friction torque as a percentage of the full-load torque of the motor. Explain how this result justifies any approximations that had to be made in order to answer question 9.

APPENDIX
INTRODUCTION TO CLOSED-LOOP CONTROL

A.1 REASONS FOR ADOPTING A SIMPLIFIED APPROACH

The aim of this Appendix is to help the readers who are not familiar with closed-loop control and feedback to feel confident when they meet such ideas in the drives context. In line with the remainder of the book, the treatment avoids mathematics where possible, and in particular it makes only passing reference to transform techniques. This approach is chosen deliberately, despite the limitations it imposes, in the belief that it is more useful for the reader to obtain a sound grasp of what really matters in a control system, rather than to become adept at detailed analysis or design.

The writer's experience when first coming into contact with control was perhaps typical in that it soon became clear how one was required to perform certain mathematical procedures to arrive at the 'right answer'. But when asked exactly what 'the answer' really meant and why it mattered, or what features of the system were critically important and what were not, some yawning gaps in understanding were revealed. Reflecting back, there are perhaps two main reasons why this experience is not uncommon. Firstly, in the majority of student textbooks, the topic of control is introduced at the same time as the Laplace transform technique, which many readers will have not met before. And secondly, the casual use of jargon that seems to be a particular characteristic of 'control' specialists can be bewildering to newcomers.

The Laplace transform approach to the analysis and design of linear systems (i.e. those that obey the principle of superposition) is unchallenged for the very good reason that it works wonderfully well. It allows the differential equations that describe the dynamic behaviour of the systems we wish to control to be recast into algebraic form. Instead of having to solve differential equations in the time domain, we are able to transform the equations and emerge in a parallel universe in which algebraic equations in the complex frequency or 's' domain replace differential equations in time. For modelling control systems this approach is ideal because we can draw simple block diagrams containing algebraic 's-domain transfer functions' that fully represent both the steady-state and dynamic (transient) relationships of elements such as motors, amplifiers, filters, transducers, etc. We can assess whether the performance of a proposed system will be satisfactory by doing simple analysis in the s-domain, and decide how to improve matters if necessary. When we have completed our studies in the s-domain, we have a set of inverse transforms to allow us to 'beam back' to our real time domain.

Despite the undoubted power of the transform method, the fact that the newcomer has first to learn what amounts to a new language presents a real challenge. And in the process of concentrating on developing transfer functions, understanding how system order influences transient response, and other supporting matters, some of the key ideas that are central to successful control systems can easily be overlooked.

In addition to having to come to grips with new mathematical techniques the newcomer has to learn many new terms (e.g. bandwidth, integrator, damping factor, dB/octave), and is expected to divine the precise meaning of other casually used terms (e.g. gain, time-constant) by reference to the particular context in which they are used.

Our aim here is to avoid these potential pitfalls by restricting our analytical exploration to steady-state performance only. In the context of a speed-controlled drive, for example, this means that although our discussion will reveal the factors that determine how well the actual steady-state speed corresponds with the target speed, we will have to accept that it will not tell us how the speed gets from one value to another when the speed reference is changed (i.e. the transient response). By accepting this limitation we can avoid the need to use transform techniques, and concentrate instead on understanding what makes the system tick.

A.2 CLOSED-LOOP (FEEDBACK) SYSTEMS

A closed-loop system is one where there is feedback from the output, feedback being defined by the Oxford English Dictionary as 'the modification of a process by its results or effects, especially the difference between the desired and actual result'. We will begin by looking at some examples of feedback systems and identify their key features.

A.2.1   *Error-activated feedback systems*

An everyday example of a feedback system is the lavatory cistern (Figure A.1), where the aim is to keep the water level in the tank at the full mark. The valve through which water is admitted to the tank is controlled by the position of the arm carrying a ball that floats in the water. The steady-state condition is shown in diagram (a), the inlet valve being closed when the arm is horizontal.

When the WC is flushed (represented by the bottom valve being opened – see diagram (b)), the water level falls rapidly. As soon as the water level falls the ball drops, thereby opening the inlet valve so that fresh water enters the tank in an effort to maintain the water level. Given that the purpose of the system is to maintain the water level at the target level, it should be clear from diagram (b) why, in a control systems context, the angle θ_e is referred to as the 'error angle': when the water is at the desired level, the error is zero.

This is an example of an 'error-activated' system, because any error automatically initiates corrective action. The fact that the water level has fallen is communicated or fed back to the valve, which responds by admitting more water to combat the 'error'. In traditional cisterns the inlet water rate is proportional to the angle θ_e, so when the tank is empty it fills at a high rate and the water level rises rapidly. As the level rises the valve begins to close and the rate at which the tank refills reduces until finally when the target level is reached the valve is fully closed and the water level is restored to its target value (i.e. full). In this particular example the 'feedback' of the ball position takes the form of a direct mechanical connection between the water level detector (the floating ball) and the inlet valve. Most of the control systems that we meet in a drives context are also error-activated, though in the majority of cases the feedback is less direct than in this example.

Alert readers will have spotted that although the aim of the system is to maintain a constant water level, the level must change every time corrective action is required. The only time that the system can have no

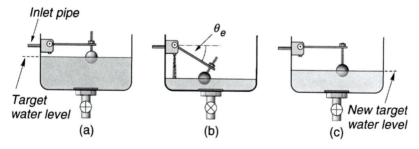

Figure A.1   *Simple closed-loop regulating system*

error is when it reaches the steady state, with no water going out and the inlet valve is fully closed. If there was a leak in the tank, causing water to drain away, the water level would stabilise at a lower level, at which the inlet valve was just open sufficiently wide to admit water at the same rate as it was leaking away. Clearly, an inlet valve that yields a high flow rate for a very small angle will be better able to maintain the correct level than one that needs a large angle error to yield the same flow rate. In control terms, this is equivalent to saying that the higher the 'gain' (expressed as *flow rate/error angle*), the better.

Another important observation is that although it is essential that there is a supply of water available at the inlet valve, the water pressure itself is not important as far as the target level is concerned: if the pressure is high the tank will fill up more quickly than if the pressure is low, but the final level will be the same. This is because the amount of water admitted depends on the integral with respect to time of the inlet flow rate, so the target level will ultimately be reached regardless of the inlet pressure. In fact, this example illustrates the principal advantage of 'integral control', in that the steady-state error ultimately becomes zero. We will return to this later.

Finally, it is worth pointing out that this is an example of a regulator, i.e. a closed-loop system where the target level does not change, whereas in most of the systems we meet in drives the target or reference signal will vary. However, if it was ever necessary to alter the target level in the cistern, it could be accomplished by adjusting the distance between ball and arm, as in diagram (c).

A.2.2 *Closed-loop systems*

To illustrate the origin and meaning of the term 'closed-loop' we will consider another familiar activity, that of driving a car, and in particular we will imagine that we are required to drive at a speed of exactly 50 km/h, the speed to be verified by an auditor from the bureau of standards.

The first essential is an accurate speedometer, because we must measure the output of the 'process' if we are to control it accurately. Our eyes convey the 'actual speed' signal (from the speedometer dial) to our brain, which calculates the difference or error between the target speed and the actual speed. If the actual speed was 30 km/h, the brain would conclude that we were 20 km/h too slow and instruct our foot to press much harder on the accelerator. As we saw the speed rising towards the target, the brain will decide that we can ease off the accelerator a little so that we don't overshoot the desired speed. If a headwind springs up, or

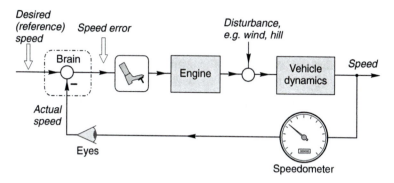

Figure A.2   *Block-diagram representation of driving a car at a target speed*

we reach a hill, any drop in speed will be seen and we will press harder on the gas pedal; conversely, if we find speed is rising due to a following wind or a falling gradient, we apply less accelerator, or even apply the brake. This set of interactions can be represented diagrammatically as in Figure A.2.

Diagrams of this sort – known as block diagrams – are widely used in control system studies because they provide a convenient pictorial representation of the interrelationships between the various elements of the system. The significance of the lines joining the blocks depends on the context: for example, the line entering the speedometer represents the fact that there is a hardware input from the wheels (perhaps a mechanical one, or more likely a train of electronic pulses whose frequency corresponds to speed), while the line leaving the speedometer represents the transmission of information via reflected light to the eyes of the driver. In many instances (e.g. an electronic amplifier) the output from a block is directly proportional to the input, but in others (e.g. the 'vehicle dynamics' block in Figure A.2) the output depends on the integral of the input with respect to time.

The lower part of the block diagram represents the speed feedback path; the upper part (known as the forward path) represents the process itself. The fact that all the blocks are connected together so that any change in the output of one affects all the others gives rise to the name *closed-loop system*.

The circles represent summing junctions, i.e. points where signals are added or subtracted. The output signal from the summing junction is the sum of the input signals, or the difference if one is negative. The most important summing junction is the one on the left (inside the driver's brain!) where the actual speed (via the eyes) is subtracted from the reference speed (stored in the brain) to obtain the speed error. The negative sign on the speed feedback signal indicates that a subtraction is required, and it is this that gives rise to the term *negative feedback*.

Factors that clearly have a bearing on whether we will be able to achieve the desired speed are the ability of the car to achieve and maintain the speed; the extent to which the driver's eyes can detect movements in the speedometer pointer; the need for the brain to subtract the actual speed from the desired speed; and the requirement for the brain to instruct the muscles of the leg and foot, and their ability to alter the pressure on the accelerator pedal. But all of these will be of no consequence if the speedometer is not accurate, because even if the driver thinks he is doing 50 km/h, the man from the standards bureau will not be impressed. A VW beetle with an accurate speedometer is potentially better than a Rolls Royce with an inaccurate one – a fact that underlines the paramount importance of the feedback path, and in particular highlights the wisdom of investing in a good transducer to measure the quantity that we wish to control. This lesson will emerge again when we look at what constitutes a good closed-loop system.

Returning to how the speed-control system operates we can identify the driver as the controller, i.e. the part of the system where the error is generated, and corrective action initiated, and it should be clear that an alert driver with keen eyesight will be better at keeping the speed constant than one whose eyesight is poor and whose actions are somewhat sluggish. We will see that in most automatic control systems expressions such as 'alert' and 'keen eyesight' translate into the 'gain' of the controller: a controller with a high gain gives a large output in response to a small change at its input. Finally, we should also acknowledge that the block diagram only accounts for some of the factors that influence the human driver: when the car approaches a hill, for example, the driver knows from experience that he will need more power and acts on this basis without waiting for the speedometer to signal a drop in speed. Although this is a fascinating aspect of control it is beyond our current remit.

A.3   STEADY-STATE ANALYSIS OF CLOSED-LOOP SYSTEMS

A typical closed-loop system is shown in block diagram form in Figure A.3. As explained in the introduction, our aim is to keep the mathematical treatment as simple as possible, so each of the blocks contains a single symbol that represents the ratio of output to input under steady-state conditions, i.e. when any transients have died out and all the variables have settled.

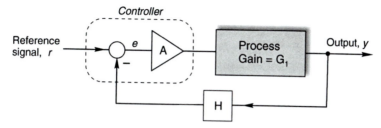

Figure A.3 *Typical arrangement of negative-feedback closed-loop control system*

A useful question to pose whenever a new block diagram is encountered is 'what is this control system for?' The answer is that the variable that is sensed and fed back (i.e. the output of the block labelled 'process' in Figure A.3) is the quantity we wish to control.

For example the 'process' block might represent say an unloaded d.c. motor, the input being the armature voltage and the output being the speed. We would then deduce that the system was intended to provide closed-loop control of the motor speed. The gain G_1 is simply the ratio of the steady-state speed to the armature voltage, and it would typically be expressed in rev/min/volt. The triangular symbol usually represents an electronic amplifier, in which case it has a (dimensionless) gain of A, i.e. the input and output are both voltages. We conclude from this that the reference input must also be a voltage, because the input and output from a summing junction must necessarily be of the same type. It also follows that the signal fed back into the summing junction must also be a voltage, so that the block labelled H (the feedback transducer) must represent the conversion of the output (speed, in rev/min) to voltage. In this case, we know that the speed feedback will indeed be obtained from a tachogenerator, with, in this case, an e.m.f. constant of H volts/rev/min.

In electronic systems where the summing junction is usually an integral part of the amplifier, the combination forms the controller, as shown by the dotted line in Figure A.3.

At this point we should pause and ask what we want of a good control system. As has already been mentioned, the output of the summing junction is known as the error signal: its perjorative name helps us to remember that we want the error to be as small as possible. Because the error is the difference between the reference signal and the fed-back signal, we deduce that in an ideal (error-free) control system the fed-back signal should equal the reference signal.

The sharp-witted reader will again wonder how the system could possibly operate if the error was zero, because then there would be no input to the forward path and unless the gain of the controller was infinite there would be nothing going into the process to produce the output that we are trying to control. So, for the present let us assume that the gain of the controller is very high, and make the justifiable assumption that we have a good system and that therefore the feedback signal is almost, but not quite, equal to the reference signal.

In terms of the symbols in Figure A.3, the almost zero-error condition is represented by approximating the reference signal (r) to the feed-back signal (Hy), i.e.

$$r \approx Hy, \quad \text{or} \quad y \approx \frac{1}{H} r \tag{A.1}$$

This equation is extremely important because it indicates that in a good control system, the output is proportional to the input, with the constant of proportionality or 'gain' of the closed-loop system depending only on the feedback factor (H). We came to the same conclusion when we looked at the example of driving a car at a given speed, when we concluded the most important aspect was that we had a reliable speedometer to provide accurate feedback.

Of course we must not run away with the idea that the forward path is of no consequence, so next we will examine matters in a little more detail to see what conditions have to be satisfied in order for the steady-state performance of a closed-loop system to be considered good. For this analysis we will assume that all the blocks in the forward path are combined together as a single block, as shown in Figure A.4, where the gain G represents the product of all the gains in the forward path. (For example, if we want to reduce Figure A.3 to the form shown in Figure A.4, we put $G = AG_1$.)

The signals in Figure A.4 are related as follows:

$$y = Ge \quad \text{and} \quad e = r - Hy \tag{A.2}$$

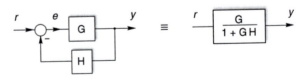

Figure A.4   *Closed-loop system (left) and its representation by a single equivalent block*

To establish the relationship between the output and input we eliminate e, to yield

$$y = \left\{ \frac{G}{1 + GH} \right\} r \qquad (A.3)$$

This shows us that the output is proportional to the input, with the closed-loop gain (output y over input, r) being given by $G/(1 + GH)$. Note that so far we have made no approximations, so we should not be surprised to see that the gain of the closed-loop system does in fact depend not only on the feedback, but also on the forward path. This result is summarised in Figure A.4, where the single block on the right behaves exactly the same as the complete closed-loop system on the left.

It is worth mentioning that if we had been using the Laplace transform method to express relationships in the s-domain, the forward and feedback paths would be written $G(s)$ and $H(s)$, where $G(s)$ and $H(s)$ represent algebraic functions describing – in the s-domain – both steady-state and dynamic properties of the forward and feedback paths, respectively. These functions are known as 'transfer functions' and although the term strictly relates to the s-domain, there is no reason why we cannot use the same terminology here, provided that we remember that we have chosen to consider only steady-state conditions, and therefore what we mean by transfer function is simply the gain of the block in the steady-state. In the context of Figure A.4, we note that the transfer function (i.e. gain) of the single block on the right is the same as the gain of the closed-loop system on the left.

Returning to equation (A.3), it is easy to see that if the product GH (which must always be dimensionless) is very much greater than 1, we can ignore the 1 in the denominator and then equation (A.3) reduces to equation (A.1), derived earlier by asking what we wanted from a good control system. This very important result is shown in Figure A.5, where the single block on the right approximates the behaviour of that on the left, which itself precisely represents the complete system shown in Figure A.4.

So now we can be precise about the conditions we seek to obtain a good system. Firstly, we need the product GH to be much greater

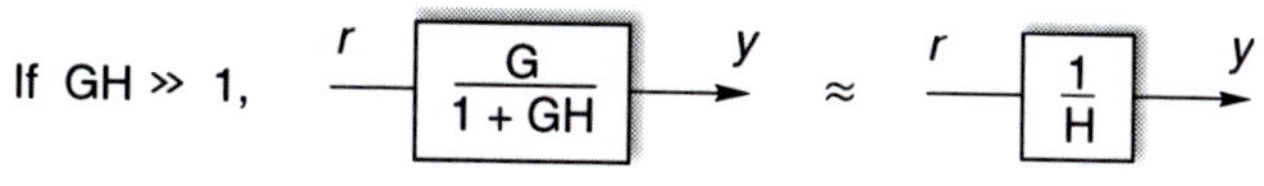

Figure A.5   *Simplification of closed-loop system model when loop gain (GH) is much greater than 1*

than 1, so that the overall performance is not dependent on the forward path (i.e. the process), but instead depends on the gain of the feedback path.

The product *GH* is called the 'loop gain' of the system, because it is the gain that is incurred by a signal passing once round the complete loop. In electronic control systems there is seldom any difficulty in achieving enough loop gain because the controller will be based on an amplifier whose gain can be very high. In drive systems (e.g. speed control) it would be unusual for the overall loop gain (error amplifier, power stage, motor, tacho feedback) to be less than 10, but unlikely to be over a thousand. (We will see later, however, that if we use integral control, the steady-state loop gain will be infinite.)

Secondly, having ensured, by means of a high loop gain, that the overall gain of the closed-loop system is given by $1/H$, it is clearly necessary to ensure that the feedback gain (H) can be precisely specified and guaranteed. If the value of H is not correct for the closed-loop gain we are seeking, or if it is different on a hot day from a cold day, we will not have a good closed-loop system. This again underlines the importance of spending money on a good feedback transducer.

A.4 IMPORTANCE OF LOOP GAIN – EXAMPLE

To illustrate the significance of the forward and feedback paths on the overall gain of a closed-loop system we will look at an operational amplifier of the type frequently used in analogue control. In their basic form these amplifiers typically have very high gains, but the gain can vary considerably even amongst devices from a single batch. We will see that as normally employed (with negative feedback) the unpredictability of the open-loop gain is not a problem.

The op-amp itself is represented by the triangle in Figure A.6(a). Voltages applied to the non-inverting input are amplified without sign change, while voltages applied to the inverting input are amplified and the sign is changed. If the open-loop gain of the amplifier is A, and the input voltages on the positive and negative inputs are v_1 and v_2, respectively, the output voltage will be $A(v_1 - v_2)$. The two inputs therefore perform the differencing function that we have seen is required in a negative feedback system.

In Figure A.6(a), we can see that a potential divider network of two high-precision resistors has been connected across the output voltage, with resistors chosen so that a fraction (0.2) of the output voltage can be fed-back to the inverting input terminal. In block diagram terms the circuit can therefore be represented as shown in Figure A.6(b), where r is

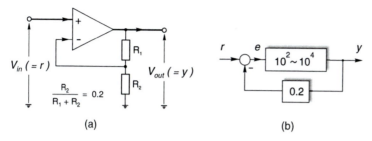

Figure A.6 *Operational amplifier with negative feedback:* (a) *circuit diagram,* (b) *equivalent block diagram*

the input or reference voltage, y is the output voltage and e represents the error signal.

The gain of the forward path (G in control systems terms) is assumed to vary between 100 and 10 000 (though for an op-amp the lower limit is unlikely to be as low as 100), and the feedback gain (H) is 0.2.

If the feedback system is good, equation (A.1) will apply and we would expect the overall (closed-loop) gain to be given by 1/0.2, i.e. 5. Table A.1 shows that the actual closed-loop gain is close to 5, the discrepancy reducing as the loop gain increases. When the loop gain is 2000 (corresponding to a forward gain of 10 000), the closed-loop gain of 4.9975 is within 0.05% of the ideal, while even a loop gain as low as 20 still gives a closed-loop gain that is only 4.8% below the ideal.

The important point to note is that the closed-loop gain is relatively insensitive to the forward path gain, provided that the loop gain remains large compared with 1, so if our op-amp happens to need replacing and we change it for one with a different gain, there will be very little effect on the overall gain of the system. On the other hand, it is very important for the gain of the feedback path to be stable, as any change is directly reflected in the closed-loop gain: for example, if H were to increase from 0.20 to 0.22 (i.e. a 10% increase), the closed-loop gain would fall by

Table A.1 Closed-loop gain is relatively insensitive to forward-path gain provided that the loop gain is much greater than 1

Forward gain, G	Feedback factor, H	Loop gain, GH	$1 + GH$	Closed-loop gain $G/(1 + GH)$
100	0.2	20	21	4.7619
1000	0.2	200	201	4.9751
10 000	0.2	2000	2001	4.9975

almost 10%. This explains why high-precision resistors with excellent temperature stability are used to form the potential divider.

We should conclude this example by acknowledging that we have to sacrifice the high inherent open-loop gain for a much lower overall closed-loop gain in the interests of achieving an overall gain that does not depend on the particular op-amp we are employing. But since such circuits are very cheap, we can afford to cascade them if we require more gain.

A.5   STEADY-STATE ERROR – INTEGRAL CONTROL

An important criterion for any closed-loop system is its steady-state error, which ideally should be zero. We can return to the op-amp example studied above to examine how the error varies with loop gain, the error being the difference between the reference signal and the feedback signal. If we make the reference signal unity for the sake of simplicity, the magnitudes of the output and error signals will be as shown in Figure A.7, for the three values of forward gain listed in Table A.1. At the highest loop gain the error is only 0.05%, rising to only 4.8% at the lowest loop gain. These figures are impressive, but what if even a very small error cannot be tolerated? It would seem that we would need infinite loop gain, which at first sight seems unlikely. So how do we eliminate steady-state error?

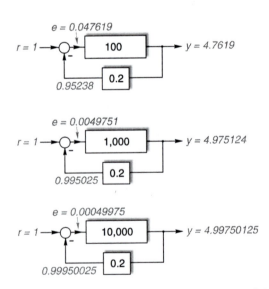

Figure A.7   *Diagram showing how the magnitude of the error signal (e) reduces as the gain of the loop increases*

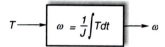

Figure A.8 *Block diagram representing the integration of torque to obtain angular velocity*

The clue to the answer lies in the observation above, that in order to have zero error we need infinite loop gain, i.e. at some point in the forward path we need a block that gives a steady output even when its input is zero. This is far less fanciful than it sounds, because all we need is for the forward path to contain a block that represents an element or process whose output depends not simply on its input at that time, but on the integrated effect of the input up to the time in question.

An example that is frequently encountered in drives is a block whose input is torque and whose output represents angular velocity. The torque (T) determines the rate of change of the angular velocity, the equation of motion being $d\omega/dt = T/J$, where J is the inertia. So the angular velocity is given by $\omega = 1/J \int T dt$, i.e. the speed is determined by the integral with respect to time of the torque, not by the value of torque at any particular instant. The block diagram for this is shown in Figure A.8. The presence of the integral sign alerts us to the fact that at any instant, the value of the angular speed depends on the past history of the torque to which it has been subjected.

To illustrate the variation in the 'effective gain' of a block containing an integrator, we can study a simple example of an inertia of $0.5 \, kg \, m^2$, initially at rest, that is subjected to a torque of 2 Nm for 1 s, followed by a torque of -1 Nm for a further second and zero torque thereafter. Plots of the input torque and the corresponding output angular speed with time are shown in Figure A.9. For the first second the acceleration is 4 rad/s/s, so the speed is 4 rad/s after 1 s. For the next second the torque is negative and the deceleration is 2 rad/s/s, so the speed after 2 s has fallen to 2 rad/s. Subsequently the torque and acceleration are zero so the speed remains constant.

If we examine the graphs, we see that the ratio of output (speed) to input (torque), i.e. the quantity that hitherto we have referred to as gain, varies. For example, just before $t = 1$, the gain is $4/2 = 2 \, rad/s/Nm$, whereas just after $t = 1$ it is $4/-1 = -4 \, rad/s/Nm$. Much more importantly, however, we see that from $t = 2$ onwards, the output is constant at 2 rad/s, but the input is zero, i.e. the gain is infinite when the steady state is reached.

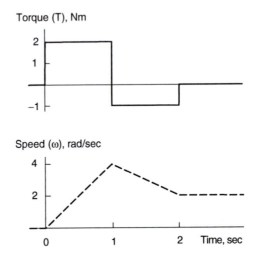

Figure A.9 *Variation of angular velocity (ω) resulting from application of torque (T)*

The fact that the gain changes until the steady state is reached means that we can only use the simple 'gain block' approach to represent an integrator when conditions have settled into the steady state, i.e. all the variables are constant. The output of an integrator can only remain constant if its input is zero: if the input is positive, the output will be increasing, and if the input is negative the output will be decreasing. So in the steady state, an integrator has infinite gain, which is just what we need to eliminate steady-state error.

The example discussed above includes an integration that reflects physical properties, i.e. the fact that the angular acceleration is proportional to the torque, and hence the angular velocity is proportional to the integral of torque. Another 'natural' integration applies to the water tank we looked at earlier, where the level of water is proportional to the integral of the flow rate with respect to time.

In many systems there is no natural integration in the forward-path process, so if it is important to achieve zero steady-state error, an integrating term is included in the controller. This is discussed in the next section.

A.6 PID CONTROLLER

We saw in Section A.3 that the simplest form of controller is an amplifier the output of which is proportional to the error signal. A control system that operates with this sort of controller is said to have 'proportional' or 'P' control. An important feature of proportional control is that as soon as there is any change in the error, proportionate action is initiated immediately.

We have also seen in Section A.5 that in order to completely eliminate steady-state error we need to have an integrating element in the forward path, so we may be tempted to replace the proportional controller in Figure A.3 by a controller whose output is the integral of the error signal with respect to time. This is easily done in the case of an electronic amplifier, yielding an 'integral' or 'I' controller.

However, unlike the proportional controller where the output responds instantaneously to changes in the error, the output of an integrating controller takes time to respond. For example, we see from Figure A.9 that the output builds progressively when there is a step input. If we were to employ integral control alone, the lag between output and input signals may well cause the overall (closed-loop) response to be unacceptably sluggish.

To obtain the best of both worlds (i.e. a fast response to changes and elimination of steady-state error), it is common to have both proportional and integral terms in the controller, which is then referred to as a PI controller. The output of the controller ($y(t)$) is then given by the expression

$$y(t) = Ae(t) + k \int e(t) \mathrm{d}t$$

where $e(t)$ is the error signal, A is the proportional gain and k is a parameter that allows the rate at which the integrator ramps up to be varied. The latter adjustment is also often – and rather confusingly – referred to as integrator gain.

We can easily see that because raising the proportional gain causes a larger output of the controller for a given error, we generally get a faster transient response. On the other hand the time lag associated with an integral term generally makes the transient response more sluggish, and increases the likelihood that the output will overshoot its target before settling.

Some controllers also provide an output term that depends on the rate of change or differential of the error. This has the effect of increasing the damping of the transient response, an effect similar to that demonstrated in Figure 4.16. PI controllers that also have this differential (or D) facility are known as PID controllers.

A.7   STABILITY

So far we have highlighted the benefits of closed-loop systems, but not surprisingly there is a potential negative side that we need to be aware of. This is that if the d.c. loop gain is too high, some closed-loop systems exhibit self-sustaining oscillations, i.e. the output 'rings' – generally at a high frequency – even when there is no input to the system. When a system behaves in this way it is said to be unstable, and clearly the consequences can be extremely serious, particularly if large mechanical elements are involved. (It is worth mentioning that whereas in control systems instability is always undesirable, if we were in the business of designing electronic oscillators we would have an entirely different perspective, and we would deliberately arrange our feedback and loop gain so as to promote oscillation at the desired frequency.)

A familiar example of spontaneous oscillation is a public address system (perhaps in the village hall) that emits an unpleasantly loud whistle if the volume (gain) is turned too high. In this case the closed loop formed by sound from the loudspeaker feeding back to the microphone is an unwanted but unavoidable phenomenon, and it may be possible to prevent the instability by shielding the microphone from the loudspeakers (i.e. lowering the loop gain by reducing the feedback), or by reducing the volume control.

Unstable behaviour is characteristic of linear systems of third or higher order and is well understood, though the theory is beyond our scope. Whenever the closed-loop system has an inherently oscillatory transient response, increasing the proportional gain and/or introducing integral control generally makes matters worse. The amplitude and frequency of the 'ringing' of the output response may become larger, and the settling time may increase. If the gain increases further the system becomes unstable and oscillation grows until limited by saturation of one or more elements in the loop.

As far as we are concerned it is sufficient to accept that there is a potential drawback to raising the gain in order to reduce error, and to take comfort from the fact that there are well-established design criteria that are used to check that a system will not be unstable before the loop is closed. Stability of the closed-loop system can be checked by examining the frequency response of the open-loop system, the gain being adjusted to ensure (by means of design criteria known as gain and phase margins) that, when the loop is closed for the first time, there is no danger of instability.

A.8   DISTURBANCE REJECTION – EXAMPLE USING D.C. MACHINE

We will conclude our brief look at the benefits of feedback by considering an example that illustrates how a closed-loop system combats the influence of inputs (or disturbances) that threaten to force the output of the system from its target value. We already referred to the matter qualitatively in Section A.2.2, when we looked at how we would drive a car at a constant speed despite variations in wind or gradients.

Throughout this book the self-regulating properties of electric motors have been mentioned frequently. All electric motors are inherently closed-loop systems, so it is fitting that we finish by revisiting one of our first topics (the d.c. machine), but this time we take a control-systems perspective to offer a fresh insight as to why the performance of the machine is so good.

The block diagram of a separately excited d.c. machine (with armature inductance neglected) is shown in Figure A.10.

This diagram is a pictorial representation of the steady-state equations (3.5), (3.6) and (3.7) developed in Chapter 3, together with the dynamic equation relating resultant torque, inertia and angular acceleration. This set of equations are repeated here for convenience.

Motor torque, $T_\mathrm{m} = kI$

Motional e.m.f, $E = k\omega$

Armature circuit, $V = E + IR$

Dynamic equation, $T_\mathrm{m} - T_\mathrm{L} = T_\mathrm{res} = J\dfrac{d\omega}{dt}$ or $\omega = \dfrac{1}{J}\displaystyle\int T_\mathrm{res}dt$

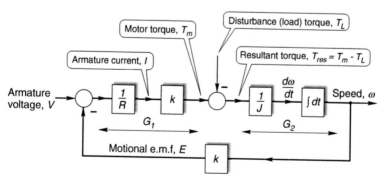

Figure A.10   *Block diagram of separately excited d.c. motor*

Each section of the diagram corresponds to one of the equations: for example, the left-hand summing junction and the block labelled $1/R$ yield the armature current, by implementing the rearranged armature circuit equation, i.e. $I = (V - E)/R$. The summing junction in the middle of the forward path allows us to represent a 'disturbance' entering the system: in this case the disturbance is the load torque (including any friction), and is shown as being subtracted from the motor torque as this is usually what happens.

A control person who was unfamiliar with motors would look at Figure A.10 and describe it as a closed-loop speed control system designed to make the speed ω track the reference voltage, V. Examination of the forward path (and ignoring the disturbance input for the moment) would reveal an integrating element, and this would signal that the loop gain was infinite under steady-state conditions. The control specialist would then say that when the loop gain is infinite, the closed-loop gain is $1/H$, where, in this case, $H = k$. Hence the speed at any voltage V is given by $\omega = (1/k)V$.

We obtained this result in Chapter 3 by arguing that if there was no friction or load torque, and the speed was steady (i.e. the acceleration was zero), the motor torque would be zero, and therefore the current would have to be zero, and hence the back e.m.f. must equal the applied voltage. We then equated $V = E = k\omega$ to obtain the result above.

Turning now to the effect of the load torque (the 'disturbance') we will focus on the steady-state condition. When the speed is steady, the signal entering the integrator block must be zero, i.e. the resultant torque must be zero, or in other words the motor torque must be equal and opposite to the load torque. But from the diagram the motor torque is directly proportional to the error signal (i.e. $V - E$). So we deduce that as the load torque increases, the steady-state error increases in proportion, i.e. the speed (E) has to fall in order for the motor to develop torque. It follows that in order to reduce the drop in speed with load, the gain of section G_1 in Figure A.10 must be as high as possible, which in turn underlines the desirability of having a high motor constant (k) and a low armature resistance (R). As expected, this is exactly what we found in Chapter 3.

We can see that the mechanism whereby the closed-loop system minimises the effect of disturbances is via the feedback path: as the speed begins to fall following an increase in the disturbance input, the feedback reduces, the error increases and more motor torque is produced to compensate for the load torque.

To quantify matters, we can develop a model that allows us to find the steady-state output due to the combined effect of the reference input (V) and the load torque (T_L), using the principle of superposition. We find the outputs when each input acts alone, then sum them to find the output when both are acting simultaneously.

If we let G_1 and G_2 denote the steady-state gains of the two parts of the forward path shown in Figure A.10, we can see that as far as the reference input (V) is concerned the gain of the forward path is G_1G_2, and the gain of the feedback path is k. Hence, using equation (A.3), the output is given by

$$\omega_V = \left\{ \frac{G_1G_2}{1+G_1G_2k} \right\} V.$$

From the point of view of the load torque, the forward path consists only of G_2, with the feedback consisting of k in series with G_1. Hence, again using equation (A.3) and noting that the load torque will usually be negative, the output is given by

$$\omega_L = \left\{ \frac{G_2}{1+G_2kG_1} \right\} (-T_L) = -\left\{ \frac{G_2}{1+G_1G_2k} \right\} T_L$$

Hence the speed (ω) is given by

$$\omega = \left\{ \frac{G_1G_2}{1+G_1G_2k} \right\} V - \left\{ \frac{G_2}{1+G_1G_2k} \right\} T_L.$$

The second term represents the influence of the disturbance – in this case the amount that the speed falls due to the load torque, T_L. Looking at the block diagram, we see that in the absence of feedback, the effect on the output of an input T_L would simply be G_2T_L. Comparing this with the second term in the equation above we see that the feedback reduces the effect of the disturbance by a factor of $1/(1+G_1G_2k)$, so if the loop gain (G_1G_2k) is high compared with 1, the disturbance is attenuated by a factor of 1/loop gain.

We can simplify these expressions in the case of the d.c. motor example by noting that $G_1 = k/R$, and, because of the integration, G_2 is infinite. Hence the steady-state speed is given by

$$\omega = \frac{V}{k} - \frac{R}{k^2} T_L,$$

This is the same result as we obtained in equation (3.10): the first term confirms that when the load torque is zero the speed is directly proportional to the armature voltage, while the second term is the drop in speed with load, and is minimised by aiming for a low armature resistance.

FURTHER READING

Acarnley, P.P. (2002) *Stepping Motors: A Guide to Modern Theory and Practice* (4th ed.). IEE Publishing, London. ISBN: 085296417x.
A comprehensive treatment at a level which will suit both students and users.

Beaty, H.W. and Kirtley, J.L. (1998) *Electric Motor Handbook*. New York: McGraw-Hill.
Comprehensive analytical treatment including chapters on motor noise and servo controls.

Hindmarsh, J. (1985) *Electrical Machines and their Applications* (4th ed.). Oxford: Pergamon.
Hindmarsh, J. (1984) *Electrical Machines and Drives* (2nd ed.). Oxford: Pergamon.
These two texts by Hindmarsh are popular with both students and practising engineers. The first covers transformers and generators as well as motors, while the second has many worked examples.

Kenjo, T. (1991) *Electric Motors and their Controls*. New York: Oxford Science Publications.
A general introduction with beautiful illustrations, and covering many small and special-purpose motor types.

Jordan, H.E. (1994) *Energy-Efficient Motors and their Applications* (2nd ed.). New York: Plenum Press.
Clearly written specialist text.

Moreton, P.L. (2000) *Industrial Brushless Servomotors*. Oxford: Newnes.
Comprehensive treatment with many worked examples.

Valentine, R. (1998) *Motor Control Electronics Handbook*. New York: McGraw Hill.
Includes hardware and software elements of digital motor/drive control.

ANSWERS TO NUMERICAL REVIEW QUESTIONS

Chapter 1

1) 2000 A
2) 1.26 T; 1.26 T
3) 2.52 mWb
4) 800 A; 400 A; 1 T
5) 100 W
6) 12.5%
7) 33 cm^2
8) 0.8 N; 8 N
9) 139 Nm

Chapter 2

2) 207 V
3) 57.6°
5) 8.13 kW; 3.66 kW
7) 0.364; 200 W

Chapter 3

10) 208 V; 41 A; 146%
11) 123 V; 80.5%
12) 2.63 Nm
13) 248 V
14) 1582 rev/min; 39.5 Nm
15) (a) 0.70 V/rad/s; 7.0 Nm; (b) 56.06 A; (c) 0.2 Ω; 11.5 V; (d) 1588 rev/min; 22.27 Nm; 87.3%
17) 5.8 V; 0.485 m Nm; 1.04 × 10^5 rad/s^2

23)   36 W

27)   20.5 Ω; 2050 W; 50 W

28)   0.605 V/rad/s; 7.53 s

29)   ±200 rev/min

30)   519.075 V; 17 W; 9603 W; 88.2 Nm; 1077.44 rev/min; 352.81 A; 176.3 Nm; 17.5%

Chapter 4

8)   60°

Chapter 5

1)   40 Hz

2)   2; 1.66%

3)   1728 rev/min; 2.4 Hz; 72 rev/min; 1800 rev/min

4)   10-pole

5)   75 A; 60 Nm

6)   None; 86.4%; 86.4%; 74.6%

8)   367 V

9)   30; 0.71 mm

10)   1.053; 2.217%

Chapter 6

5)   8.7%

17)   1770 rev/min; 1717 rev/min

18)   458 V; 20.8 kW; 1450 rev/min

21)   22.2 kW

Chapter 7

1)   120 V; 4 A

2)   80 Ω, resistive

3)   55.6 mA; 13.3 W

5)   1.11%; 18.8

6)   15.9 Nm

7)   62.8 Nm

9)   4.2%

10)   42.8 A; 0.90; 22.9 A; 0.89

11)   1380 rev/min, 0.08; 60 rev/min, 0.67

Chapter 8

1)   1760 rev/min; 140 rev/min

2)   60 A; 150 A; 0.67 Hz

4) $0.2 N_s$; 10%

5)   76%; 44%

Chapter 9

4) 0.15 step
5) 0.66 Nm
6) 40 Hz
7) 15°; 7.5°; 1.8°
9) 72 rev/min
10) 22.5 V; 6 Ω

Chapter 10

1) 24
2) 350 V
8) 25.4°

Chapter 11

1) 82.5 rev/min; 67.5 rev/min
2) 36
3) 14.9 min
7) 1.4 kW
8) 4-pole; 5.5 kW; 1380 rev/min
9) 12.88 kgm^2; 101.85 kJ; 100 kJ
10) 2%

INDEX